Polymer-Based Multifunctional Nanocomposites and Their Applications

Edited by

Kenan Song
Advanced Materials Advanced Manufacturing Laboratory (AMAML), Ira A. Fulton Schools of Engineering, Arizona State University, USA

Chuntai Liu
National Engineering Research Center for Advanced Polymer Processing Technology, Zhengzhou University, China

John Zhanhu Guo
Integrated Composites Laboratory (ICL), Department of Chemical & Biomolecular Engineering, University of Tennessee, USA

ELSEVIER

Higher Education Press

Elsevier
Radarweg 29, PO Box 211, 1000 AE Amsterdam, Netherlands
The Boulevard, Langford Lane, Kidlington, Oxford OX5 1GB, United Kingdom
50 Hampshire Street, 5th Floor, Cambridge, MA 02139, United States

Library of Congress Cataloging-in-Publication Data
A catalog record for this book is available from the Library of Congress

British Library Cataloguing-in-Publication Data
A catalogue record for this book is available from the British Library

ISBN: 978-0-12-815067-2

Publisher: Matthew Deans
Acquisition Editor: Edward Payne
Editorial Project Manager: Naomi Robertson
Production Project Manager: Sruthi Satheesh
Cover Designer: Mark Rogers

Typeset by SPi Global, India

Contents

Contributors

Numbers in parenthesis indicate the pages on which the authors' contributions begin.

Khalid A. Askar (111), Department of Materials Science and Engineering, Masdar Institute of Science and Technology, Abu Dhabi, United Arab Emirates

Adarsh Bafana (157), Nanobiomaterials and Bioprocessing Laboratory, Dan F. Smith Department of Chemical Engineering, Lamar University, Beaumont, TX, United States

Tracy Benson (157), Operando Spectroscopy Laboratory, Dan F. Smith Department of Chemical Engineering, Lamar University, Beaumont, TX, United States

David Bird (175), Interdisciplinary Program in Materials Science & Engineering, New Jersey Institute of Technology, Newark, NJ, United States

Linh Doan (157), Operando Spectroscopy Laboratory, Dan F. Smith Department of Chemical Engineering, Lamar University, Beaumont, TX, United States

Xuehui Ge (1), The State Key Lab of Chemical Engineering, Department of Chemical Engineering, Tsinghua University, Beijing, China

Natali Gendelberg (175), Interdisciplinary Program in Materials Science & Engineering, New Jersey Institute of Technology, Newark, NJ, United States

Haoqing Hou (201), Department of Chemistry and Chemical Engineering, Jiangxi Normal University, Nanchang, China

Clayton Jeffryes (157), Nanobiomaterials and Bioprocessing Laboratory, Dan F. Smith Department of Chemical Engineering, Lamar University, Beaumont, TX, United States

Tuo Ji (77), Intelligent Composites Laboratory, Department of Chemical and Biomolecular Engineering, The University of Akron, Akron, OH, United States

Shaohua Jiang (201), College of Materials Science and Engineering, Nanjing Forestry University, Nanjing, China

Li Jin (201), Electronic Materials Research Laboratory, Key Laboratory of the Ministry of Education & International Center for Dielectric Research, School of Electronic and Information Engineering, Xi'an Jiaotong University, Xi'an, China

Weijun Li (261), State Key Laboratory of Heavy Oil Processing, China University of Petroleum, Beijing, China

Yang Lu (157), Materials Engineering And Nanosensor [MEAN] Laboratory, Department of Chemical and Biological Engineering, The University of Alabama, Tuscaloosa, AL, United States

Nitin Mehra (77), Intelligent Composites Laboratory, Department of Chemical and Biomolecular Engineering, The University of Akron, Akron, OH, United States

Liwen Mu (77), Intelligent Composites Laboratory, Department of Chemical and Biomolecular Engineering, The University of Akron, Akron, OH, United States

Vladimir N. Popok (35), Department of Materials and Production, Aalborg University, Aalborg, Denmark

Nuggehalli M. Ravindra (175), Interdisciplinary Program in Materials Science & Engineering, New Jersey Institute of Technology, Newark, NJ, United States

Min Zhi Rong (137), Key Laboratory for Polymeric Composite and Functional Materials of Ministry of Education, GD HPPC Lab, School of Chemistry, Sun Yat-Sen (Zhongshan) University, Guangzhou, China

Kenan Song (111), Department of Manufacturing Engineering, Arizona State University, Tempe, AZ, United States

Zhenhao Tian (1), The State Key Lab of Chemical Engineering, Department of Chemical Engineering, Tsinghua University, Beijing, China

Yujun Wang (1), The State Key Lab of Chemical Engineering, Department of Chemical Engineering, Tsinghua University, Beijing, China

Suying Wei (157), Department of Chemistry and Biochemistry, Lamar University, Beaumont, TX, United States

Jia Wei (245), PCI Synthesis, Devens, MA, United States

Evan K. Wujcik (157), Materials Engineering And Nanosensor [MEAN] Laboratory, Department of Chemical and Biological Engineering, The University of Alabama, Tuscaloosa, AL, United States

Jianhong Xu (1), The State Key Lab of Chemical Engineering, Department of Chemical Engineering, Tsinghua University, Beijing, China

Quan Xu (261), State Key Laboratory of Heavy Oil Processing, China University of Petroleum, Beijing, China

Xing Yang (285), Viance LLC, Charlotte, NC, United states

Xiao Ji Ye (137), Key Laboratory for Polymeric Composite and Functional Materials of Ministry of Education, GD HPPC Lab, School of Chemistry, Sun Yat-Sen (Zhongshan) University, Guangzhou; Guangdong Provincial Public Laboratory of Analysis and Testing Technology, China National Analytical Center, Guangzhou, China

Guoqiang Yu (157), Dan F. Smith Department of Chemical Engineering, Lamar University, Beaumont, TX, United States

Ming Qiu Zhang (137), Key Laboratory for Polymeric Composite and Functional Materials of Ministry of Education, GD HPPC Lab, School of Chemistry, Sun Yat-Sen (Zhongshan) University, Guangzhou, China

Lin Zhang (201), Electronic Materials Research Laboratory, Key Laboratory of the Ministry of Education & International Center for Dielectric Research, School of Electronic and Information Engineering, Xi'an Jiaotong University, Xi'an, China

Wei Zhang (285), Sustainable Biomaterials Department, Virginia Tech, Blacksburg, VA, United States

Jiahua Zhu (77), Intelligent Composites Laboratory, Department of Chemical and Biomolecular Engineering, The University of Akron, Akron, OH, United States

Yong Zhu (137), Key Laboratory for Polymeric Composite and Functional Materials of Ministry of Education, GD HPPC Lab, School of Chemistry, Sun Yat-Sen (Zhongshan) University, Guangzhou, China

Preface

Composite materials are those with multiple phases that display improved mechanical performance or functional properties. Polymer-based composites are based on polymer matrices and have applications in fields such as aerospace, automobiles, construction, biomedicine, catalysis, and electronics, among others. Despite all the potential these materials hold, a lack of development in novel fabrication methods, understanding of microstructural control, and bridging between theoretical predictions and experimental measurements remain significant hurdles to their widespread application.

In this book, we show novel approaches for preparing polymer nanocomposites (i.e., microfluidic technology in Chapter 1, cluster beam method in Chapter 2), property explorations (i.e., thermal conductivity in Chapter 3, multifunctional property in Chapter 4), microstructural creations (i.e., self-healing paths in Chapter 5), and novel applications (i.e., sensors for environmental monitoring in Chapter 6, biomedical applications in Chapter 7, electronics fields in Chapter 8, fuel cells in Chapter 9, adhesive systems in Chapter 10, and fire retardant materials in Chapter 11). This book aims to showcase recent advances in the field of polymer-based nanocomposites used for structural, thermal, electrical, optical, and other functional applications.

In Chapter 1, Xu's Group comprehensively reviews microfluidic technology in synthesizing nanoparticles and fabricating nanocomposites. The advantages of using this technology include the precise control of nanoparticle size and composition, temperature and mass transportation, and in situ screening and optimization of the manufacturing procedure. The authors cover the composition and structural management in nanoparticle and nanocomposite processing, as well as current applications and future perspectives.

In Chapter 2, Popok presents a state-of-the-art review of synthesizing metal and metal-oxide-included polymer composites. The author compares other widely used approaches with the cluster beam technique. Topics such as physical principles of cluster nucleation, growth and formation, polymer-nanoparticle interfacial interactions and immersions, and different methods in the use of cluster beams are provided. Following is a discussion of practical issues such as the application of polymer/metal and polymer/metal-oxide composites and future perspectives on the cluster beam technique.

In Chapter 3, Zhu's Group focuses on thermal conductivity between polymers and its composites. Study of the phonon is most fundamental to describe any heat conduction at the microscopic level. Nanoscale thermal conductivity depends on the development of efficient phonon transport pathways. Therefore, the improvement of thermal conductivity will correspond to phonons transportations without heat barriers or scattering. Understanding of thermal conductivity in polymer chains and across interface poses a challenge when carbon fillers, ceramic fillers, and metallic fillers are introduced in polymer matrices. Strategies to enhance thermal conduction are filler alignment, filler surface modifications, filler hybridizations, and so on.

Khalid's Group in Chapter 4 describes a multifunctional nanocomposite coating system that can be wear-resistant, optically transparent, self-cleaning, and self-healing. The coating materials obtained from simple spray-coating methods and their multiple functions rely on nanoparticle orientation modifications. Mechanical reinforcement and wear-resistance improvements are primarily correlated with the out-of-plane alignment of nanoparticles. Self-cleaning and self-healing functions are results of compatible thermoset-thermoplastic composites.

The Zhang Group discusses the self-pressurized healing system and mechanism in Chapter 5. The material system depends on the self-healing agent contained in micropipelines. In this way, the healing agent can flow without manual intervention. Epoxy/mercaptan healing agents and foaming agent filling in polypropylene tubes embeds in glass fiber/epoxy composites. Internal pressure is built up via the decomposition of the foaming agent and gas inflation. The pressurized healing fluid will burst out upon damage to tubes and initiate epoxy/hardener crosslinking. Healing efficiency designed in this way is significantly improved, and it may suggest many practical applications.

In Chapter 6, the Wujcik Group applied nanoscience and nanoengineering to sensors for environmental monitoring. The traditional environmental monitoring methods are UV-Vis, LC, GC, ICP-AES, F-AAS, and XRF. These analytical methods have a high cost, are time-consuming, and require pretraining; nanocomposite-based sensors, on the other hand, provide an alternative that is simple, cost-efficient, determined on-site, highly sensitive, and selective, and responds quickly. The group reviews monitoring of inorganic gases, carcinogenic gases, and organic gases in the air, soil monitoring, and organic pollutants and heavy metals in water. The chapter concludes with challenges and critical elements in developing nanocomposite sensors.

In Chapter 7, Ravindra's Group gives a brief overview of progress in nanocomposites in biomedical applications. Self-healing, biocompatibility, and mechanical durability, along with other properties, play critical roles in nanocomposites used in this field. Smart materials including biopolymers, shape memory polymers, and self-healing materials with different working mechanisms are comprehensively reviewed.

In Chapter 8, Jiang's Group concentrates on polymer-based nanocomposites with high dielectric permittivity. Polymer materials do not display dielectric properties, and dielectric/conductive fillers are introduced for high permittivity behaviors. This chapter presents the fundamentals of dielectrics, high permittivity in nanocomposites due to dielectric fillers, conductive fillers, and hybrid fillers, and fabrication methods of these composites. The chapter is concluded with outlooks, challenges, and future perspectives of nanocomposites with high dielectric permittivity.

Wei in Chapter 9 discusses the specific application of proton-conducting materials in polymer electrolyte members used in fuel cells. Starting from the fundamental proton transport mechanism in fuel cells and mechanisms of proton conducting, the author also reviews different types of fuel cells and fuel-cell-based electrolyte membranes. Current development of heterocycle-polymer systems for high-temperature polymer electrolyte membranes is introduced in detail.

Xu's Group in Chapter 10 reveals the advancement of smart adhesion surfaces, especially in gecko toe pad and mussel thread adhesion systems. The recent development of biomimicry adhesion systems, including polymers and other molecules, inspired by gecko toe pads and mussel threads, and their biomimicry applications are summarized.

Zhang in Chapter 11 reviews wood-polymer composites and their flame retardancy. The author starts from the fabrication techniques and characterization methods in making wood-polymer composites, followed by a review of phosphorus-based flame retardants, boron-based flame retardants, metal hydroxide flame retardants, graphite-based flame retardants, and multifiller-based flame retardants. The summary is concluded with challenges such as understanding the synergy effects of multiple flame retardants, burning behavior of wood-polymer composites, and requirement of regulatory fire tests.

This book is an attempt to collate information from a group of internationally known researchers and professors, and demonstrate the state-of-the-art of polymer-based nanocomposites regarding their design, fabrication, characterization, and applications. We would like to thank all the contributors for their willingness to participate in this exercise and being patient during the compilation work. The editor also wishes to thank the National Engineering Research Center for Advanced Polymer Processing Technology in Zhengzhou University, Integrated Composites Laboratory at University of Tennessee, and Advanced Materials Advanced Manufacturing Laboratory at Arizona State University for all the support.

Kenan Song
Chuntai Liu
Zhanhu Guo

Chapter 1

Nanoparticles and Nanocomposites With Microfluidic Technology

Zhenhao Tian, Xuehui Ge, Yujun Wang and Jianhong Xu
The State Key Lab of Chemical Engineering, Department of Chemical Engineering, Tsinghua University, Beijing, China

Chapter Outline

1.1 INTRODUCTION

Over the past few decades, chemically or physically synthesized nanoparticles and nanocomposites have drawn attention both in fundamental researches, and for practical applications. Compared to macro-size particles, nanoparticles and nanocomposites have unique physical and chemical properties, such as luminescent properties, photoelectric properties, catalytic properties, magnetic response, biocompatibility, and so on. Therefore, nanoparticles and nanocomposites have a wide range of applications in the fields of optical materials [1], semiconductor materials [1,2], catalysts [3,4], energy [5,6], bio-labels [7–10], drug deliveries [10], and many others. Therefore, it is

Polymer-Based Multifunctional Nanocomposites and Their Applications
https://doi.org/10.1016/B978-0-12-815067-2.00001-9

important to develop appropriate methods for the synthesis of nanoparticles and nanocomposites.

A wide variety of nanoparticles could be prepared by flask reactors or batch reactors [11–35]. In batch reactors, many methods and strategies have been developed to synthesize different kinds of nanoparticles and nanocomposites, such as reduction methods [13–16], sol-gel methods [17–20], hydrothermal methods [21–24], hot injection methods [25–30], multistep methods [31–35], and so on. For example, noble metal nanoparticles are synthesized via reduction methods. Au nanoparticles could be synthesized from $HAuCl_4$ reduced by sodium borohydride in the presence of surfactants [13]. Metal oxide nanoparticles are usually synthesized via sol-gel methods or hydrothermal methods. Fe_3O_4 could be obtained by sol-gel methods [19]. Under vigorous stirring, iron sources, $FeCl_3$ and $FeCl_2$, and a precipitator, aqueous ammonia, are added together to form Fe_3O_4 powders. In addition, the morphologies of ZnO nanoparticles can be controlled by changing the temperature and reaction time of hydrothermal reactions in batch reactors [23]. Quantum dots can be obtained via hot injection methods. Typically [25], Se precursors are injected into preheated Cd precursors, and then, the temperature is kept at around 260°C for nucleation and growth of CdSe quantum dots. Nanocomposites are often synthesized step by step according to different compositions and structures. For example, in order to synthesize ZnO/Ag core-shell nanoparticles [35], first, ZnO cores are synthesized through hydrothermal methods. Then, $AgNO_3$ is reduced to form silver layers. Therefore, the chemical system and research methods provided by the batch reactors provide a good example from which to learn.

No matter which methods or strategies for synthesis of nanoparticles in flask reactors are used, researchers often control the temperatures, residence times, precursor concentrations, ligands, feeding methods, and so forth to synthesize nanoparticles with different sizes or shapes. However, in conventional flask reactors, the temperature field and the concentration field are homogenized by means of rapid injection and stirring in a small volume [36]. Thus, the precise control of temperatures and concentrations is difficult, which can lead to the irreproducibility of sizes, size distributions, shapes, and quality of the nanomaterials from batch to batch [37]. In addition, for scaling up the yield of nanoparticles, simply expanding the volume of the reactor could make it difficult to maintain a homogeneous solvent environment [38]. So there are some drawbacks in the synthesis of nanoparticles via batch reactors.

As the mass and heat transfer efficiency in microfluidic devices are much higher than that in batch reactors [39–41], many methods based on microfluidic devices for nanoparticles and nanocomposites synthesis have been developed. In recent years, microfluidic devices have been widely used in chemical and biological fields. Due to efficient heat and mass transfers and spatial limitations in microfluidic devices, the size distribution of nanoparticles become more uniform [42]. Through convenient parameter adjustments in microreactors, nanoparticles with different sizes, shapes, and compositions could be synthesized reproducibility. In addition, because of the sealability in microreactors, many synthesis processes no longer needed inert gas protection, and solvent toxicity could be confined. In situ process monitoring, combined with transparent microchannels [43,44] could control the properties of nanoparticles and study the formation mechanism of nanoparticles conveniently.

In this review, we will firstly, introduce the types and fabrication techniques of microfluidic devices and the flow patterns commonly used for the preparation of nanoparticles. Then, the preparation methods and typical examples of organic nanoparticles, inorganic nanoparticles, and hybrid nanocomposites by using microreactors will be introduced, respectively.

1.2 MICROFLUIDIC PLATFORMS FOR NANOPARTICLES AND NANOCOMPOSITES SYNTHESIS

1.2.1 The Types and Fabrication Techniques of Microfluidic Platforms

The synthesis of nanoparticles by solvothermal methods, whether it is with a conventional flask reactor or microreactor, is necessary to control the reaction parameters, growth parameters, and crystallization parameters of the crystal system.

A uniform preparation environment, including uniformity of concentration and temperature, is critical for the synthesis of nanoparticles. In microreactors, the flow is laminar, and diffusion is the main pattern for mass and heat transfer. Due to the small size of the microreactors, the reactor volume is around a nanoliter or microliter, and the short diffusion distance and high specific surface area lead to efficient heat transfer and mass transfer. Thus, it is easy to obtain a uniform preparation environment in microreactors. The design of suitable microreactors and continuous preparation of controllable nanoparticles, depending on the preparation conditions of different nanoparticles, has been a popular topic in recent years.

Capillary reactors were first used to synthesize nanoparticles. Commonly used capillary microreactors were made of glass capillaries [45], PTFE tubes [46], stainless steel tubes [47,48], and so on. These materials are commercially available for easy fabricating and connection. In the glass capillary [45] (Fig. 1.1A), researchers achieved the preparation of the emulsion [52–54]. In addition, because the glass capillary is transparent, online characterization and online light curing facilitated the preparation of nanoparticles. Temperature and pressure for the glass capillary micro-reaction have some restrictions; generally the reaction processes take place under normal temperature and pressure. Polytetrafluoroethylene tubes [46] with excellent chemical stability, that are soft and easy to bend, and have a certain heat resistance, are commonly used

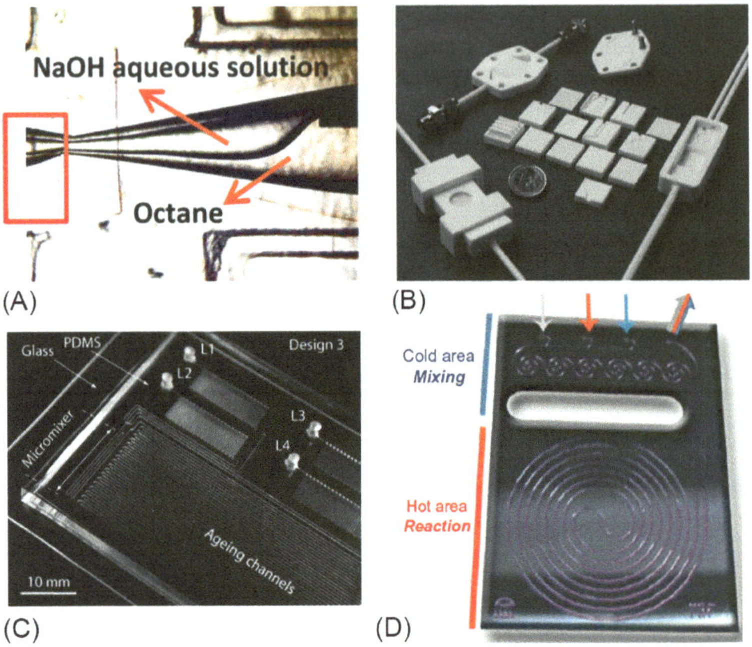

FIG. 1.1 Microfluidic platforms for nanoparticles and nanocomposites synthesis. (A) Glass capillary. Reproduced from Ref. [45] with permission of The Royal Society of Chemistry. (B) Ceramic microreactions. Reproduced from Ref. [49] with permission of The Royal Society of Chemistry. (C) PDMS/Glass chip [50]. Copyright 2004 American Chemical Society. (D) Silicon/Pyrex chip [51]. Published by The Royal Society of Chemistry.

in the preparation of a long-residence-time microreactors. De Mello et al. [46] synthesized quantum dots by the use of PTFE tubes as a microreactor. However, the PTFE tube is not resistant to high pressure, and the upper limit operating temperature is only about 200°C. Stainless steel tubes [47,48] are suitable for high temperature and high pressure operation; Tian et al. [48] used stainless steel tube-synthesized CdSe quantum dots at 270°C. But stainless steel tubes are not suitable for the operation under strong acid system. Besides, stainless steel tubes are opaque, and the formation of nanoparticles cannot be observed and characterized online. Moreover, ceramic microreactors [49,55] (Fig. 1.1B) can withstand extremely high temperatures. Kartavya et al. [55] synthesized Si_3N_4 nanocrystals at 1000°C by ceramic microreactors prepared by abrasives.

Compared with the capillary reactors, microreactors on chips are more easily integrated. Processes of injection of precursors, mixing, reaction, nucleation, growth, and the addition of precursors and quenching could be achieved in a chip. PDMS chip microreactors [50] (Fig. 1.1C), which are transparent, biocompatible, and have mature fabrication technology, are the most widely used, and many kinds of nanoparticles have been synthesized in PDMS chips. Commonly, PDMS chip microreactors are fabricated by soft lithography methods

[56,57]. However, the PDMS chip microreactors also have some shortcomings, such as intolerance of high temperature, channel swelling in organic solvents, and so on. In addition, glass, silicon, metal, resin, and other materials are also used to fabricate chip microreactors. Glass and silicon chip microreactors are fabricated by etching. Glass chips with a transparent feature are suitable for online characterization. Additionally, silicon chips are tolerant of high temperatures, and some semiconductor nanocrystals can be synthesized by silicon chips. Jensen's group used a Silicon/Pyrex chip microreactor (Fig. 1.1D) to synthesize many kinds of inorganic nanoparticles [51]. It is worth mentioning that, as the 3D printing process has matured, the use of 3D printed resin chip microreactors have generated more and more attention in the field of microsystems.

1.2.2 Flow Patterns

The microreactors can be divided into two types: continuous laminar flow microreactors and segmented flows microreactors, according to the flow patterns in channels during the process of synthesis nanoparticles (Fig. 1.2) [58].

Continuous laminar flow microreactors are the most basic microreactors, with all the advantages of microreactors. But on the other hand, due to the laminar flow in continuous flow microreactors, the boundary layer effect is the biggest challenge to overcome in these kinds of methods [59]. The boundary layer effect would cause the parabolic distribution of fluid velocity, which leads to uneven mixing of axial fluid, uneven distribution of residence time, and other

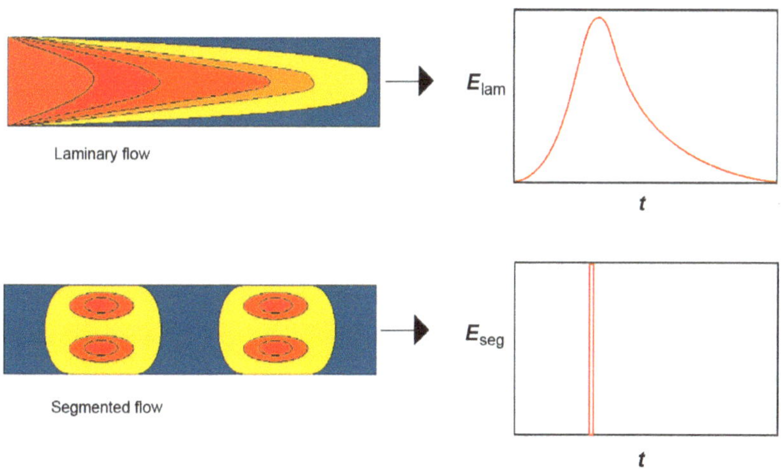

FIG. 1.2 Large residence time distribution in the case of laminar flow of homogeneous fluids in microchannels and narrow residence time distribution and circular flow in microfluid segments. *(Reproduced from J.M. Kohler, S.N. Li, A. Knauer, Why is micro segmented flow particularly promising for the synthesis of nanomaterials? Chem. Eng. Technol. 36 (2013) (6) 887–899 with permission of John Wiley and Sons.)*

issues, which greatly affect the monodispersity of nanoparticles. Additionally, the contact of the nanoparticles with the channel wall of the microreactor would cause contamination of the nanoparticles and the clogging of microreactors.

The segmented flow microreactors are formed by the use of two immiscible fluids. Commonly, the segmented flow microreactor used in the synthesis of nanoparticles is a liquid-liquid system in which droplets or liquid plugs contain reactants. As a result, each droplet or liquid plug with the volume of about a nanoliter is a microreactor. In the segmented flow microreactors, first, because droplets are independent of each other, the uneven mixing of axial fluid in the continuous laminar flow microreactor is eliminated, and the residence time distribution is uniform [58,60,61]. The particle size distribution of nanoparticles is narrower than that of those that are synthesized in the continuous laminar flow microreactors. Second, due to the different surface velocities of the droplets, the eddy current would enhance the mixing of the reactants inside the droplets. In addition, the channel wall could be wetted by continuous phase fluid, and a protective film would be formed, so that the reactants in the droplets and the nanoparticles would not have contact with the channel wall. This solves the problem of channel wall pollution. However, the challenge of segmented flow reactors is to select the appropriate continuous phase fluid. The reactants in the droplets cannot be dissolved in the continuous phase fluid, and cannot react with the continuous phase fluid. Second, the separation of the two phases may be problematic. Furthermore, in order to make the continuous phase fluid compatible with the channel wall, sometimes the surface of the channel needs to be modified, by hydrophilic modification, lipophilic modification, fluoride modification, and so forth. Finally, in the segmented flow microreactor, it is more difficult for supplement precursors to form a multi-stage and multi-step reaction.

To solve these challenges, some researchers selected a gas-liquid system to form gas slug flow microreactors [62]. The inert gas could be easily removed and the reaction between the reactants with gas could be easily avoided. In addition, the separation of the gas and liquid was also simple. It is also convenient to add liquid feeds to the gas-liquid system. However, in the gas-liquid segmented flow microreactors, because the liquid columns containing the reactants are still in contact with the channel wall, and the gas could be compressed, the microreactors are still easily plugged. Because of this, the reactants might pass through the liquid film between the gas and channel wall, leading to an uneven mixing of axial fluid, which would affect the nanoparticle size distributions.

1.3 SYNTHESIS OF ORGANIC NANOPARTICLES BY MICROREACTORS

Organic nanoparticles are usually made of biodegradable polymers to be applied in drug encapsulation and release. Relatively little work has been done to harness the benefits of nanoparticles into drugs previously because of the inability to control the mixing processes, which requires variable nanoparticle physicochemical properties [63].

The introduction of microfluidic technology into nanoparticle technologies has shown considerable potential for the development of organic nanoparticles with narrow size distribution, distinct shapes, better control over the process of nanoprecipitation, and screening of various formulation conditions; which reinforces the promise for their maximal exploitation. The biodegradable polymers, including PLGA, PEG [64], chitosan [65,66], alginate [67], liposomes [68], and so forth, are the original materials that are used in the formation of organic nanoparticles.

Karnik et al. [64] synthesized PLGA-PEG nanoparticles in a microfluidic device by rapidly mixing polymer-acetonitrile solutions and water using hydrodynamic flow, focusing on a controlled nanoprecipitation process (Fig. 1.3A). The polymer-acetonitrile solutions stream flows along the central channel, and meet two adjacent water streams flowing at higher flow rates. At low Reynolds numbers, the central stream is squeezed into a narrow stream between the two adjacent streams. The narrow width of the focused stream then enables rapid mixing through diffusion. Due to water acting as a kind of nonsolvent for PLGA-PEG, homogeneous PLGA-PEG nanoparticles with diameters typically in the range of 10–50 nm would be obtained by the nanoprecipitation process.

Chitosan nanoparticles have attracted attention as drug delivery vehicles due to their ability to protect the drug from degradation, increase loading, and release it in a slow, sustained manner. Kamat et al. [69] synthesized chitosan nanoparticles through an ionic gelation method in a microreactor with magnetic stirring (Fig. 1.3B). Chitosan dissolved in 1% acetic acid was mixed and reacted with sodium tripolyphosphate under magnetic stirring. The size of chitosan could be controlled from 250 ± 27 nm to 130 ± 9 nm via different concentrations and flow rates. Moreover, the fabrication simplicity of this method was further exploited for loading chitosan nanoparticles with amphotericin B (AmB), an antifungal drug, for drug delivery applications.

Alginate is a biopolymer with favorable pH-sensitive properties for oral delivery of peptides and proteins. Bazban-Shotorbani et al. [67] developed an on-chip hydrodynamic flow-focusing approach for synthesis of alginate nanogels with adjustable pore sizes in the range of $68-138$ nm to achieve a fine-tunable release profile of the encapsulated bioactive agents (Fig. 1.3C). A PDMS microfluidic device with three inlets and one outlet was microfabricated for synthesizing nanogels. An alginate aqueous solution was hydrodynamically focused by the lateral $CaCl_2$ streams into a narrowly focused stream, and polymer chains were cross-linked by Ca^{2+} ions to form alginate nanogels at pH 7.4. The average size and average pore size of alginate nanogels was related to synthesis parameters such as flow ratio and mixing time. The particle size shrunk in the simulated gastric fluid (pH 1.2), while in the simulated intestinal fluid (pH 6.8), the particle size had no significant difference compared with its initial size.

Liposomes, with their ability to encapsulate compounds, promote initial lipid vesicles as drug delivery vehicles. Jahn et al. [68] developed a

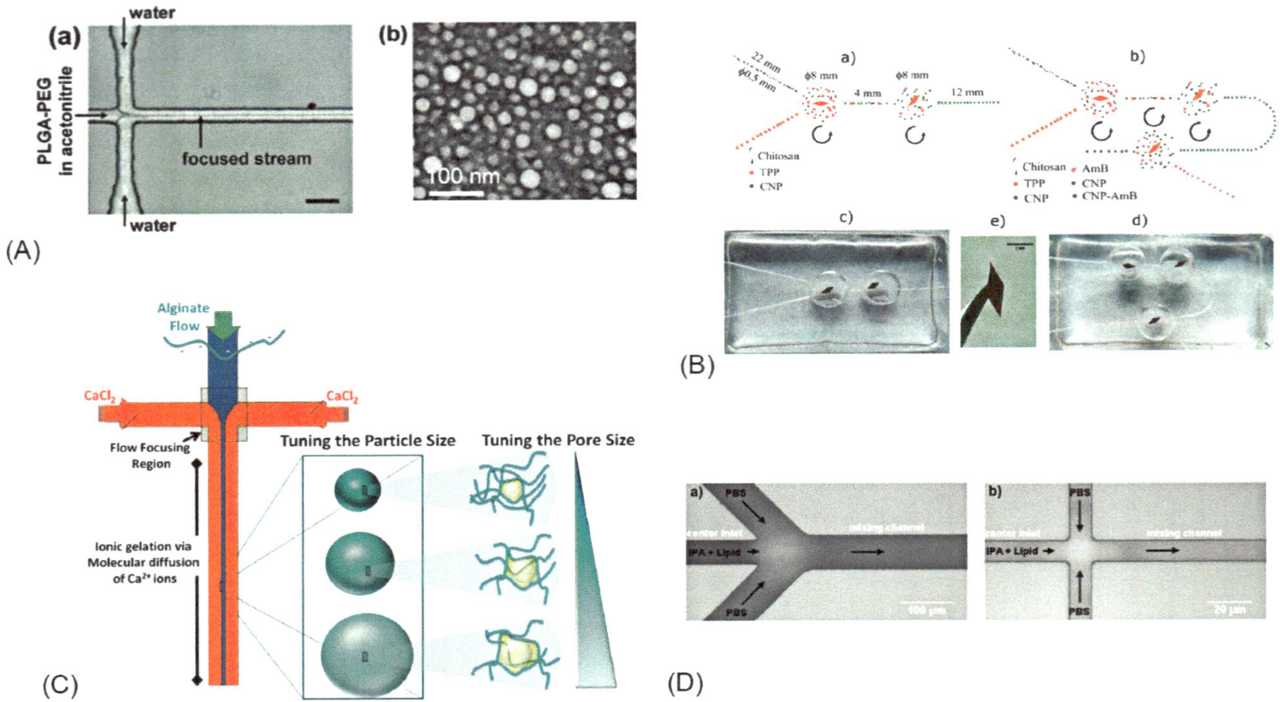

FIG. 1.3 Examples of synthesis of organic nanoparticles by microreactor. (A) A microfluidic device for hydrodynamic flow focusing of polymeric nanoparticles in water and TEM image of PLGA-PEG nanoparticles synthesized by nanoprecipitation. Reprinted with permission from Ref. [64]. Copyright 2008 American Chemical Society. (B) Schematic representation and optical image of fabricated microreactor chip for chitosan nanoparticles synthesis by two-chamber method, and drug loading by three-chamber method. Reprinted with permission from Ref. [69]. Copyright 2015 American Chemical Society. (C) Schematic representation of the microfluidic assisted approach to generate alginate nanogels with tunable size and pore size. Reprinted with permission from Ref. [67]. Copyright 2016 American Chemical Society. (D) Optical micrographs of microfluidic devices used for liposome synthesis with controlled microfluidic mixing and nanoparticle determination. Reprinted with permission from Ref. [68]. Copyright 2010 American Chemical Society.

hydrodynamic flow using a microreactor to synthesize unilamellar lipid vesicle (liposomes) nanoparticles with diameters of tens of nanometers. Apart from fluid flow parameters, they found that microfluidic device geometry (Fig. 1.3D), in conjunction with hydrodynamic flow focusing, strongly influences vesicle size distributions. Further numerical simulations of the mixing process expanded the understanding of the microfluidic environment that controls liposome self-assembly, and yields several technological advances for the on-chip synthesis of nanoscale lipid vesicles.

1.4 SYNTHESIS OF INORGANIC NANOPARTICLES BY MICROREACTOR

Inorganic nanoparticles have a wide range of catalytic, energy, and biological applications, among others. Many types of inorganic nanoparticles can be synthesized using a flask reactor. However, the size and size distribution of the nanoparticles are not reproducible by batch methods. In recent years, there have been many methods of synthesizing inorganic nanoparticles by using microreactors. The size, size distribution, and morphology of inorganic nanoparticles can be controlled by changing the flow parameters and reaction parameters in microreactors. We will introduce the preparation research necessary for metals, oxides, and semiconductor nanoparticles in microreactors.

1.4.1 Metal Nanoparticles

A variety of metal nanoparticles could be prepared by microreactors, such as Au [62,70–73], Ag [70,73–75], Cu [75–77], Pd [71,78], Pt [79,80], Ni [81], and so on. The principle of preparation was to reduce the metal salt solution to obtain metal nanoparticles via a reduction reaction. The commonly used reducing agents were $NaBH_4$, sodium citrate, anti-cyclic acid, a mixture of lithium hydrotriethyl borate (Li [Bet3H]) with 3 (*N,N*-dimethyldodecylammonia) propane sulfonate (SB12), N_2H_4, and so forth. Stabilizers such as CTAB, PVP, and fatty amines would prevent nanoparticles from producing macroparticles and improve the monodispersity of nanoparticles.

Sharada et al. [78] synthesized palladium nanoparticles by a continuous flow reactor (Fig. 1.4A). $PdCl_2$, as a precursor of Pd, and the reducing agent $NaBH_4$ were injected through the peristaltic pump, respectively, through the Y-channel mixing. And Pd nanoparticles were synthesized at 25°C. The *N*-cetyl-*N*, *N*, *N*-tri-methyl ammonium bromide (CTAB) was added to the precursor for the stabilizing reaction system. By controlling the precursor concentration and flow rates, Pd nanoparticles with diameters from 5 to 200 nm could be obtained.

Wagner et al. [70] synthesized Au nanoparticles and Ag nanoparticles by means of multi-step feeding in continuous flow (Fig. 1.4B). In the case of synthesis of Au nanoparticles, the Au precursor solutions made from $HAuCl_4$ and K_2CO_3 were mixed in the first mixer. The concentration of precursor was

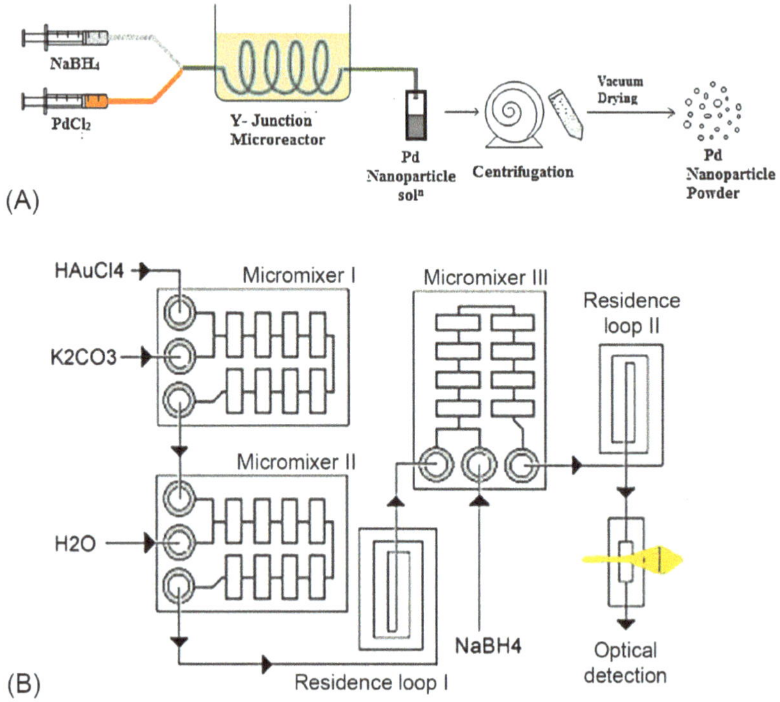

FIG. 1.4 Metal nanoparticles synthesized in a continuous flow pattern. (A) Experimental setup for stable palladium (Pd) nanoparticle (NP) synthesis using continuous flow microreactor. Reprinted from Ref. [78], Copyright 2016 with permission from Elsevier. (B) Modular microreactor arrangement for flow-through process for synthesis of gold nanoparticles. Reprinted from Ref. [70], Copyright 2007 with permission from Elsevier.

adjusted by adding water in a second mixer, and the reducing agent $NaBH_4$ was added to the third mixer. The Au nanoparticles were obtained in the residence loop II.

The method of continuous flow synthesis for metal nanoparticles is simple, and the nucleation and growth stage of nanocrystals could be carried out through step-by-step feeding. However, the nonuniformity of the residence time leads to size distribution of the nanoparticles that is a bit wide, and the microchannels might be blocked for a long time during synthesis. Thus, in recent years, lots of methods for synthesizing metal nanoparticles in segmented flow reactors have been developed.

Zhang et al. [71] used a droplet microreactor to synthesize noble metal nanoparticles such as Pd, Au, and core-shell hybrid nanoparticles (Fig. 1.5A). In the case of synthesis for Pd nanoparticles, Na_2PdCl_4 was a Pd precursor, and ascorbic acid was the reducing agent. They were injected and mixed from R1 and R2 channels, respectively, followed by the addition of silicone oil as a continuous phase to generate droplets. Each droplet with a volume of 0.25 microliters was a

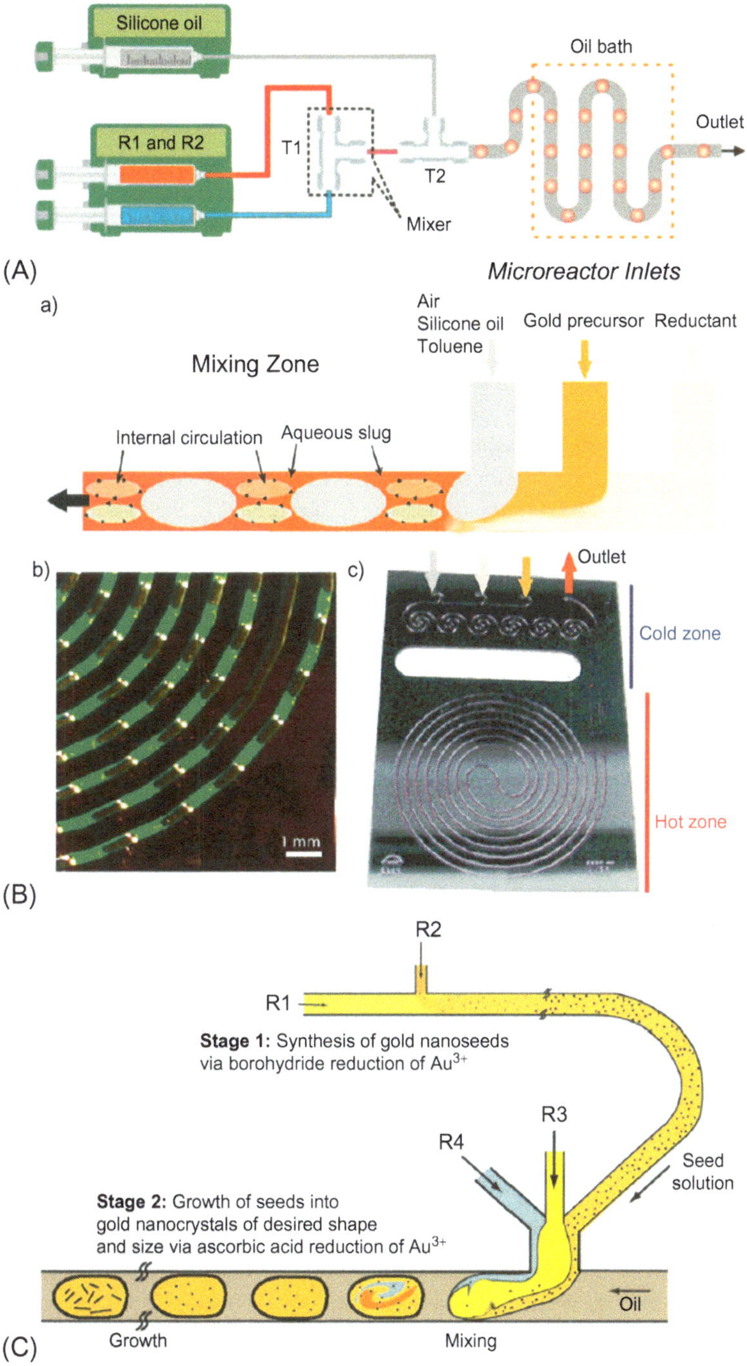

FIG. 1.5 Metal nanoparticles synthesized in segmented flow pattern. (A) Schematic illustration of the device used for generating milliliter-sized droplets for Pd nanoparticles synthesis. The device is assembled from commercially available syringe pumps, T connectors, and PTFE tubes with different inner diameters. Reprinted with permission from Ref. [71]. Copyright 2014 American Chemical Society. (B) Schematic of the segmented flow generation in a hydrophilic MFR. Detail of segmented slugs generated at a residence time of 10 s. Spiral silicon/Pyrex microfluidic reactor designed for gold nanocrystal synthesis. Reprinted with permission from Ref. [62]. Copyright 2012 American Chemical Society. (C) Schematic of the two-stage microfluidic method showing the synthesis of gold nanoparticle seeds in Stage 1 and the growth of gold nanocrystals of desired shape and size in Stage 2. Reproduced from Ref. [72] with permission of John Wiley and Sons.

micro vessel reactor and reacted at 80 ° C for 10 minutes to obtain Pd nanoparticles. In order to obtain nanoparticles with different sizes and morphologies, the previously prepared Pd nanoparticles were injected from the R1 channel as the nuclei. Ascorbic acid, and Pd or Ag precursors were added to the R2 channel to prepare different particle sizes and morphologies Pd nanoparticles or core-shell hybrid Pd/Ag nanoparticles. The nanoparticles' yield was 1 ~ 10 g/h by using this method.

Jensen's group used the plug stream to synthesize Au nanoparticles [62] (Fig. 1.5B). Au precursor HAuCl4 and reducing agent NaBH4 aqueous solution were injected into the microchannel and mixed. Unlike other segmented flow reactors, the aqueous reactant in the reaction was used as a continuous phase fluid. Meanwhile, air or silicone oil or toluene were used as the dispersed phase fluid, so that the nanoparticles would be formed in the continuous phase between segments. Studies have found that the viscosity of different inert components affects the final Au nanoparticle properties. The smaller the viscosity of the inert component, the smaller the particle size of the final synthesized Au nanoparticles.

Duraiswamy et al. [72] combined continuous flow and segmented flow reactors to prepare anisotropic Au nanoparticles with different morphologies (Fig. 1.5C). First, Au precursors, reducing agents $NaBH_4$ and CTAB, were added through R1 and R2 channels, and Au nanoparticles with small particle size were prepared by continuous flow reactors as seeds for further growth. After the addition of Au precursors, ascorbic acid and water from R3 and R4 channels, silicone oil was used as a continuous phase to form segmented flow reactors. In the segmented flow reactors, further growth of the Au seeds took place. By controlling the flow rate of each stream, the crystal morphology of Au nanoparticles could be controlled, such as spherical gold nanoparticles, oblate particles, and gold nanorods with varying aspect ratios.

1.4.2 Metal Oxide Nanoparticles

Metal oxide nanoparticles are widely used in the fields of catalysts, photocatalysis, sensors, and so on. Many kinds of metal oxides, such as TiO_2 [82–84], SiO_2 [50,82,85], iron oxide [82,86–89], ZnO [90–93], Al_2O_3 [94], were synthesized from corresponding metal salt solutions through sol-gel reactions or hydrothermal reactions in microreactors.

Frenz et al. [89] synthesized magnetic iron oxide through droplet reactors at room temperature (Fig. 1.6A). The precursor used was a mixture of $FeCl_2$ and $FeCl_3$ (molar ratio of 2:1) aqueous solution, and the precipitant was ammonia. Droplet pairs containing precursors were created based on hydrodynamic coupling of two spatially separated nozzles. The iron oxide nanoparticles were synthesized via a very fast (millisecond) and reproducible reaction after fusion of droplet pairs by electro-coalescence.

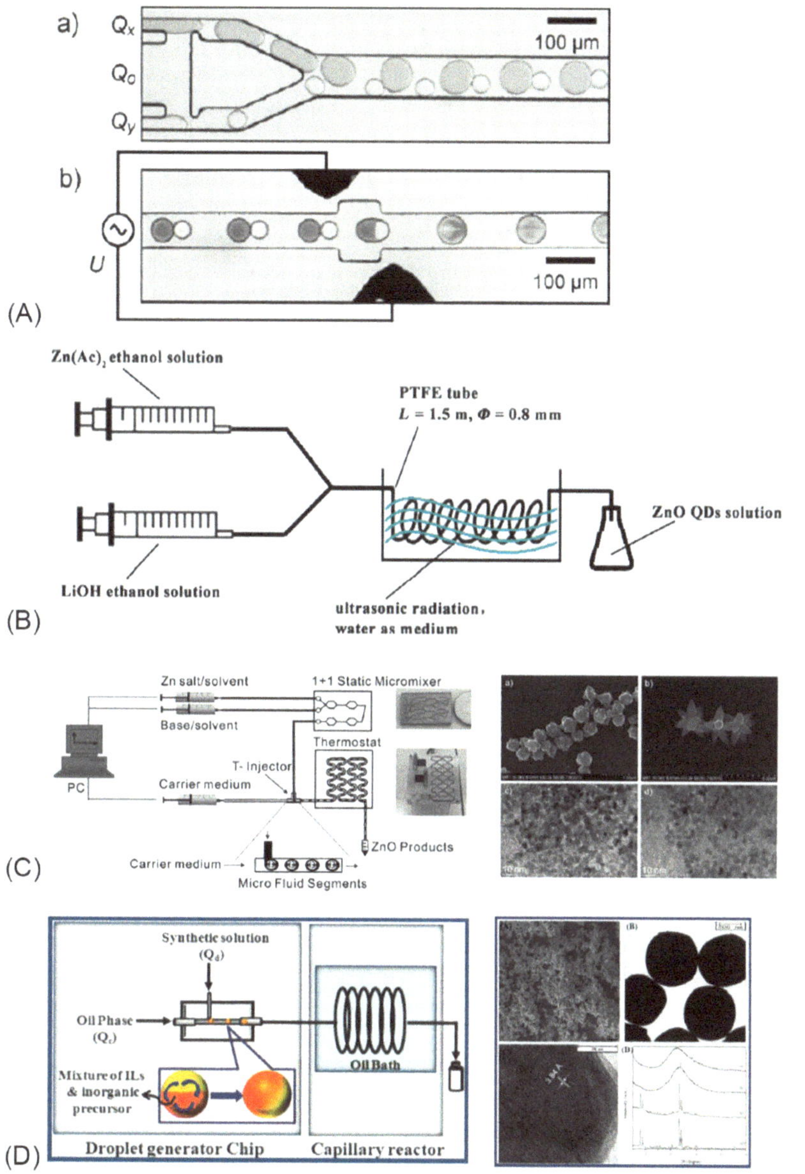

FIG. 1.6 Metal oxide nanoparticles synthesized in microreactors. (A) Two aqueous phases are injected by the outer channels and are synchronously emulsified by the central oil channel. Paired droplets can be coalesced by applying an electrical voltage U between the two electrodes. Reproduced from Ref. [89] with permission of John Wiley and Sons. (B) Experimental setup of ultrasonic microreactor for ZnO nanoparticles synthesis. Reprinted from Ref. [91], Copyright 2016 with permission from Elsevier. (C) Experimental setup for synthesis of ZnO particles. SEM images of ZnO microparticles via hydrothermal method and TEM images of ZnO nanoparticles via sol-gel method. Reproduced from Ref. [93] with permission of John Wiley and Sons. (D) Illustration of the DIM system for synthesis of various inorganic nanomaterials and ZSM-5 characterization. Reprinted with permission from Ref. [94]. Copyright 2011 American Chemical Society.

Yang et al. [91] synthesized ZnO nanoparticles by an ultrasonic microreactor (Fig. 1.6B). Ultrasonic radiation brought bubbles through ultrasonic cavitation. These bubbles built a microreactor inside the microreactor. The photoluminescence properties of ZnO nanoparticles were controlled through different flow rates, ultrasonic power, and temperature. In particular, ultrasonic power changed the ultrasonic cavitation intensity, which affected the reaction energy and the separation of the solution. With the increasing of ultrasonic power, the sizes of ZnO nanoparticles first decreased, then increased, while the quantum yields kept increasing.

Li et al. [92,93] synthesized ZnO nanocrystals by using a droplet flow reactor. Zinc acetate aqueous solution was used as a zinc precursor, and two different base solutions, NaOH and LiOH, were used as precipitants, respectively. First, the effect of water content on growth and optical properties of ZnO nanoparticles was studied by using NaOH. Water/EG binary solvent were used to vary to water content [92]. The diameters of ZnO nanoparticles varied from 18 to 436 nm when the water content ranged between 15 v% and 90 v% at lower pH (pH 12.3, 100°C). The diameters of ZnO nanoparticles varied from 33 to 168 nm when the water content ranged between 15 v% and 60 v% at higher pH (pH 12.7, 90°C). Then, they compared two different methods to synthesize ZnO nanoparticles [93] (Fig. 1.6C). Flower-shaped particles could be observed with a flower branch changed from 0.15 to 1.38 mm when using NaOH/H_2O at 150°C via a hydrothermal method. Dot-like particles could be observed with the diameter changed from 3.9 to 4.6 nm when using LiOH/EtOH at room temperature via a sol-gel method.

The experimental condition of hydrothermal reactions, which needs high temperature and pressure, is a critical challenge in microfluidic devices. Hoang et al. [94] reported droplet- and ionic liquid-assisted microfluidic devices to solve the problem (Fig. 1.6D). ZSM-5, γ-AlOOH, and β-FeOOH nanorods were synthesized through this method. For example, the zeolite precursor solution was prepared by adding sodium aluminate ($NaAlO_2$) to a homogeneous mixture of 1-butyl-3-methylimidazolium chloride (C_4MimCl) and deionized water. The addition of ionic liquids increased the boiling point of the solution. And ZSM-5 could be synthesized in 150°C at normal pressure. Surprisingly, the use of microchannel synthesis ZSM-5 only took 15 min, while the traditional hydrothermal method will take 1 day. This was mainly because the crystallization described was a diffusion controlled system, rather than a reaction controlled system, due to the fast reaction rates of precursors used in these syntheses of inorganic nanomaterials, the crystallization in the nanoliter-scale droplet was facilitated by the enhanced mass transfer with a short diffusion distance.

1.4.3 Quantum Dots

Semiconductor nanoparticles (quantum dots), which can emit monochrome and tunable fluorescent light based on their sizes and compositions, have undergone many fundamental studies for their potential applications.

Because the optical properties of the quantum dots are related to the size and size distribution, in the flask reactors, the synthesis process of quantum dots exhibits nonrepeatability from batch to batch as restricted on the applications of quantum dots. On the other hand, by the microreactor synthesis of quantum dots, the size can be easily controlled through flow rates and temperatures, and high mass and heat transfer efficiency would narrow the size distribution of quantum dots. Additionally, in-situ optical detection can acquire the property parameters of quantum dots conveniently, which contributes to the optimization of reactions and the understanding of the mechanism.

A great deal of research has been done on synthesizing quantum dots in continues flow microreactors [95–99]. Luan et al. [96,97] synthesized monodisperse CdSe quantum dots via continuous flow microreactors that are open to air (Fig. 1.7A). By varying temperature, residence time, and concentration ratio between Cd and Se, CdSe quantum dots with different sizes could be synthesized. In a typical synthesis process, Cadmium oleate and TOP=Se dissolved in 1-octodecene were injected to a micromixer at room temperature, and then the flow was heated in a tube furnace at 260°C. At last, the product was collected and purified. Tian et al. [98] synthesized CdSe quantum dots with 27.6 nm fluorescent FWHM by using mixed oleylamine and dodecyl amine as ligands via a continuous flow reactor. In the preceding methods, the nucleation stage and growth stage of quantum dots were exposed to the same reaction

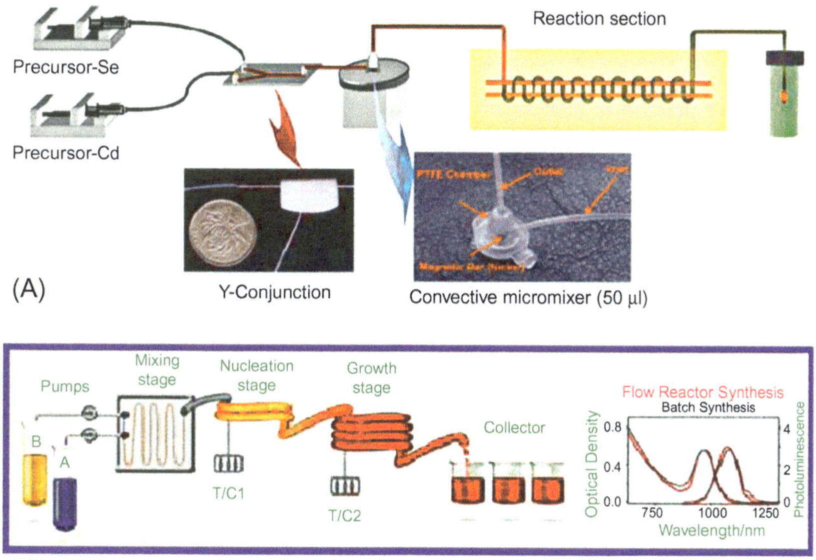

FIG. 1.7 Quantum dots synthesized in a continuous flow pattern. (A) Schematic graph of the microfluidic capillary reactor for CdSe quantum dots synthesis. Reprinted from Ref. [97], with the permission of Springer Nature. (B) Schematic graph of the microfluidic devices separated the nucleation and growth stage for PbS quantum dots. Reprinted with permission from Ref. [99]. Copyright 2013 American Chemical Society.

temperature. Thus, an additional nucleus would form during the growth stage, which would cause a wider size distribution. In order to separate the nucleation stage and growth stage, Sargent et al. [99] developed a dual-temperature-stage continuous flow reactor for the synthesis of PbS quantum dots (Fig. 1.7B). The mixed precursors were first heated up to a 120°C for the nucleation of PbS quantum dots, and then the flow was led to a 90°C oil bath for the growth of PbS quantum dots. A narrower size distribution of PbS quantum dots was obtained compared with the batch reactor and single-stage continuous flow reactor.

Due to the high viscosity in the quantum dots synthesis system, the velocity distribution in the continuous flow leads to a wider distribution of the quantum dots' size. The accumulation of solids on the wall of the microchannel for a long time could lead to the inaccurate control of the residence time and even the channel clogging. The segmented flow reactor solves this problem by introducing another inert fluid. During the synthesis of quantum dots, an elevated reaction temperature was often required to improve the crystallinity of nanocrystals. Thus, the inert fluid not only cannot react or dissolve the precursors, but also needs to withstand high temperatures. The gas and perfluoropolyether (PFPE) are often used as inert fluids during research.

Jensen's group [100] synthesized the oil-soluble CdSe quantum dots by a gas column reactor (Fig. 1.8A). The microreactor was similar to the synthesized metal nanoparticles, and finally the monodisperse CdSe quantum dots with FWHM of about 30 nm were obtained. Chan et al. [101] developed a nanoliter-volume droplet reactor to the high-temperature synthesis of CdSe quantum dots (Fig. 1.8B). A flow-focusing nanojet structure with a step increase in channel height reproducibly generated octadecene droplets in PFPE. Cadmium and selenium precursors flowing in octadecene droplets through a high-temperature (240–300°C) glass microreactor produced high-quality CdSe nanocrystals. DeMello's group [46] used a droplet flow reactor to synthesize the CdSe quantum dots (Fig. 1.8C). This feed method is worth noting. Unlike the use of the sudden expansion of the channel, the two streams were mixed directly at the same point and sheared by PFPE to form liquid droplet reactors. In order to verify the universality of the reactor, they also used this reactor to synthesize nanocrystals such as TiO_2 and Ag.

In the synthesis of oil-soluble quantum dots, the reaction temperature is high, which limits the selection of solvent. The most commonly used solvent is ODE with a high boiling point. The precursors composition of the preceding methods are the same as, or similar to, the typical use of the batch reactor's synthesis quantum dots. However, ODE will lead to a high viscosity system, while the purification process after the synthesis process would become more complex. On the other hand, at sufficiently high pressure, virtually any common solvent, precursor, and ligand will either remain liquid or become supercritical at the temperatures required for QD synthesis. In the supercritical regime, properties of the fluid can be tuned from liquid-like to gas-like, displaying the high miscibility and rapid diffusion rates typical of gases, but with sufficiently high

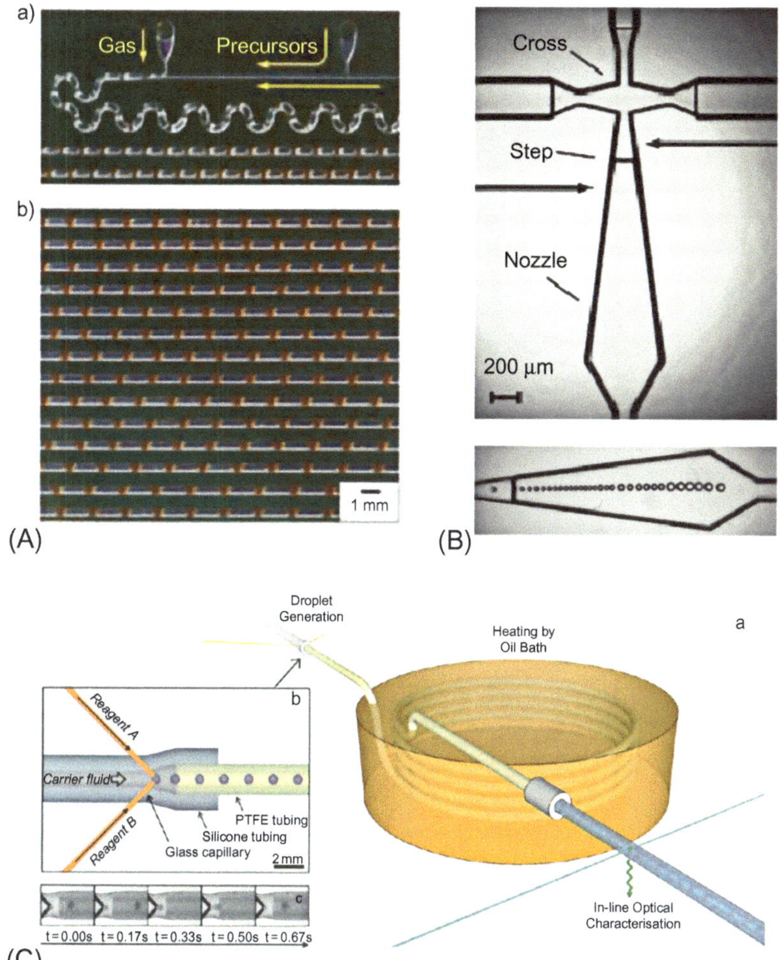

FIG. 1.8 Quantum dots synthesized in a segmented flow pattern. (A) A gas column reactor for synthesis of the CdSe at a high temperature. Reproduced from Ref. [100] with permission of John Wiley and Sons. (B) Microreactor channel design with droplet jet injector. The optical micrograph of droplet injection cross and the optical micrographs of droplet formation. Reprinted with permission from Ref. [101]. Copyright 2005 American Chemical Society. (C) Schematic of full system for CdSe nanoparticle synthesis and close-up of droplet generation stage. Reproduced from Ref. [46] with permission of The Royal Society of Chemistry.

densities for solubilizing a wide range of compounds of low to medium polarity, typical of liquids. Jensen's group [102] synthesized CdSe quantum dots under high temperature and high pressure conditions by the use of a low boiling point solvent (*n*-hexane) to form supercritical fluid (Fig. 1.9). Compared with a conventional high-boiling-point solvent, squalane, CdSe quantum dots synthesized

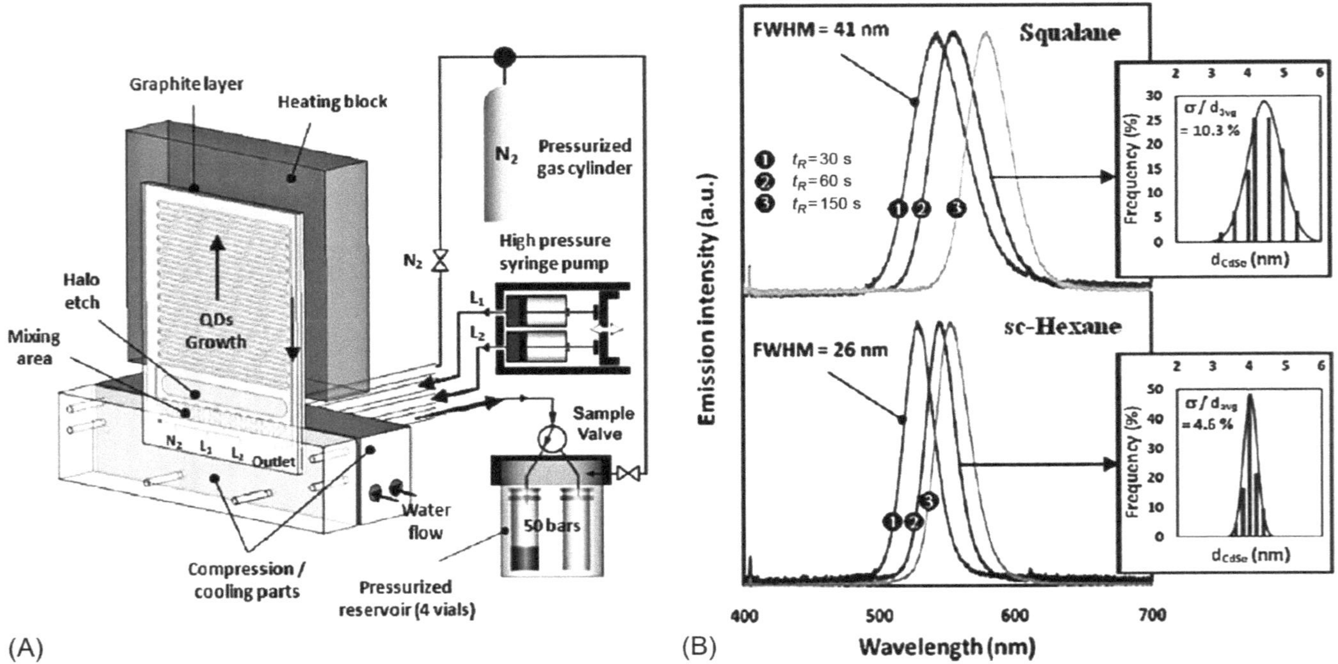

FIG. 1.9 Synthesis of CdSe quantum dots via sc-hexane. (A) Experimental setup including a high-pressure high-temperature microreactor, a compression-cooling aluminum part, a high-pressure syringe pump, a 5-way high-pressure valve, and a high-pressure reservoir containing four vials. (B) PL spectra at different residence times by the use of sc-hexane and squalene. *(Reproduced from Ref S. Marre, et al., Supercritical continuous-microflow synthesis of narrow size distribution quantum dots. Adv. Mater. 20 (2008) (24) 4830 with permission of John Wiley and Sons)*

via supercritical hexane (sc-hexane) have significantly narrower size distribution, which FWHM of fluorescence spectroscopy is 26nm.

Apart from oil-soluble quantum dots, the synthesis of water-soluble quantum dots has also been achieved in microchannels. Hung et al. [103] used a droplet flow reactor to synthesize water-soluble CdS quantum dots. $Cd(NO_3)_2$ and NaS aqueous solutions were respectively sheared by silicone oil to produce droplets containing precursors. A tapered chamber is designed to generate a velocity gradient and allow droplet approaching and fusing, and CdS quantum dots were synthesized at room temperature (see Fig. 1.10A). Due to the toxicity of cadmium, cadmium-free quantum dots have been studied extensively in recent years. Jensen's group [104] developed a continuous three-stage silicon-based microfluidic system consisting of mixing, aging, and sequential injection stages to synthesize InP quantum dots. Indium myristate $[In(MA)_3]$ was an indium precursor, and tris(trimethylsilyl) phosphine $[(TMS)_3P]$ was a phosphorus precursor. In most cases of InP nanocrystal synthesis, both the aging and sequential growth reactors are maintained at 320°C and 65 bars to utilize supercritical octane. Octane was selected as the solvent in order to provide excellent mixing, fast diffusivity, and sufficient density for the solubilization of all reagents. This is different from II-VI CdSe quantum dots, because InP molecular precursors are rapidly depleted regardless of temperature. Jensen's group speculated that InP nanocrystal growth is dominated by nonmolecular processes, such as the coalescence of particles and interparticle ripening (Fig. 1.10B). In further work [106], they found that trace amounts of water impurities in indium myristate precursors can negatively impact InP quantum dots' growth by limiting their size tunability because of the behavior to the hydroxylation effect of water. DeMello's group [105] reported the use of a droplet-based microfluidic platform for the synthesis of $CsPbX_3$ quantum dots. Full-color emitting $CsPbX_3$ quantum dots were obtained by control the molar ratios of Cs, Pb, and halide precursors, reaction temperatures, and reaction times (see Fig. 1.10C).

1.5 INORGANIC HYBRID NANOPARTICLES AND NANOCOMPOSITES

Hybrid nanoparticles and nanocomposites are composed of two or more materials. As the nanoparticles have small particle size, large surface area, high surface energy, high chemical activity and instability; surface atoms on nanoparticles have many unsaturated electrons or empty orbits, and will cause surface defects affecting the nanoparticles' nature. Hybrid nanoparticles and nanocomposites can mask surface defects, and offer higher nanoparticle stability. In addition, many hybrid rice nanoparticles and nanocomposites have the properties of two nanomaterials, and can be used as a multifunctional material. Therefore, hybrid nanoparticles and nanocomposites have great potential in applications. Microchannels can be easily added to the material, and can be

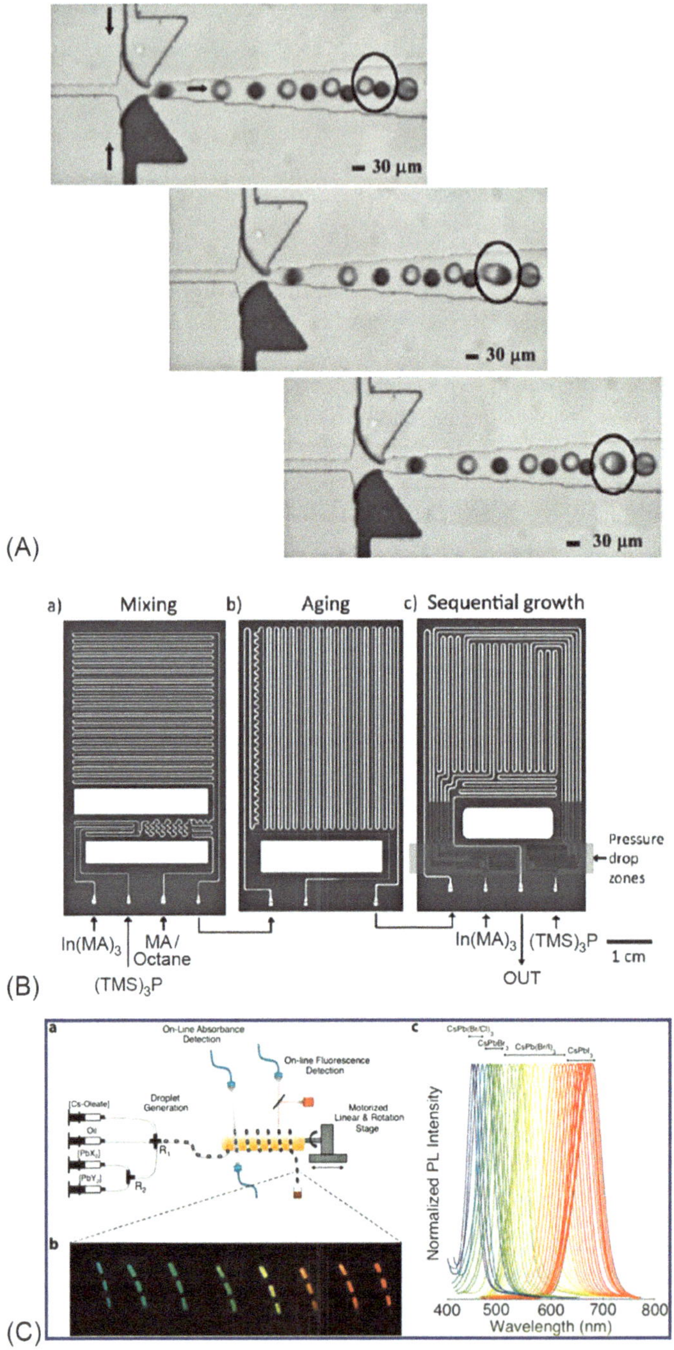

FIG. 1.10 Synthesis of water-soluble quantum dots and cadmium-free quantum dots in microreactors. (A) Dynamic fusion of alternating droplets in the synthesis of water-soluble CdS quantum dots. Reproduced from Ref. [103] with permission of The Royal Society of Chemistry. (B) Three-stage high-temperature and high-pressure microfluidic system of InP quantum dots with (a) a mixing stage, (b) an aging stage, and (c) a sequential injection microreactor with six additional injection channels. Reproduced from Ref. [104] with permission of John Wiley and Sons. (C) Illustration of the droplet-based microfluidic platform integrated with online absorbance and fluorescence detection for the synthesis and real time characterization of $CsPbX_3$ perovskite NCs. Reprinted with permission from Ref. [105]. Copyright 2016 American Chemical Society.

prepared in different hydraulics' structure of the emulsion. The use of microchannel synthesis of hybrid nanoparticles and nanocomposites has been extensively studied.

1.5.1 Metal Alloy Materials

Similar to the method of synthesizing different sizes of metal nanoparticles, researchers often synthesize a metal nanoparticle, and then add another metal precursor solution to prepare the core-shell structure of metal alloy materials [107].

Based on previous research, Knauer et al. [107] synthesized Au/Ag and Au/Ag/Au core-shell nanoparticles through a two-step droplet microreactor (Fig. 1.11A). In a typical synthesis process, the prepared Au nanocrystals were injected through channel A, and $AgNO_3$ was added to the B channel. Au/Ag nanoparticles could be obtained by reaction at 80°C. Then, Au/Ag nanoparticles were injected through channel A, and $HAuCl_4$ was added to channel B, and the Au/Ag/Au nanoparticles could be synthesized.

Jensen's group [51] synthesized a wide library of metallic nanostructures: monometallic, bimetallic, anisotropic growth, and heterostructures in a segmented microreactor (Fig. 1.11B). Some of the nanostructures had not reported, such as Pt-Ru, Pt-Ni and Pt-Conanodendrites, Pt-Pd heterostructures, Ag-Pd core-shell NPs, Au-Pd nanodumbbells and Au-Pd Nanosheets were obtained. The main strategy for preparing alloy nanoparticles with different morphologies and heterostructures was to control the reduction potential and nucleation rate through reaction temperatures, flow rates, concentrations and ratios of precursors, activities of reductants, and ligands. Taking Pt-Pd heterostructures, for example, Pt-Pd core–shell nanoparticles were obtained by using $NaBH_4$ at 130°C with residence time as short as 5 s. Pt with a reduction potential higher than Pd nucleated faster and Pd grew heterogeneously on the Pt cores. When ethanol was added to the system, ethanol would slow down the platinum nucleation, and Pt and Pd agglomerates were obtained. On the other hand, bimetallic single crystal PtPd nanodendrites could be synthesized by changing the reductant from $NaBH_4$ to a mild reducing agent, ascorbic acid.

1.5.2 Core-Shell Quantum Dots

Due to the presence of surface atom dangling bonds, quantum dots' surface defects can affect quantum yield. Preparation of core-shell quantum dots can passivate quantum dot surface defects, and improve the quantum yield and chemical stability. Meanwhile, cadmium-free shell material could reduce the toxicity of quantum dots. Therefore, most of the current applications were based on core-shell quantum dots.

Wang et al. [108] synthesized CdSe/ZnS core-shell quantum dots via a continuous flow reactor (Fig. 1.12A). Diethylzinc and bis(trimethylsilyl) sulfide

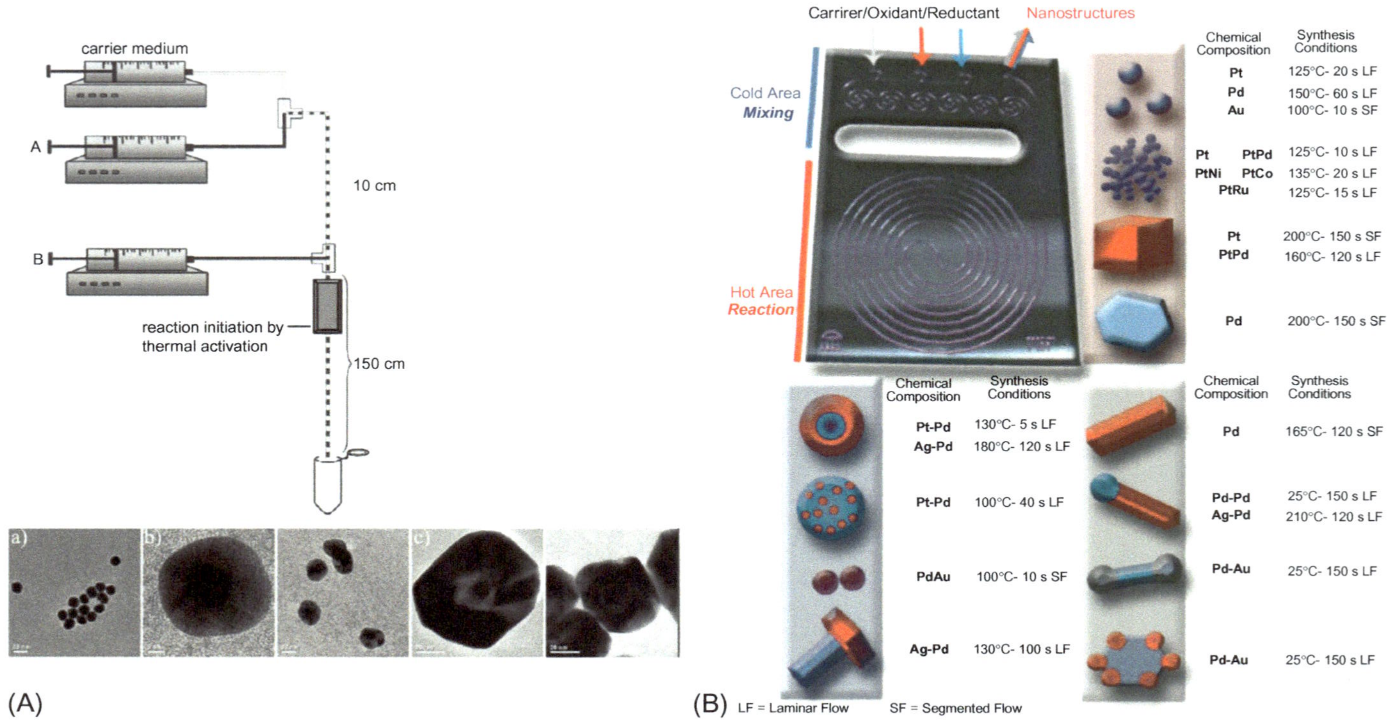

FIG. 1.11 Synthesis of metal alloy materials in microreactors. (A) General setup for micro-segmented flow synthesis of bimetallic multi-shell nanoparticles. HRTEM pictures of the three different particle classes: (a) Au seeds; (b) Au/Ag core/shell particles obtained by flow-through synthesis; (c) Au/Ag/Au double shell particles from flow-through synthesis. Reprinted from Ref. [107], Copyright 2010 with permission from Elsevier. (B) Silicon-Pyrex microfluidic reactor device used to synthesize the library of nanostructures. Schematic of the library of metallic nanostructures obtained and their synthesis conditions: monometallic and bimetallic spherical nanoparticles, nanodendrites, nanocubes, core-shell nanoparticles, Janus nanostructures, dumbbells, nanorods and nanosheets [51]. Published by The Royal Society of Chemistry.

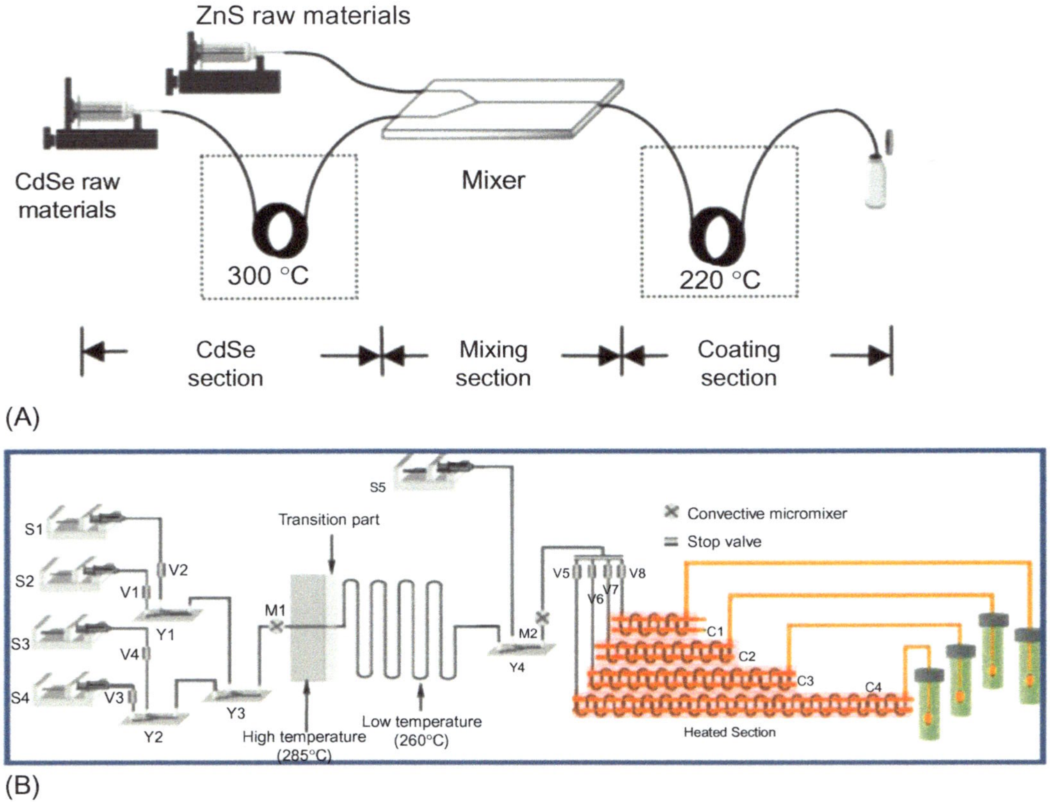

FIG. 1.12 Synthesis of core-shell quantum dots in microreactors. (A) Schematic illustration of the multi-step continuous synthesis system for core-shell CdSe/ZnS quantum dots. Reproduced from Ref. [108] with permission of The Royal Society of Chemistry. (B) Schematic of the microreaction system for the continuous synthesis of CdSe/ZnS and CdS/ZnS NCs. Reprinted with permission from Ref. [109]. Copyright 2009 American Chemical Society.

dissolved in TOP and TOPO were used as ZnS raw materials. There were three sections whose functions included the CdSe synthesis, mixing (with ZnS raw materials), and ZnS coating. The particle size and layer thickness can be directly adjusted by the residence time. Similarly, Yang et al. [109] used zinc diethyldithiocarbamate as ZnS precursors to continuous synthesis CdSe/ZnS and CdS/ZnS quantum dots (Fig. 1.12B). The best capping temperature was 140°C in their research.

1.6 ORGANIC HYBRID FUNCTIONAL NANOPARTICLES SYNTHESIS AND THEIR APPLICATIONS FOR DRUG DELIVERY

As described herein, compared with flask reactors, nanoparticles could be controllable and synthesized in microreactors due to their high mass and heat transfer efficiency, precise parameter adjustments, and many other advantages in microreactors. Many kinds of functional nanoparticles with different applications have been successfully synthesized in microreactors. In this section, we will briefly introduce the microfluidic synthesis methods of functional nanoparticles and their applications for drug delivery. In the Synthesis of Organic Nanoparticles by Microreactors section, we have introduced research on PLGA, alginate, chitosan, liposomes, and nanoparticles, and their potential applications in drug delivery. Compared with these non-shell nanoparticles, hybrid nanoparticles are more suitable for drug delivery, because the physical and chemical properties could be fine-tuned to fit different kinds of applications, such as tunable sizes, structures, surface modification, rigidity, and so on.

Lipid-polymer hybrid nanoparticles, which can deliver a wide range of drugs in a controlled manner, have been widely studied. Compared with liposome nanoparticles, lipid-polymer hybrid nanoparticles are more stable in blood and have a higher loading efficacy [110,111]. Size control and wide size distributions are the main challenges for lipid-polymer hybrid nanoparticles synthesis in flask reactors by emulsion methods. Kim et al. [112] synthesized lipid-polymer hybrid nanoparticles with superior reproducibility and homogeneity in a pattern-tunable microvortex platform (Fig. 1.13A). A three-inlet microfluidic channel, into which the organic solution of PLGA was injected to the central inlet, and the aqueous solution containing lecithin and DSPE – PEG was injected to the outer inlet, which generates microvortices at a relatively higher Reynolds number regime ($Re \sim 75$) than those used in conventional microfluidics ($Re \sim 1$). After rapid mixing of the streams, lipid-polymer hybrid nanoparticles were synthesized. The sizes of nanoparticles could be controlled by varying flow rates (i.e., Reynolds number). Feng et al. [113] developed a two-stage microfluidic chip for synthesizing the lipid-PLGA hybrid NPs. In the first stage, the PLGA core was formed by precipitation through a hydrodynamic flow focusing approach (Fig. 1.13B). In the second stage, a lipid solution was introduced and the lipid shell was adsorbed onto the surface of the PLGA core through hydrophobic interaction.

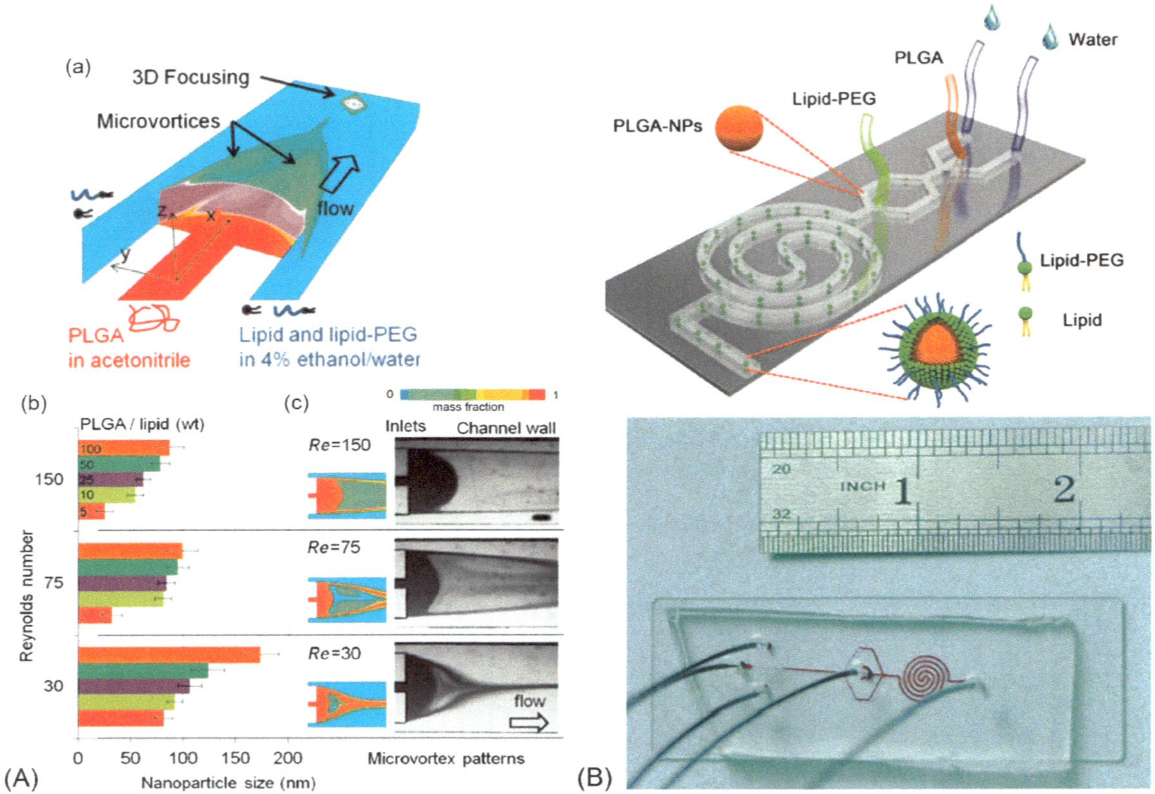

FIG. 1.13 Synthesis of lipid-polymer hybrid nanoparticles in microreactors. (A) Size-controllable nanoparticle syntheses with variation of the Reynolds number in controlled microvortex patterns. Reprinted with permission from Ref. [112]. Copyright 2012 American Chemical Society. (B) Schematic and photograph of the two-stage microfluidic chip for synthesizing the lipid-PLGA hybrid NPs. Reprinted from Ref. [113], with the permission of AIP Publishing.

Drug-polymer core-shell structure nanoparticles, which have superior colloidal stability, minimal off-target effects, as well as precisely controlled drug-release profiles, are widely used for drug delivery. Many kinds of drug-polymer core-shell structure nanoparticles have been synthesized through multistep-nanoprecipitation microfluidic devices. Liu et al. [114] successfully fabricated an optimal nano-in-nano vector, consisting of a drug (sorafenib, SFN, or itraconazole, ICZ) nanocrystal core with a polymer (folic acid conjugated spermine-functionalized acetalated dextran, ADSFA) shell (Fig. 1.14A). For the preparation of drug nanocrystal cores, the drug organic solution was pumped into the outer inlets and the nonsolvent (i.e., water) for drug molecules were added into inner inlet. The fast diffusion and mixing between two streams in the microreactor led to the formation of drug nanocrystals. After purification to remove the organic solvents and excess stabilizers, the drug nanocrystals were dispersed in an ethanol water solution (50%, v/v) containing ADSFA. The nonsolvent for both drug and polymer molecules was added into the outer inlet. Finally, the ADS-FA precipitated and deposited onto the surface of drug nanocrystals. Liu et al. [115] then developed a superfast sequential microfluidic nanoprecipitation device for fabrication of nanocomposites, consisting of a drug (sorafenib, SFN, or paclitaxel, PTX) nanocrystal core with a polymer (enteric coating polymer hypromellose acetate succinate, H grade fine powders, HF) shell (Fig. 1.14B). They used different solubility of drugs and polymers at different pHs to form drug core nanoparticles and drug-polymer core-shell hybrid nanoparticles. Specifically, in basic water (pH 10.5), the mixture fraction (ξ, $\xi = 1$ for the nonsolvent, and $\xi = 0$ for the solvent) for both PTX and SFN was around 0.5, while the mixture fraction for HF was around 0. In acidic water (pH 3.0), the mixture fraction value was 0.6. At first, a drug-acetone solution containing HF (Inner fluid 1) was mixed with a basic aqueous solution (pH 10.5, Inner fluid 2) and drug nanocrystal cores formed. Then, another acidic aqueous solution (pH 4, Outer fluid) was added for the HF precipitation to form drug-polymer core-shell hybrid nanoparticles.

1.7 CONCLUSIONS AND OUTLOOKS

Microfluidic devices have great potential to synthesize nanoparticles and nanocomposites, due to the efficient heat and mass transfers and spatial limitations in microfluidic devices, convenient parameter adjustments, and in situ process monitoring in microreactors. Many kinds of nanoparticles with controllable sizes, size distributions, and morphologies, such as organic nanoparticles, inorganic nanoparticles, and hybrid nanoparticles have been successfully synthesized in microfluidic devices. Both continuous flow reactors and segmented flow reactors have their own advantages and shortcomings in the synthesis of nanoparticles.

The microreactor offers an important means of studying nanoparticles. Depending on the conditions of the reactions, designing a suitable microreactor

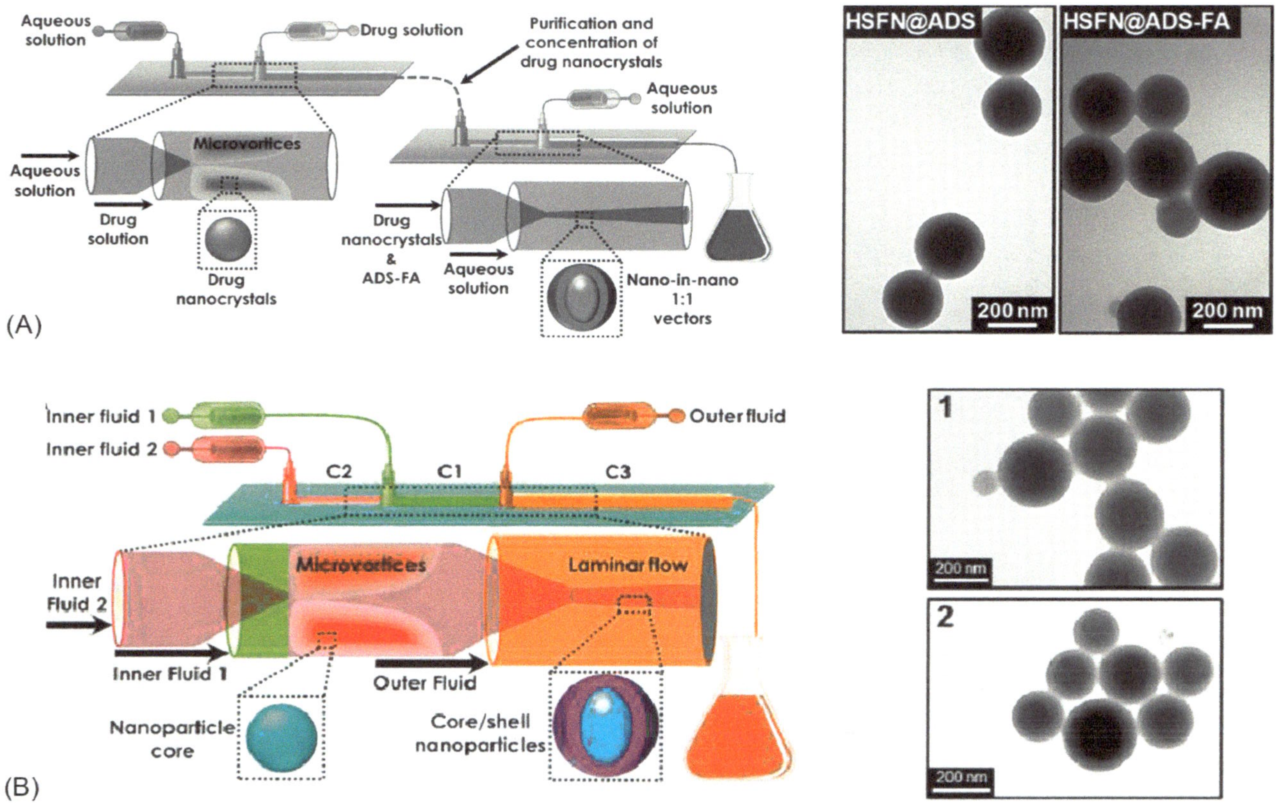

FIG. 1.14 Synthesis of drug-polymer hybrid nanoparticles in microreactors. (A) The multi-step microfluidic nanoprecipitation for HSFN@ADS-FA nanoparticles and the TEM images of HSFN@ADS and HSFN@ADS-FA nanovectors. Reproduced from Ref. [114] with permission of John Wiley and Sons. (B) A 3D glass capillary device to prepare structured core/shell nanocomposites. TEM images of PTX@HF (1) and SFN@HF (2). Reprinted with permission from Ref. [115]. Copyright 2017 American Chemical Society.

is the big challenge. In the course of subsequent research, separation of the nucleation and growth stage, studying the mechanism of nanoparticles' nucleation and growth mechanisms, synthesis of functional nanoparticles, in-situ purification and separation of nanoparticles, and other aspects of using microfluidic devices will have significant research value.

ACKNOWLEDGMENTS

The authors gratefully acknowledge the support of the National Natural Science Foundation of China (Grant Number: 21322604, 21476121) for this work.

REFERENCES

[1] Y. Shirasaki, et al., Emergence of colloidal quantum-dot light-emitting technologies, Nat. Photonics 7 (1) (2013) 13–23.
[2] J.A. Smyder, T.D. Krauss, Coming attractions for semiconductor quantum dots, Mater. Today 14 (9) (2011) 382–387.
[3] H. Zhang, et al., Shape-controlled synthesis of Pd nanocrystals and their catalytic applications, Acc. Chem. Res. 46 (8) (2013) 1783–1794.
[4] A.C. Chen, C. Ostrom, Palladium-based nanomaterials: synthesis and electrochemical applications, Chem. Rev. 115 (21) (2015) 11999–12044.
[5] P.V. Kamat, Quantum dot solar cells. semiconductor nanocrystals as light harvesters, J. Phys. Chem. C 112 (48) (2008) 18737–18753.
[6] R.S. Selinsky, et al., Quantum dot nanoscale heterostructures for solar energy conversion, Chem. Soc. Rev. 42 (7) (2013) 2963–2985.
[7] K. Saha, et al., Gold nanoparticles in chemical and biological sensing, Chem. Rev. 112 (5) (2012) 2739–2779.
[8] S.S. Agasti, et al., Nanoparticles for detection and diagnosis, Adv. Drug Deliv. Rev. 62 (3) (2010) 316–328.
[9] J. Wang, W. Gao, Nano/microscale motors: biomedical opportunities and challenges, ACS Nano 6 (7) (2012) 5745–5751.
[10] P.K. Jain, et al., Noble metals on the nanoscale: optical and photothermal properties and some applications in imaging, sensing, biology, and medicine, Acc. Chem. Res. 41 (12) (2008) 1578–1586.
[11] Y.N. Xia, et al., Shape-controlled synthesis of metal nanocrystals: simple chemistry meets complex physics? Angew. Chem. Int. Ed. 48 (1) (2009) 60–103.
[12] J. Park, et al., Synthesis of monodisperse spherical nanocrystals, Angew. Chem. Int. Ed. 46 (25) (2007) 4630–4660.
[13] M. Giersig, P. Mulvaney, Preparation of ordered colloid monolayers by electrophoretic deposition, Langmuir 9 (12) (1993) 3408–3413.
[14] M.J. Hostetler, et al., Alkanethiolate gold cluster molecules with core diameters from 1.5 to 5.2 nm: core and monolayer properties as a function of core size, Langmuir 14 (1) (1998) 17–30.
[15] P.J.G. Goulet, R.B. Lennox, New insights into Brust-Schiffrin metal nanoparticle synthesis, J. Am. Chem. Soc. 132 (28) (2010) 9582–9584.
[16] R. Sardar, J.S. Shumaker-Parry, 9-BBN induced synthesis of nearly monodisperse omega-functionalized alkylthiol stabilized gold nanoparticles, Chem. Mater. 21 (7) (2009) 1167.

[17] Z.L. Zhang, L. Guo, W.D. Wang, Synthesis and characterization of antimony oxide nanoparticles, J. Mater. Res. 16 (3) (2001) 803–805.
[18] Y.D. Wang, et al., Preparation of nanocrystalline metal oxide powders with the surfactant-mediated method, Inorg. Chem. Commun. 5 (10) (2002) 751–755.
[19] Y.H. Zheng, et al., Synthesis and magnetic properties of Fe3O4 nanoparticles, Mater. Res. Bull. 41 (3) (2006) 525–529.
[20] Z.X. Tang, et al., Size-dependent Curie-temperature in nanoscale Mnfe2o4 particles, Phys. Rev. Lett. 67 (25) (1991) 3602–3605.
[21] S.J. Chen, et al., Preparation and characterization of nanocrystalline zinc oxide by a novel solvothermal oxidation route, J. Cryst. Growth 252 (1–3) (2003) 184–189.
[22] G.Q. Zhang, X.W. Lou, General solution growth of mesoporous NiCo2O4 nanosheets on various conductive substrates as high-performance electrodes for supercapacitors, Adv. Mater. 25 (7) (2013) 976–979.
[23] D.R. Chen, X.L. Jiao, G. Cheng, Hydrothermal synthesis of zinc oxide powders with different morphologies, Solid State Commun. 113 (6) (1999) 363–366.
[24] Y.Y. Liang, et al., Covalent hybrid of spinel manganese-cobalt oxide and graphene as advanced oxygen reduction electrocatalysts, J. Am. Chem. Soc. 134 (7) (2012) 3517–3523.
[25] C.B. Murray, D.J. Norris, M.G. Bawendi, Synthesis and characterization of nearly monodisperse Cde (E = S, Se, Te) semiconductor nanocrystallites, J. Am. Chem. Soc. 115 (19) (1993) 8706–8715.
[26] C.B. Murray, C.R. Kagan, M.G. Bawendi, Synthesis and characterization of monodisperse nanocrystals and close-packed nanocrystal assemblies, Annu. Rev. Mater. Sci. 30 (2000) 545–610.
[27] L.H. Qu, X.G. Peng, Control of photoluminescence properties of CdSe nanocrystals in growth, J. Am. Chem. Soc. 124 (9) (2002) 2049–2055.
[28] W.W. Yu, X.G. Peng, Formation of high-quality CdS and other II-VI semiconductor nanocrystals in noncoordinating solvents: tunable reactivity of monomers, Angew. Chem. Int. Ed. 41 (13) (2002) 2368–2371.
[29] Z.A. Peng, X.G. Peng, Formation of high-quality CdTe, CdSe, and CdS nanocrystals using CdO as precursor, J. Am. Chem. Soc. 123 (1) (2001) 183–184.
[30] L.H. Qu, Z.A. Peng, X.G. Peng, Alternative routes toward high quality CdSe nanocrystals, Nano Lett. 1 (6) (2001) 333–337.
[31] D. Aherne, et al., From Ag nanoprisms to triangular AuAg nanoboxes, Adv. Funct. Mater. 20 (8) (2010) 1329–1338.
[32] L.L. Feng, et al., Well-controlled synthesis of Au@Pt nanostructures by gold-nanorod-seeded growth, Chem Eur J 14 (31) (2008) 9764–9771.
[33] Y.C. Tian, et al., Coupled composite CdS-CdSe and core-shell types of (CdS)CdSe and (CdSe) CdS nanoparticles, J. Phys. Chem. 100 (21) (1996) 8927–8939.
[34] X.B. Zhang, et al., Magnetically recyclable Fe@Pt core-shell nanoparticles and their use as electrocatalysts for ammonia borane oxidation: the role of crystallinity of the core, J. Am. Chem. Soc. 131 (8) (2009) 2778.
[35] F. Li, et al., Synthesis and characterization of ZnO-Ag core-shell nanocomposites with uniform thin silver layers, Appl. Surf. Sci. 256 (20) (2010) 6076–6082.
[36] M.S. Jin, et al., Synthesis of Pd nanocrystals enclosed by {100} facets and with sizes $<$ 10 nm for application in CO oxidation, Nano Res. 4 (1) (2011) 83–91.
[37] A.R. Rathmell, et al., The growth mechanism of copper nanowires and their properties in flexible, transparent conducting films, Adv. Mater. 22 (32) (2010) 3558.

[38] M. Rasheed, T. Wirth, Intelligent microflow: development of self-optimizing reaction systems, Angew. Chem. Int. Ed. 50 (2) (2011) 357–358.
[39] G.T. Liu, et al., Liquid-liquid microflows and mass transfer performance in slit-like microchannels, Chem. Eng. J. 258 (2014) 34–42.
[40] G.M. Whitesides, The origins and the future of microfluidics, Nature 442 (7101) (2006) 368–373.
[41] P. Yager, et al., Microfluidic diagnostic technologies for global public health, Nature 442 (7101) (2006) 412–418.
[42] J.B. Edel, et al., Microfluidic routes to the controlled production of nanoparticles, Chem. Commun. 10 (2002) 1136–1137.
[43] A.J. deMello, Control and detection of chemical reactions in microfluidic systems, Nature 442 (7101) (2006) 394–402.
[44] S. Krishnadasan, et al., On-line analysis of CdSe nanoparticle formation in a continuous flow chip-based microreactor, J. Mater. Chem. 14 (17) (2004) 2655–2660.
[45] K. Xu, et al., A region-selective modified capillary microfluidic device for fabricating water-oil Janus droplets and hydrophilic-hydrophobic anisotropic microparticles, RSC Adv. 5 (58) (2015) 46981–46988.
[46] A.M. Nightingale, et al., A stable droplet reactor for high temperature nanocrystal synthesis, Lab Chip 11 (7) (2011) 1221–1227.
[47] J.X. Ju, et al., Continuous synthesis of zeolite NaA in a microchannel reactor, Chem. Eng. J. 116 (2) (2006) 115–121.
[48] Z.H. Tian, et al., Intensification of nucleation stage for synthesizing high quality CdSe quantum dots by using preheated precursors in microfluidic devices, Chem. Eng. J. 302 (2016) 498–502.
[49] R. Knitter, M.A. Liauw, Ceramic microreactors for heterogeneously catalysed gas-phase reactions, Lab Chip 4 (4) (2004) 378–383.
[50] S.A. Khan, et al., Microfluidic synthesis of colloidal silica, Langmuir 20 (20) (2004) 8604–8611.
[51] V. Sebastian, K.F. Jensen, Nanoengineering a library of metallic nanostructures using a single microfluidic reactor, Nanoscale 8 (33) (2016) 15288–15295.
[52] R.K. Shah, et al., Designer emulsions using microfluidics, Mater. Today 11 (4) (2008) 18–27.
[53] L.Y. Chu, et al., Controllable monodisperse multiple emulsions, Angew. Chem. Int. Ed. 46 (47) (2007) 8970–8974.
[54] A.S. Utada, et al., Dripping, jetting, drops, and wetting: the magic of microfluidics, MRS Bull. 32 (9) (2007) 702–708.
[55] K. Jain, et al., Synthesis of nanoparticles in high temperature ceramic microreactors: design, fabrication and testing, Int. J. Appl. Ceram. Technol. 6 (3) (2009) 410–419.
[56] J.C. Love, J.R. Anderson, G.M. Whitesides, Fabrication of three-dimensional microfluidic systems by soft lithography, MRS Bull. 26 (7) (2001) 523–528.
[57] Y.N. Xia, G.M. Whitesides, Soft lithography, Annu. Rev. Mater. Sci. 28 (1998) 153–184.
[58] J.M. Kohler, S.N. Li, A. Knauer, Why is micro segmented flow particularly promising for the synthesis of nanomaterials? Chem. Eng. Technol. 36 (6) (2013) 887–899.
[59] C.X. Zhao, et al., Nanoparticle synthesis in microreactors, Chem. Eng. Sci. 66 (7) (2011) 1463–1479.
[60] A.M. Nightingale, J.C. de Mello, Microscale synthesis of quantum dots, J. Mater. Chem. 20 (39) (2010) 8454–8463.
[61] A. Gunther, K.F. Jensen, Multiphase microfluidics: from flow characteristics to chemical and materials synthesis (vol 6, pg 1487, 2006), Lab Chip 7 (7) (2007) 935.

[62] V.S. Cabeza, et al., Size-controlled flow synthesis of gold nanoparticles using a segmented flow microfluidic platform, Langmuir 28 (17) (2012) 7007–7013.
[63] X. Xie, et al., Microfluidic fabrication of colloidal nanomaterials-encapsulated microcapsules for biomolecular sensing, Nano Lett. 17 (3) (2017) 2015–2020.
[64] R. Karnik, et al., Microfluidic platform for controlled synthesis of polymeric nanoparticles, Nano Lett. 8 (9) (2008) 2906–2912.
[65] F.S. Majedi, et al., On-chip fabrication of paclitaxel-loaded chitosan nanoparticles for cancer therapeutics, Adv. Funct. Mater. 24 (4) (2014) 432–441.
[66] F.S. Majedi, et al., Microfluidic assisted self-assembly of chitosan based nanoparticles as drug delivery agents, Lab Chip 13 (2) (2013) 204–207.
[67] S. Bazban-Shotorbani, et al., Microfluidic directed synthesis of alginate nanogels with tunable pore size for efficient protein delivery, Langmuir 32 (19) (2016) 4996–5003.
[68] A. Jahn, et al., Microfluidic mixing and the formation of nanoscale lipid vesicles, ACS Nano 4 (4) (2010) 2077–2087.
[69] V. Kamat, et al., Synthesis of monodisperse chitosan nanoparticles and in situ drug loading using active microreactor, ACS Appl. Mater. Interfaces 7 (41) (2015) 22839–22847.
[70] J. Wagner, T.R. Tshikhudo, J.M. Koehler, Microfluidic generation of metal nanoparticles by borohydride reduction, Chem. Eng. J. 135 (2008) S104–S109.
[71] L. Zhang, et al., Continuous and scalable production of well-controlled noble-metal nanocrystals in milliliter-sized droplet reactors, Nano Lett. 14 (11) (2014) 6626–6631.
[72] S. Duraiswamy, S.A. Khan, Dual-stage continuous-flow seedless microfluidic synthesis of anisotropic gold nanocrystals, Part. Part. Syst. Charact. 31 (4) (2014) 429–432.
[73] L.L. Lazarus, et al., Two-phase microfluidic droplet flows of ionic liquids for the synthesis of gold and silver nanoparticles, ACS Appl. Mater. Interfaces 4 (6) (2012) 3077–3083.
[74] A. Knauer, et al., Microsegmented flow-through synthesis of silver nanoprisms with exact tunable optical properties, J. Phys. Chem. C 116 (16) (2012) 9251–9258.
[75] L. Xu, et al., Synthesis of copper nanocolloids using a continuous flow based microreactor, Appl. Surf. Sci. 355 (2015) 1–6.
[76] S. Biswas, et al., Developing a millifluidic platform for the synthesis of ultrasmall nanoclusters: ultrasmall copper nanoclusters as a case study, Small 8 (5) (2012) 688–698.
[77] Y.J. Song, et al., Investigations into sulfobetaine-stabilized cu nanoparticle formation: toward development of a microfluidic synthesis, J. Phys. Chem. B 109 (19) (2005) 9330–9338.
[78] S. Sharada, et al., Synthesis of palladium nanoparticles using continuous flow microreactor, Colloids Surf. A Physicochem. Eng. Asp. 498 (2016) 297–304.
[79] S.K. Lee, et al., Synthesis, assembly and reaction of a nanocatalyst in microfluidic systems: a general platform, Lab Chip 12 (20) (2012) 4080–4084.
[80] M.J. Hossain, et al., High-yield synthesis of PVP-stabilized small Pt clusters by microfluidic method, Catal. Today 183 (1) (2012) 101–107.
[81] L. Xu, et al., Synthesis of nickel nanoparticles by aqueous reduction in continuous flow microreacto(r), Chem. Eng. Process. 93 (2015) 44–49.
[82] S.A. Khan, K.F. Jensen, Microfluidic synthesis of titania shells on colloidal silica, Adv. Mater. 19 (18) (2007) 2556−+.
[83] H.Z. Wang, et al., Preparation of titania particles utilizing the insoluble phase interface in a microchannel reactor, Chem. Commun. (14) (2002) 1462–1463.
[84] M. Takagi, et al., Production of titania nanoparticles by using a new microreactor assembled with same axle dual pipe, Chem. Eng. J. 101 (1-3) (2004) 269–276.
[85] N.J. Carroll, et al., Droplet-based microfluidics for emulsion and solvent evaporation synthesis of monodisperse mesoporous silica microspheres, Langmuir 24 (3) (2008) 658–661.

[86] A. Abou Hassan, et al., Synthesis of iron oxide nanoparticles in a microfluidic device: preliminary results in a coaxial flow millichannel, Chem. Commun. (15) (2008) 1783–1785.
[87] T. Miyake, et al., Synthesis of Fe2O3 in the capillary-tube reactor, J. Mater. Sci. 40 (18) (2005) 5011–5013.
[88] W.B. Lee, et al., Biomedical microdevices synthesis of iron oxide nanoparticles using a microfluidic system, Biomed. Microdevices 11 (1) (2009) 161–171.
[89] L. Frenz, et al., Droplet-based microreactors for the synthesis of magnetic iron oxide nanoparticles, Angew. Chem. Int. Ed. 47 (36) (2008) 6817–6820.
[90] C.H. Choi, J.B. Levin, C.H. Chang, Continuous formation of a seed layer and vertical ZnO nanowire arrays enabled by tailored reaction kinetics in a microreactor, CrystEngComm 18 (44) (2016) 8645–8652.
[91] W.M. Yang, et al., High quantum yield ZnO quantum dots synthesizing via an ultrasonication microreactor method, Ultrason. Sonochem. 33 (2016) 106–117.
[92] S. Li, S. Meierott, J.M. Kohler, Effect of water content on growth and optical properties of ZnO nanoparticles generated in binary solvent mixtures by micro-continuous flow synthesis, Chem. Eng. J. 165 (3) (2010) 958–965.
[93] S.N. Li, et al., Local structure of ZnO micro flowers and nanoparticles obtained by microsegmented flow synthesis, ChemPhysChem 13 (6) (2012) 1557–1561.
[94] P.H. Hoang, H. Park, D.P. Kim, Ultrafast and continuous synthesis of unaccommodating inorganic nanomaterials in droplet- and ionic liquid-assisted microfluidic system, J. Am. Chem. Soc. 133 (37) (2011) 14765–14770.
[95] H. Nakamura, et al., Preparation of CdSe nanocrystals in a micro-flow-reactor, Chem. Commun. (23) (2002) 2844–2845.
[96] W.L. Luan, et al., Open-to-air synthesis of monodisperse CdSe nanocrystals via microfluidic reaction and its kinetics, Nanotechnology 18 (17) (2007).
[97] H.W. Yang, et al., Synthesis of monodisperse nanocrystals via microreaction: open-to-air synthesis with oleylamine as a coligand, Nanoscale Res. Lett. 4 (4) (2009) 344–352.
[98] Z.H. Tian, et al., Microfluidic synthesis of monodispersed CdSe quantum dots nanocrystals by using mixed fatty amines as ligands, Chem. Eng. J. 285 (2016) 20–26.
[99] J. Pan, et al., Automated synthesis of photovoltaic-quality colloidal quantum dots using separate nucleation and growth stages, ACS Nano 7 (11) (2013) 10158–10166.
[100] B.K.H. Yen, et al., A microfabricated gas-liquid segmented flow reactor for high-temperature synthesis: the case of CdSe quantum dots, Angew. Chem. Int. Ed. 44 (34) (2005) 5447–5451.
[101] E.M. Chan, A.P. Alivisatos, R.A. Mathies, High-temperature microfluidic synthesis of CdSe nanocrystals in nanoliter droplets, J. Am. Chem. Soc. 127 (40) (2005) 13854–13861.
[102] S. Marre, et al., Supercritical continuous-microflow synthesis of narrow size distribution quantum dots, Adv. Mater. 20 (24) (2008) 4830.
[103] L.H. Hung, et al., Alternating droplet generation and controlled dynamic droplet fusion in microfluidic device for CdS nanoparticle synthesis, Lab Chip 6 (2) (2006) 174–178.
[104] J. Baek, et al., Investigation of indium phosphide nanocrystal synthesis using a high-temperature and high-pressure continuous flow microreactor, Angew. Chem. Int. Ed. 50 (3) (2011) 627–630.
[105] I. Lignos, et al., Synthesis of cesium lead halide perovskite nanocrystals in a droplet-based microfluidic platform: fast parametric space mapping, Nano Lett. 16 (3) (2016) 1869–1877.
[106] L.S. Xie, et al., Effect of trace water on the growth of indium phosphide quantum dots, Chem. Mater. 27 (14) (2015) 5058–5063.
[107] A. Knauer, et al., Au/Ag/Au double shell nanoparticles with narrow size distribution obtained by continuous micro segmented flow synthesis, Chem. Eng. J. 166 (3) (2011) 1164–1169.

[108] H.Z. Wang, et al., Continuous synthesis of CdSe-ZnS composite nanoparticles in a microfluidic reactor, Chem. Commun. (1) (2004) 48–49.

[109] H.W. Yang, et al., Continuous synthesis of full-color emitting core/shell quantum dots via microreaction, Cryst. Growth Des. 9 (11) (2009) 4807–4813.

[110] J.M. Chan, et al., PLGA-lecithin-PEG core-shell nanoparticles for controlled drug delivery, Biomaterials 30 (8) (2009) 1627–1634.

[111] Y. Zhang, W.Y. Kim, L. Huang, Systemic delivery of gemcitabine triphosphate via LCP nanoparticles for NSCLC and pancreatic cancer therapy, Biomaterials 34 (13) (2013) 3447–3458.

[112] Y. Kim, et al., Mass production and size control of lipid-polymer hybrid nanoparticles through controlled microvortices, Nano Lett. 13 (10) (2013) 4997.

[113] Q. Feng, et al., Microfluidic based high throughput synthesis of lipid-polymer hybrid nanoparticles with tunable diameters, Biomicrofluidics 9 (5) (2015).

[114] D.F. Liu, et al., A nano-in-nano vector: merging the best of polymeric nanoparticles and drug nanocrystals, Adv. Funct. Mater. 27 (9) (2017).

[115] D.F. Liu, et al., Core/shell nanocomposites produced by superfast sequential microfluidic nanoprecipitation, Nano Lett. 17 (2) (2017) 606–614.

Chapter 2

Cluster Beam Synthesis of Polymer Composites with Nanoparticles

Vladimir N. Popok
Department of Materials and Production, Aalborg University, Aalborg, Denmark

Chapter Outline

2.1 INTRODUCTION

2.1.1 Functionalities of Polymers with Nanoparticles

In recent years, nanoparticles (NPs) of metals and semiconductors, as well as their oxides, have been widely used in research and industry for a variety of applications. New functionalities can be obtained if NPs are deposited on or embedded in a matrix, thus forming a nanocomposite. Polymers, as low-cost, flexible and easily processed materials, are very attractive hosting media for NPs [1–3]. By varying the metal filling factor, it is possible to tune electric

Polymer-Based Multifunctional Nanocomposites and Their Applications
https://doi.org/10.1016/B978-0-12-815067-2.00002-0

conductivity from electron tunnelling and variable range hopping (VRH) between individual NPs to metal-like conductance under particle percolation conditions, thus, achieving an insulator-to-metal transition (IMT) [4–9]. Due to the flexibility of polymers, it is possible to control the conductance in the films with fixed filling factor of NPs by simple stretching or contracting [10]. On the other hand, applying mechanical forces allows to change the conductance, and the thin metal-polymer films can be used in strain gauges [11].

Another attractive area of application is related to plasmonic properties of coinage metal NPs embedded into an organic matrix [12–16]. This gives rise to characteristic absorption and strong local field enhancement. These specific properties are widely used for optical sensing of gases, environmental species, various bio-objects, and food elements, as well as for surface enhanced Raman scattering (SERS) applications [17–22]. Recently, metal-polymer nanocomposites with stable plasmonic tuning under cyclic strain conditions have been reported [23]. Utilizing the localized surface plasmon resonance (LSPR) of metal NPs in polymers also allows designing the metamaterials with either perfect absorption of visible light [24], or full transparency in the visible interval of wavelengths [25]. The latter phenomenon can also be combined with high conductivity provided by the metal doping, thus representing a polymer-based composite enabling substitution of indium tin oxide. Utilization of optical properties of polymer composites with NPs is not solely based on LSPR. Adding NPs of Si, ZnS, and iron sulphides into polymer materials allows increasing the refractive index, thus making such composites promising for the formation of optical elements such as filters, reflectors, and waveguides [26,27]. Mixing the semiconductor nanocrystals representing quantum dots into polymers gives rise to photoluminescence for photonics applications [28–30]. NPs of ZnO and TiO_2 embedded into organic films demonstrate excellent photo-protective properties, including absorption of UV light [26].

The materials absorbing electromagnetic waves of GHz frequencies are of high interest for microwave technologies. For such applications, polymer coatings filled by composite NPs containing ferromagnetic materials demonstrate excellent properties, also adding such advantages as lightweight antioxidation and anticorrosion [26]. It is also found that polymer media with metal NPs of ferromagnetic materials can exhibit the magnetic properties that can be tuned between superparamagnetic and ferromagnetic ones, depending on the metal filling factor [31–35].

Recently, successful processes have been developed to add NPs to polymers in the synthesis of membranes used in filtration and separation processes. Incorporation of nano-sized materials leads to synergistic effects influencing permeability, selectivity, hydrophilicity, conductivity, mechanical strength, and thermal stability; as well as the antiviral and antibacterial properties of the polymeric membranes, thus allowing significant improvement in performance [36,37]. Embedding of metal NPs into polymers can also be used as a tool to study polymer properties; in particular, to measure glass transition temperature at the surface [38].

This overview on the different functionalities and applications of polymers with embedded NPs provides clear evidence that the composites can be tuned to very different properties, depending on particular needs. However, the fabrication requires specific methodological and technological approaches, as well as a thorough understanding of the physics and chemistry at the particle/polymer interfaces. Possible methods for particle preparation and deposition, or embedment into host organic matrixes, cover quite a number of different physical and chemical approaches.

2.1.2 Synthesis of Polymers with NPs

Polymer composites with NPs are often prepared using methods of wet chemistry [1,26,39]. In an ex situ approach, first the particles are synthesized by chemical methods; for example, through reduction of metal salts or decomposition of metal complexes, and then dispersed in a polymer solution [40,41]. To avoid aggregation, NPs can be functionalized or stabilized by organic coatings. An advantage of ex situ approaches is in the possibility of control of particle size and shape that is often required for optical and bio-related applications [42]. For in situ, NPs are directly synthesized in an organic matrix. One of the popular approaches is a sol-gel method. Both metal and metal oxide NP synthesis can be performed in parallel with the formation of the organic polymer [26].

Among the physical methods, vapor phase deposition is the most common approach, allowing particle formation on a substrate surface. This can be realized either by simple evaporation, or sputtering of a metal using magnetron techniques or laser ablation [3,43–45]. If NPs need to be spread in a volume, the polymer can be co-deposited by evaporation of monomers and polycondensation on the substrate, thermal cracking and repolymerization, or formation of so-called plasma polymers by RF sputtering [1,2,46,47]. Another possible approach is to thermally anneal a polymer with deposited NPs that favors the particle immersion [43,48]. The physics of this process will be described in more detail below. Recently it has been shown that the particle immersion can be also stimulated by post-implantation [49]. It is worth mentioning that the annealing and post-implantation allow embedding NPs only into a relatively thin sub-surface layer.

Ion implantation is another widely used approach to form NPs in dielectric media [4,12,50–53]. At high fluences, concentration of the embedded impurity overcomes the solubility limit in a polymer, thus leading to nucleation of NPs. With fluence increase, the implantation of metal ions can even result in a granular film formation in the near-surface layer [54]. Using ion implantation, one can embed practically any chemical element into any type of substrate, which is a great advantage. However, there are a couple of drawbacks to this method. One of them is in statistical distribution of the implanted impurity with depth and formation of NPs within a wide size range. Another problem that is, in particular, very significant for polymers, is radiation damage leading to structural and compositional evolution of material upon ion stopping. It can cause scission of polymer chains and cross-linking [55,56]. Breakage of chemical bonds can be followed by degassing of volatile compounds, thus leading to carbonization of the matrix [57,58]. Atmospheric oxygen can penetrate into the implanted layer and decorate the radiation damaged areas by making chemical bonds with radicals; for example, carbonyl groups [58,59]. Thus, the initial structure and composition of the polymer is significantly changed, and it can lose some of the properties attractive for practical applications.

The cluster beam technique has been known for a few decades as a versatile tool with a number of unique capabilities for materials synthesis and manipulation on the nano-scale [60–67]. The advantages of the method are in (i) a very good control of cluster composition because the particles are formed in a vacuum from ultra-pure targets, (ii) a possibility of cluster size (mass) selection, and (iii) tuning of cluster kinetic energy to adjust the mechanisms of cluster-matter interaction [68]. Recently, cluster beams have been applied for the synthesis of metal-polymer composites [13,69,70], and demonstrated great capabilities in the formation of materials with advanced properties [10,16,18,23,71–74].

The following sections present an overview on the main methods and techniques of cluster formation, as well as analysis on fundamental and applied aspects of the polymer composites with NPs synthesized by gas aggregation approach.

2.2 FORMATION OF CLUSTER BEAMS

2.2.1 Cluster Nucleation and Growth

Cluster formation requires a thermodynamic non-equilibrium condition that is realized in a cluster source. A cluster originates from an atomic or molecular vapor (atoms or molecules in a gas phase), and follows nucleation (initial condensation to form a cluster nucleus), growth (enlargement of a cluster by adding atoms or molecules), coagulation (merging of small clusters to form larger ones), and in some cases, coalescence (growth of larger clusters at the expense of small ones, Ostwald ripening) stages [75–77]. One can see the main steps in the schematic drawing in Fig. 2.1.

To describe nucleation, it is necessary to consider three-body collisions in a gas phase leading to the dimer formation.

$$A + A + B(E_1) \rightarrow A_2 + B(E_2 > E_1). \tag{2.1}$$

The third partner, B, has the role of fulfilling energy and momentum conservation; in particular it removes an excess energy (E), thus stabilizing the dimer, which serves as a nucleus for further cluster growth. To make the nucleation step more efficient, atoms of an inert carrier (cooling) gas typically play the role of the third partner. Further growth occurs by an incorporation of atoms,

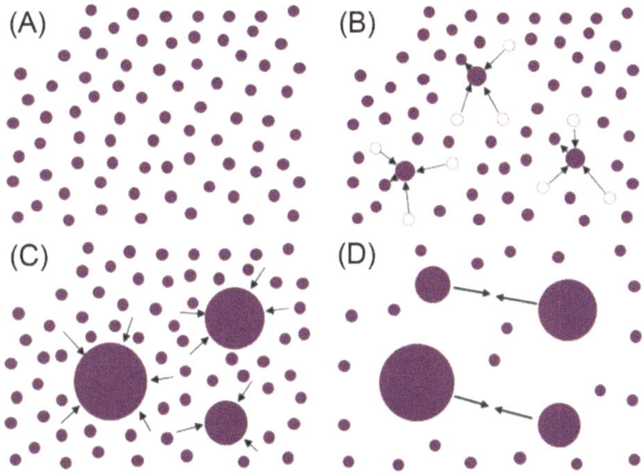

FIG. 2.1 Schematics of the main stages of cluster generation: (A) initial atomic or molecular vapor; (B) formation of cluster nuclei; (C) cluster growth; and (D) coagulation.

one at a time. Then, collisions of small clusters can lead to coagulation, yielding larger particles. This stage can further be followed by coalescence. A competing process to the generation is evaporation (loss of one or more atoms), or cluster decay, due to the excess energy (heat of condensation). Hence, for the formation of a stable beam, the clusters need to be cooled. More details about possible cooling mechanisms can be found elsewhere [78].

2.2.2 Cluster Sources

There are a number of approaches for a cluster beam formation [61,75,79–83]. The preferred method can be chosen depending on the type of cluster species, desired size of the particles, and other requirements of the beam properties. The most widely used techniques of cluster production are briefly overviewed in the following sections.

2.2.2.1 Evaporation Sources

The gas aggregation method, in which a solid or liquid is evaporated into a carrier gas and the atoms and molecules are forming clusters, is one of the simplest approaches [84,85]. After the aggregation, the clusters expand through a nozzle into the next vacuum chamber, forming either a subsonic or supersonic beam, depending on nozzle configuration and pressure difference between the regions separated by the nozzle. More details about supersonic (free-jet) expansion will be given in Section 2.2.2.3. Aggregation sources typically produce continuous beams of clusters made of materials with a relatively low melting point (<2000 K), usually metals. They can also be used to produce composite clusters where one of the components is a metal. The cluster size distribution depends on the physical dimensions of the cluster cell, and can reach many thousands of atoms [86,87]. This type of source has considerable flux, and it is useful for controllable deposition and growth of nanostructured thin films [85].

2.2.2.2 Surface Erosion Sources

In this type of approach, the material is vaporized from solid state targets using *heavy-ion beams*, *laser ablation*, *arc discharge*, or *magnetron sputtering*, and the clusters are formed by cooling the vapor using a carrier gas. Among these methods, the latter three are of most practical use, and will be briefly overviewed herein. For more details about ion sputtering cluster sources, one can look in [68,78] and the references therein.

Laser ablation sources are typically used to produce clusters of semiconductors and metals, especially those with high melting points that are hard to evaporate by heating. The vapor is formed by pulsed laser ablation of the material [79,88]. Typically, a beam of a solid state or excimer laser is focused on the rod or plate, which is driven in a slow motion so that a fresh area of its surface continues to be exposed to the laser. The ablated material (plume of plasma with temperature of around 10^4 K) is typically entrained in a pulse of the carrier gas for cooling and cluster nucleation [89]. The gas/cluster mixture is ejected from a nozzle forming a beam. The use of a laser leads to partial cluster ionization, thus, this source generates neutral and charged clusters. The cluster size distribution depends on the source conditions, and typically ranges from a few atoms up to a few hundred of them per cluster; see, for example, Refs. [90,91]. One of the advantages of laser ablation sources is an easy way to produce binary clusters (consisting of two different chemical elements) using binary alloy targets [92] or a dual-target configuration [93] or even a dual-target dual-laser technique [94]. One of the disadvantages of these sources is that the beam intensity is typically low.

One more approach is to vaporize material from the surfaces of a cathode and anode, between which an intense electrical discharge is ignited, and then provides conditions for cluster formation, which is realised in *pulsed-arc (arc discharge) cluster ion sources* [95]. The vapor is cooled down by means of a carrier gas similar to the preceding cases. Typically, about 10% of the emitted material is charged. In general, the cluster beams produced by such a source tend to be more intensive compared with laser ablation sources. However, the cluster production is limited to only well-conductive materials. A cluster deposition rate of about 5 nm/min over an area of ca. 50 mm^2 has been reported for a so-called pulsed microplasma cluster apparatus representing a modified arc discharge source [96]. Use of aerodynamic focusing coupled to supersonic expansion allows good control of cluster mass and spatial distribution [97].

Magnetron sputtering sources are currently the most widely used setups for the formation of metal and semiconductor clusters due to their applicability to a wide range of materials, and additional advantages that are outlined as follows.

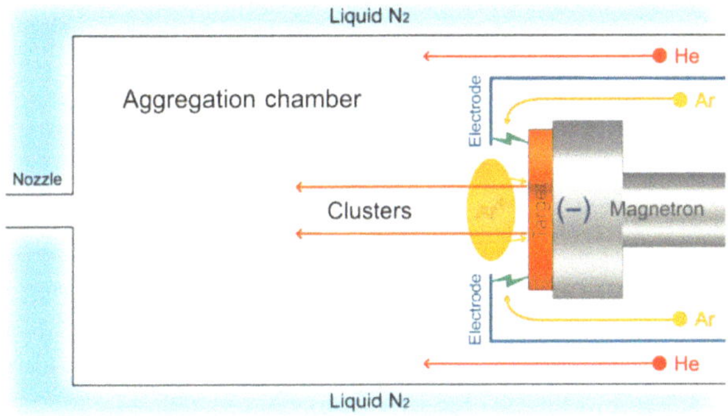

FIG. 2.2 Schematic overview of an aggregation chamber in a typical magnetron sputtering cluster source.

Therefore, this type of source is of special interest for synthesis of metal-polymer nanocomposites. The schematic picture is shown in Fig. 2.2. The sputtering is carried out by plasma that is ignited in an inert gas (usually argon) over a target surface by applying direct current or radio frequency (RF) potentials. Presence of a magnet behind the target helps to confine electron trajectories near the surface that promotes the ionization of plasma and, hence, intensifies the sputtering [98]. The cluster condensation is facilitated by a carrier gas (He in Fig. 2.2) that can be cooled down by collisions with the chamber walls where liquid nitrogen is supplied. The particle sizes can vary within a very wide range, from a few tens of atoms, to more than a million. A very significant fraction of clusters (which can reach 80% for certain species) is ionized [80]. The aggregated clusters are expanded through a nozzle. The source can provide a rather high cluster beam flux of up to 10^{12} clusters/cm^2. Combining the magnetron sputtering source with a novel time-of-flight mass filter demonstrated a capability to produce intense cluster beams with a constant mass resolution over a broad range of cluster sizes (from 2 to 70000 atoms) [99]. Another approach utilizing the electrostatic quadrupole mass selection showed filtering efficiency within 7%–10% of a relative standard deviation of particle diameter, which are very good values for many practical applications [16,100].

2.2.2.3 Supersonic (Free-Jet) and Matrix Assembly Sources

The aggregation of clusters can also be based on condensation of volatile precursors, such as atoms or molecules of gases or vapors in the adiabatically expanding nozzle flow [101]. This approach is often used to form clusters of inert gases [102,103]. Molecular clusters of N_2, O_2, CO_2, CF_6, and so forth can also be produced and used for collision or sputtering applications [102,104]. Free-jet expansion can be combined with the above-mentioned gas aggregation or sputtering approaches. In this case, metal or semiconductor clusters are seeded into the carrier gas at a stagnation pressure of several bars and expanded through a nozzle of appropriate shape, providing conditions for the supersonic beam formation [61].

At the end of this section, it is worth mentioning the very recently introduced new approach based on the assembly of metal clusters within a condensed matrix (*matrix assembly cluster source*). It uses an ion beam to enhance collisions between metal atoms in the matrix, and to sputter clusters forming a beam [105]. The source is capable of tuning the cluster sizes from a few to a few thousand atoms. In the prototype, a flux equivalent to nA current is reached, but it is claimed to be scalable to mA, giving great promise to promote cluster deposition systems toward industrial applications.

For many practical reasons, narrowing the mass distribution is desirable, and clusters of selected sizes are required. There are different approaches that can be used for mass (size) filtering, depending on the charge state of particles. Description of these methods is outside the scope of this chapter. Exhaustive information on commonly used mass-selecting techniques can be found, for example, in [61,63,68,75].

For convenience of use in the experiments, as well as for deposition on or implantation in the substrates, the clusters are typically collimated into beams that can be steered using either aerodynamic approaches, in the case of neutral particles, or ion lenses, together with electrostatic deflectors, in the case of cluster ions.

2.3 CLUSTER DEPOSITION/EMBEDMENT ON/IN POLYMERS

2.3.1 Fundamental Aspects of Nanoparticle Interaction with Polymer Surfaces

Atoms located at surfaces are characterized by reduced coordination numbers. They are missing a number of neighboring atoms, compared with the bulk case. This situation leads to additional energy, which is known as a surface energy, or surface tension (γ). It defines how the internal energy of the system is changed (∂U), if the surface area is increased or decreased (∂A), while keeping constant other parameters such as volume, temperature, and total number of atoms:

$$\gamma = \frac{\partial U}{\partial A}. \tag{2.2}$$

Common polymers have relatively low values of γ, typically a few tens of mJ/m^2, while the surface energy of bulk metals and semiconductors is one to two orders of magnitude higher [106–108]. Information about surface energies of metal and semiconductor nanostructures is a bit controversial. Theoretical models predict that γ of small clusters (fewer than approximately 100 atoms) of alkali and transition metals is considerably smaller than that of the bulk, while for larger particles, it should quickly approach the bulk values [109]. However, a number of experimental studies show that the surface energy of metal NPs can be much greater compared with the bulk, reaching values of a few J/m^2 [110–112]. Both assumptions that the surface energy of NP is either similar to that of bulk, or much greater, lead us to the conclusion that γ for the clusters with sizes on the scale of a few or few tens of nm should be much larger than that of typical polymers. This large difference in the surface energies can lead to two possible scenarios for NP deposited on a polymer: wetting by a thin layer of polymer, and immersion into the organic matrix. Both cases were found experimentally [16,113], as can be seen in Fig. 2.3, showing high resolution transmission microscopy images (HRTEM) for gold and silver NPs embedded into polymers.

The driving force for the immersion of an NP into polymer has a thermodynamics nature. In this approximation, it is considered that a particle is spherical in shape, and rigid compared with a host matrix, which is soft. Thus, the particle shape does not change under the embedding. It is worth noting that gravitational forces are disregarded in this case as a negligible factor. These conditions effectively describe a real situation of a metal NP on polymer. The NP immerses if the following condition is fulfilled [114]

$$\gamma_{NP} > \gamma_{pol} + \gamma_{NP/pol} \tag{2.3}$$

where γ_{NP}, γ_{pol} and $\gamma_{NP/pol}$ are the tensions at the interfaces of NP/air, polymer/air and NP with the polymer, respectively. The immersion is continued until the equilibrium condition is reached [115]

$$\gamma_{NP} = \gamma_{NP/pol} + \gamma_{pol}\, cos\varphi, \tag{2.4}$$

where φ is the contact angle as illustrated in Fig. 2.4. Thus, depending on the values of γ for the interfaces in Eq. (2.4), the particle can become either fully or partly embedded (to the depth z as shown in Fig. 2.4). More details on the process dynamics can be found in [115].

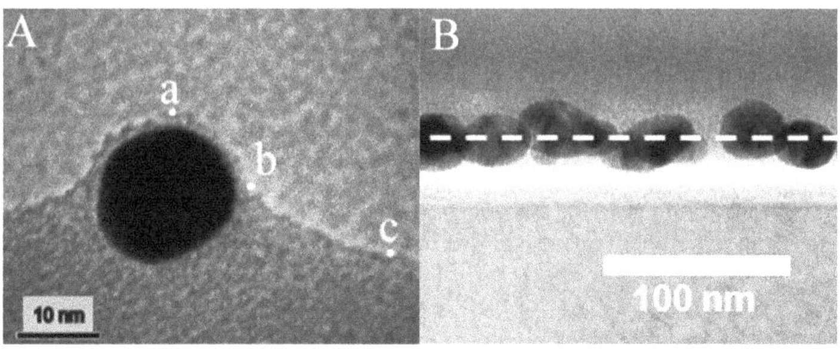

FIG. 2.3 HRTEM images of (A) gold NP embedded into polystyrene melt with wetting layer covering the particle and (B) silver NPs embedded into solid poly(methyl methacrylate) (PMMA) layer (bright stripe) using thermal annealing. Dashed line just provides visual guidance for the PMMA surface. *(Image (A) is reprinted with permission from R.D. Deshmukh, R.J. Composto, Direct observation of nanoparticle embedding into the surface of a polymer melt. Langmuir 23 (2007) 13169–13173. Copyright (2007) American Chemical Society.)*

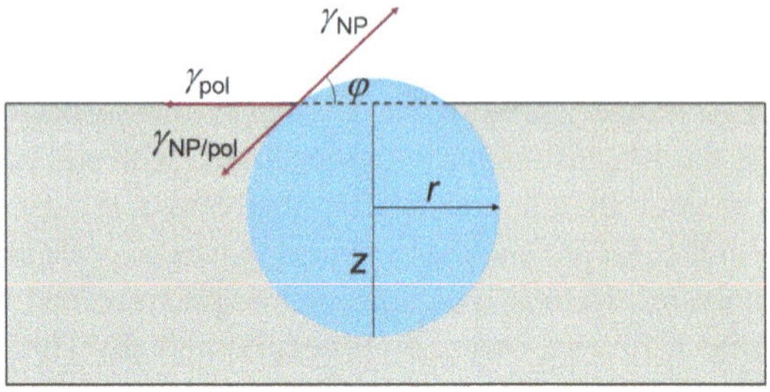

FIG. 2.4 Schematic image of a rigid nanoparticle embedded into soft material. See text for the details.

In practice, the particle embedment into a polymer requires creation of conditions for the immersion. The polymers must provide a cavity for NP. For that, the polymer chain mobility should be increased, which can be realized by heating at temperatures close to or above the glass-transition point, or even melting the polymer, as in the case described in [113] and shown in Fig. 2.3A. For the latter situation, the wetting layer creates capillary pressure on an NP. It was also shown in the corresponding model that the layer thickness is independent of the polymer molecular weight. Using this model, and the measured layer thickness, the energy at the NP/polystyrene interface was calculated to be as high as 8.7 J/m^2.

Embedding of metal NPs into polymers was also studied using X-ray and atomic force microscopy (AFM). In X-ray scattering experiments (see Fig. 2.5), the nominal gold layer thickness line was found to shift with the temperature increase. Since the depth distribution of gold clusters was consistent with the electron density profiles, the dependences allowed measurement of the particle immersion as a function of the polymer (polystyrene, PS) temperature [116,117]. It can be seen in Fig. 2.5B that the NPs embedment starts at a temperature of approximately 345 K, which is below the glass transition point of bulk PS, due to the increased mobility of the PS segments at the surface. This process was found to be strongly dependent on the amount of gold NPs. It was suggested that a short-range metal-polymer interaction can exceed the van der Waals forces and notably reduce the chain mobility, thus requiring higher temperatures for the embedment in the case of high cluster coverage.

AFM allows direct measurement of the cluster height towered above the surface. According to the study presented in [48], the start of the immersion process required a characteristic coating time to make a wetting layer covering the particles, which varied between approximately 3500–4800 s (ca. 60–80 min.) for different temperatures and cases of Ag or Au NPs either in PS or PMMA. Thereafter, the embedding occurred according to linear law with time. The velocity of immersion was observed to increase with temperature following the Arrhenius type of dependence, allowing calculation of the activation energies for particular combinations of metals and polymers, indicating that this energy primarily depends on the metal species (see Fig. 2.6) [48].

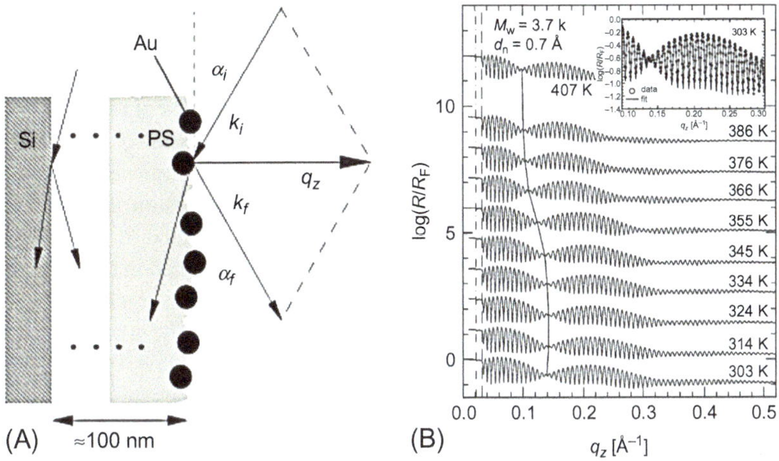

FIG. 2.5 (A) Sketch of a sample with Au NPs on PS layer covering a silicon substrate and X-ray scattering geometry for reflectivity measurements. k_i and k_f denote the incident and the final wave vector, respectively. q_z refers to the vertical component of the scattering vector. (B) X-ray reflectivity data of PS layer with gold NPs corresponding to nominal thickness of 0.7 Å at different temperatures. The inset shows magnification of the curve at 303 K. *(Reprinted with permission from R. Weber, I. Grotkopp, J. Stettner, M. Tolan, W. Press, Embedding of gold nanoclusters on polystyrene surfaces: influence of the surface modification on the glass transition. Macromolecules 36 (2003) 9100–9106. Copyright (2003) American Chemical Society.)*

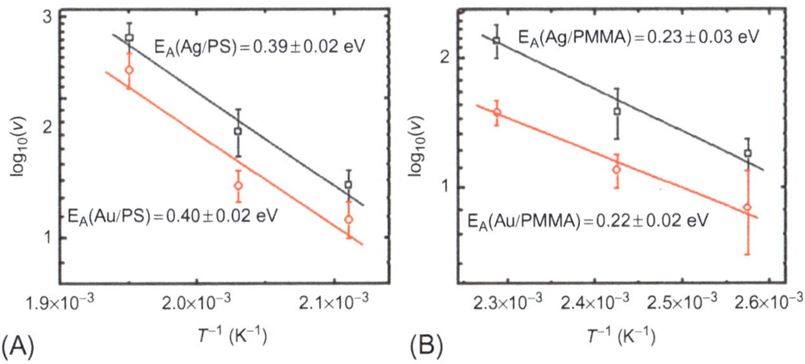

FIG. 2.6 Arrhenius plots of the embedding velocity as a function of annealing temperature for (A) Ag/Ps and Au/PS and (B) Ag/PMMA and Au/PMMA systems. The solid lines are exponential fits of the experimental data. *(Reprinted with permission from F. Ruffino, V. Torrisi, G. Marletta, M.G. Grimaldi, Effects of the embedding kinetics on the surface nano-morphology of nano-grained Au and Ag films on PS and PMMA layers annealed above the glass transition temperature. Appl. Phys. A Mater. Sci. Process. 107 (2012) 669–683.)*

However, quite different results were obtained for embedding of Au NPs (spheres of 10 and 20nm in diameter) into PS reported in [118]. These experiments showed that metal NPs started immersing into polymers to the depth of 3–4nm at temperatures below the bulk glass transition point, and in much faster regimes compared with those reported in [48]. It was suggested that the thin near-surface layer of polymer is characterized by higher molecular mobility and lower glass transition temperature. Later, experiments with immersion of NPs and relaxation of holes on the polymer surface confirmed this suggestion [119,120]. By other researchers [121], it has been demonstrated that for approximately 50nm thick spin-coated PMMA and PS films, the thickness of the mobile layer was around 6 and 14nm, respectively, near the glass transition point. Thus, embedding of a particle into polymer under annealing at $T > T_g$ can occur to the depth of approximately half of the diameter within just 10min (see Fig. 2.7). The results on immersion of particles deposited using the cluster beam technique, which are discussed in the next sections, also demonstrate fast embedding dynamics. The origin of the difference in the embedding time scales is not entirely clear. One of the reasons could be in the methods used for preparing the metal NPs. Silver and gold particles reported in [48] were formed using a sputter coater, thus, the clusters self-aggregated on the polymer surface in a variety of sizes, and they were not necessarily spherical in shape, while in the other experiments [118–120], the geometry and size of particles were well controlled.

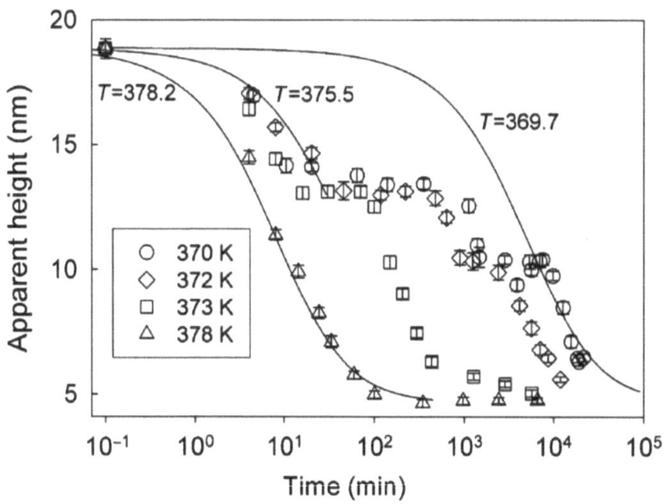

FIG. 2.7 Height of 20nm in diameter gold NP on PS film (100nm thick) versus annealing time at different temperatures. 378K is above T_g, 373K is close to T_g while the other temperatures are below, thus demonstrating different dynamics of embedding. *(Reprinted with permission from D. Qi, M. Ilton, J.A. Forrest, Measuring surface and bulk relaxation in glassy polymers. Eur. Phys. J. E 34 (2011) 56.)*

In the experiments on the investigation of surface glass transition temperature by embedding metal NPs into polymers, it was shown that the immersion strongly depends on the cluster radius [122]. Since the surface tension is a reciprocal function of the particle radius, smaller clusters are characterized by higher surface energy, thus increasing the driving force for embedding into the polymer. On the other hand, small clusters have less effect on the mobility of polymer chains. However, this effect needs to be considered carefully, because with an increase of surface coverage, even by small NPs, the properties do not reflect the free polymer surface, but rather a new composite material.

2.3.2 Deposition and Implantation of Clusters

The first experiments on the direct embedment (implantation) of metal clusters into polymers using the cluster beam technique were reported by the P. Milani's group [69]. Using the supersonic beam of gold clusters, they produced a shallow layer of PMMA filled with the metal NPs. TEM images of the polymer cross-sections with the clusters implanted at room temperature and 95°C are shown in Fig. 2.8. Despite the relatively low impact energy of 0.5 eV/atom (velocity of 1000 m/s), the particles were found to be dispersed in the layers to the depth of 50–70 nm. The authors suggested a couple of possible mechanisms capable of explaining the high penetration depth. One of them is the so-called clearing the way effect, according to which, the front atoms of the cluster push away the matrix atoms in collisions, thus reducing the stopping power (loss of energy per unit pass) for the rest of the atoms of the cluster, and increasing the penetration depth [123,124]. However, considering only this mechanism does not explain the deep cluster penetration for such low energy (0.5 eV/atom). Apart from a momentum driven effect, the cluster impact induces a local thermal and pressure spike in the target due to the multiple collision effect [64,125]. The used clusters had a mean diameter of 2.7 nm that corresponds to approximately 650

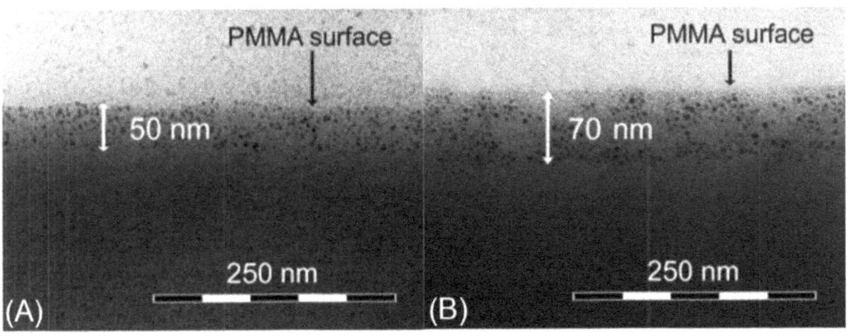

FIG. 2.8 TEM cross-sectional micrographs of PMMA with Pd nanoparticles deposited at (A) room temperature and (B) at 95°C. *(Reprinted with permission from L. Ravagnan, G. Divitini, S. Rebasti, M. Marelli, P. Piseri, P. Milani, Poly(methyl methacrylate)–palladium clusters nanocomposite formation by supersonic cluster beam deposition: a method for microstructured metallization of polymer surfaces. J. Phys. D. Appl. Phys. 42 (2009) 082002.)*

atoms in the spherical cluster approximation. Indeed, despite the low energy per atom, the cluster delivers quite a high total energy, very locally, and on the time scale of only a few tens of fs. The kinetic energy is partly converted into the thermal one heating the polymer matrix, thus facilitating the particle embedding by a local increase in chain mobility. At the same time, the reduced stopping power allows the cluster to penetrate to the significant depth. Hence, the thermal processes should play the key role in the cluster embedding, which is confirmed by the implantation experiments into PPMA films heated up to 95°C. The penetration depth of the palladium cluster increased from 50 to 70 nm [69] (see Fig. 2.8). It is very important to stress that due to the low impact energy per atom, which is below the cohesive energy for the cluster, the cluster constituents stay intact, thus, the particles preserve under the implantation.

This qualitative model has recently been confirmed by the simulations of gold cluster implantation into polydimethylsiloxane (PDMS) [126,127]. It was found that the impact of gold clusters of 3 nm in radius with energy of 2.0 eV/atom causes local heating along the track up to 600 K (see Fig. 2.9). This temperature wave is dumped while propagating from the track inside PDMS, but under a high-fluence cluster deposition (under conditions of high track density and short time intervals between successive impacts), the overall temperature of the near-surface layer can increase for about 50 degrees; decreasing the substrate density for about 5% [127]. These thermally-induced phenomena greatly facilitate metal cluster embedding into polymers. As shown experimentally, an exposure of the substrate to the nanoparticle beam with energy of 0.5 eV/atom for 3 s at a deposition rate of 0.02 nm/s (surface power density is of a few $\mu W/cm^2$) led to the gold cluster implantation into PDMS to the depth of up to 180 nm.

Recent experiments on implantation of metal NPs from supersonic cluster beams into various polymers have shown great capabilities in the formation of metal/polymer micropatterns; stretchable electrodes with high sustainability

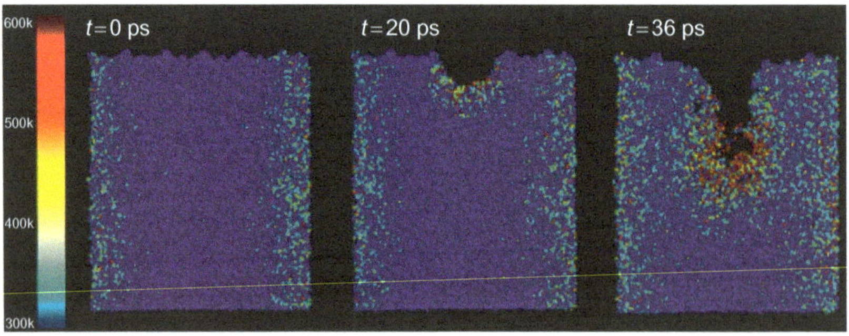

FIG. 2.9 Time evolution of temperature in PDMS during the implantation of Au cluster (radius of 3 nm) with energy of 2.0 eV/atom. *(Reprinted with permission from R. Cardia, C. Melis, L. Colombo, Neutral-cluster implantation in polymers by computer experiments. J. Appl. Phys. 113 (2013) 224307.)*

against large deformations; composite materials with controllable wrinkling, showing good perspectives to be used for bio-interfaces and sensors; passive electrical components on paper, as well as stretchable and tunable plasmonic films [10,23,73,74,128]. These application-oriented studies will be discussed in more detail in the next section.

Another possibility for the formation of organic composites with embedded NPs is co-deposition of clusters and polymer material. This can be realized by a combination of a cluster source providing NPs and a technique for a simultaneous plasma polymerization process. Thus, the clusters are deposited into the plasma polymer grown on a substrate by RF magnetron sputtering in a vacuum chamber [70,129–131]. Different polymers can be synthesized. For example, a C:H plasma polymer can be formed by sputtering of graphite in an Ar/*n*-hexane gas mixture, or C:H:N:O films can be synthesized using a nylon target and nitrogen as a carrier gas, or a $SiO_xC_yH_z$ polymer matrix can be produced by cracking hexamethyldisiloxane [13,132,133]. There are more routes for plasma polymer synthesis that can, for example, be found in [71], demonstrating a spectrum of possible polymer media for the formation of composites. Roughness of plasma polymers can be varied by controlling the deposition angle, allowing formation of smooth, granular, or even nanocolumnar films (see Fig. 2.10) [133]. Samples with increased roughness are of great interest in biomedical applications and sensor technologies.

Composite films of plasma polymers with different metal species were synthesized, and their properties studied. Deposition of Ti NPs simultaneously with particles of C:H polymer allowed formation of granular films with variable roughness, depending of the deposition times of both titanium and the polymer [70]. Because Ti is known to form oxides at ambient conditions very quickly, it was not a surprise to find that the films exposed to the atmosphere contain different TiO_x compounds with dominating fractions of TiO_2 (ca. 60% as found by X-ray photoelectron spectroscopy, XPS). Formation of Ti-C compounds (reaching approximately 10%) was also observed, evidencing titanium reactions

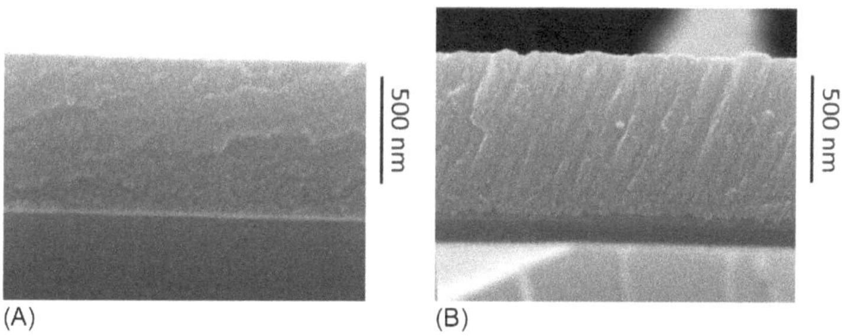

FIG. 2.10 Cross-sectional SEM images of RF magnetron sputtered nylon films deposited at angles of (A) 0 degree and (B) 80 degree. *(Reprinted with permission from O. Kylian, A. Shelemin, P. Solar, A. Choukourov, J. Hanus, M. Vaidulych, A. Kuzminova, H. Biederman, Plasma polymers: from thin films to nanocolumnar coatings. Thin Solid Films 630 (2017) 86–91.)*

with the elements composing the polymer. Aluminum and copper particles were found to be less reactive with oxygen, but they also gradually converted into oxides, or formed oxide shells under long-term exposure to ambient conditions [72,132]. For example, the Cu/O ratio reached 1:1 saturation after 7 days in air. This gradual oxidation typically changes electric and optical properties of the composite material, as will be discussed in more detail below. On the other hand, the embedded metal NPs typically do not show any considerable change of the polymer composition, except the cases of very high metal filling factors.

It is worth mentioning that NPs of different metals may have different geometrical shapes. It is well known that small clusters follow specific geometries yielding polyhedrons, which allow high packing densities. Some of these figures are close to spheres in shape; for example, icosahedrons having only hexagonal faces, while others may follow different polyhedral types, such as truncated decahedrons, or cuboctahedrons having square facets [134]. With an increasing number of atoms in the cluster, there is a tendency to change packing to that which is typical for bulk. There is no strict size limit for this transition, and it is very dependent on chemical elements. It can occur in the range between approximately a few hundred to a few thousands of atoms. Many studies showed that clusters of silver and copper on the scale larger than 4–5 nm typically represent the geometries close to spheres, and have fcc structure [13,72,100,135]. However, recent experiments with silver have shown that the clusters can still be represented by exotic 3D morphologies such as icosahedrons, decahedrons, and so forth, even if they reach a billion atoms [136]. The experiments on the deposition of Al and Ti NPs gave evidence that a high fraction of these particles have morphologies with hexagonal and rectangular facets, or can even be cubic (see Fig. 2.11). Cubic shapes were also found for iron clusters [137]. After embedding into polymers, the particles typically preserve these shapes.

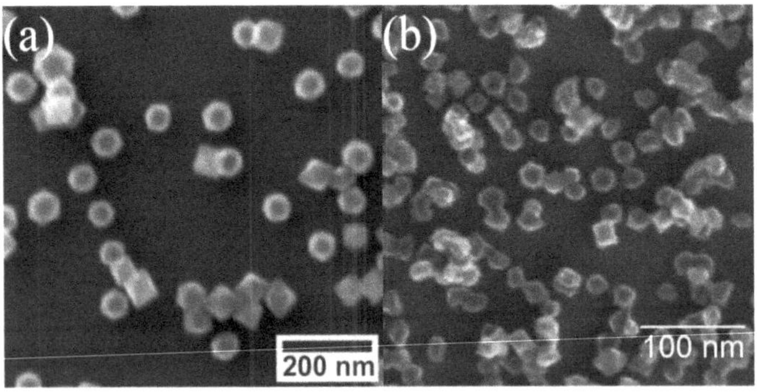

FIG. 2.11 SEM images of clusters produced by magnetron sputtering and gas aggregation: (A) Al NPs and (B) Ti NPs on Si substrates. *(Panel (A) is reprinted with permission from O. Polonskyi, O. Kylián, M. Drábik, J. Kousal, P. Solař, A. Artemenko, J. Čechvala, A. Choukourov, D. Slavínská, H. Biederman, Deposition of Al nanoparticles and their nanocomposites using a gas aggregation cluster source, J. Mater. Sci. 49 (2014) 3352–3360.)*

The last approach on synthesis of polymer films with embedded NPs, which is described in this chapter, is soft landing of particles from cluster beams followed by thermal annealing favoring immersion of NPs. It has been shown in the experiments with size-selected Ag and Cu clusters (mean sizes of approximately 13 and 15 nm, respectively) obtained by magnetron sputtering that after deposition on polymer films, they preserve almost spherical shapes and do not immerse at room temperature into the organic matrix, if it is properly solidified [16,138]. However, if the polymer film is prepared by a standard spin-coating procedure, and not annealed, the medium is of low density and inhomogeneous. For the deposition carried out on such a type of polymer matrix, the NPs become partly embedded already at room temperature [16] due to the driving forces described in the section about the fundamental aspects of immersion and easiness for NPs to find a cavity in a disordered structure. AFM images of silver NPs with corresponding height histograms for the cases of annealed and non-annealed PMMA films prepared by spin-coating are shown in Figs. 2.12 and 2.13, respectively. Post-deposition annealing at temperatures

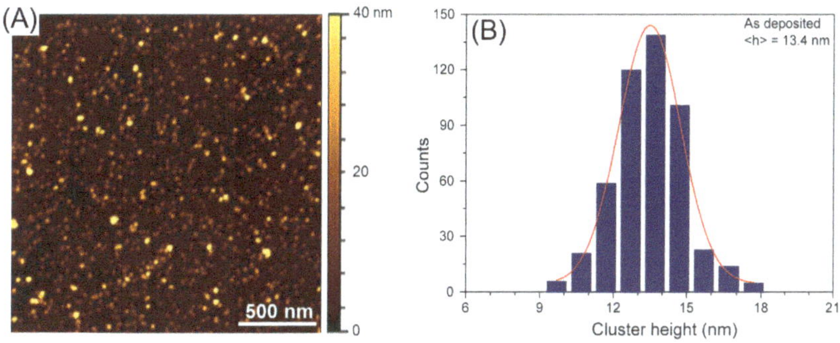

FIG. 2.12 AFM image and height histogram for Ag clusters deposited on PMMA annealed at 100°C for 10 min.

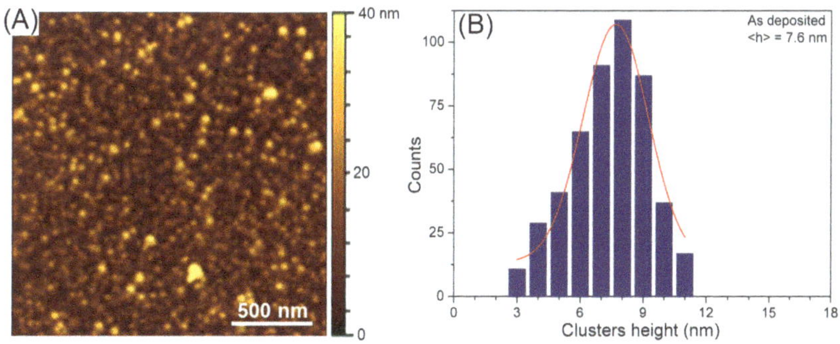

FIG. 2.13 AFM image and height histogram for Ag clusters deposited on non-annealed PMMA. The polymer film was only dried at room temperature for 12 h to evaporate the solvents.

close to or slightly above T_g leads to rapid (on the time scale of a few minutes) embedding of NPs. This immersion is proved not only by AFM measurements, but also by cross-sectional TEM imaging (see Fig. 2.14). Very similar short embedding times are also found for copper clusters deposited on the 50 nm thick PMMA films spin-coated on silicon or quartz [138].

Immersion of individual NPs can be monitored using special surface markers. Such experiments were carried out for size-selected copper clusters on PMMA [139]. Height analysis of the NPs by AFM (see examples of images in Fig. 2.15), indicated embedding for about 70%–75% of the diameter after 5 min of annealing at 125°C (T_g = 105°C). Longer annealing led to deeper immersion. However, the majority of particles were never fully embedded (even after about 2 h of annealing), having a small top part above the PMMA surface. Very similar trends of immersion are obtained on the size-selected Cu NPs deposited on the spin-coated PS films.

The preceding examples of relatively rapid immersion of NPs are quite similar to those described in [119], and controversial, compared with those presented in [48], where the embedding took around 1 h just to start. A possible

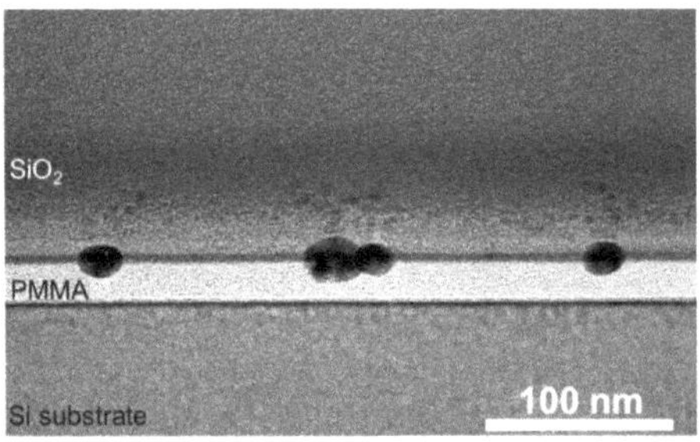

FIG. 2.14 TEM cross-sectional image of PMMA layer (bright stripe) with silver clusters after the thermal annealing at 125°C for 10 min.

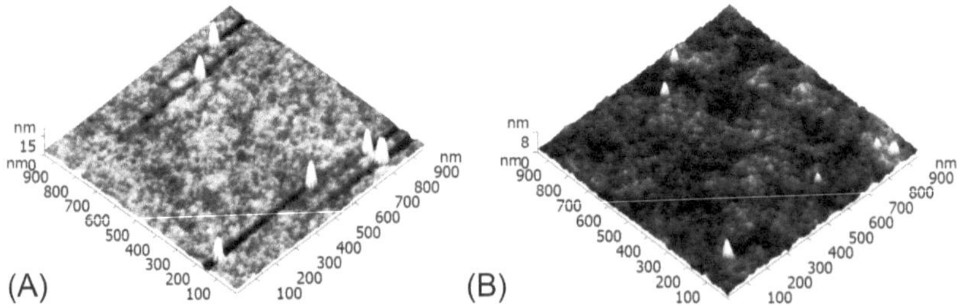

FIG. 2.15 AFM images of the same area of PMMA with (A) as deposited Cu NPs of mean height 17 nm and (B) after thermal annealing at 125°C for 5 min. Embedding of individual clusters can be clearly monitored by comparing panels (A) and (B).

reason could be related to the good control of size and spherical shape of particles prepared and deposited using the cluster beam technique.

An important point for a number of practical applications of polymer composites is a possibility to form arrays or patterned configurations of NPs. One simple approach is a patterning through a stencil mask. In [127], metal grids and meshes were used in front of PDMS films as masks during the cluster deposition, allowing formation of arrays of different geometrical configurations on the micrometer scale, as can be seen in Fig. 2.16. This approach can be used for the patterned cluster deposition in order to produce electrodes, electronic components, or optically active areas of desired configurations composed of individual clusters. These topics will be addressed in more detail in the next section.

To obtain arrays of NPs on the nanometer scale, cluster beam deposition can be combined with electron beam lithography (EBL) capable of writing patterns in thin polymer films. The approach is realized in [139] and schematically shown in Fig. 2.17, demonstrating (i) the EBL writing in the top polymer film to form a pattern that can be of a different geometrical configuration, (ii) cluster deposition across the entire surface, (iii) thermal annealing for partial embedding of clusters into polymers, and (iv) the chemical lift-off to remove the polymer between the patterned areas. This approach was recently applied, and linear stripes (ca. 200 nm in width) of size-selected (15 nm in diameter) Cu NPs were obtained on the polymer surface (see Fig. 2.18). The produced structures represent approximately two layers of clusters, with the bottom one partially embedded into the polymer film. These stripes show great stability against wet chemistry applied in the lift-off procedure, removing the rest of the polymer.

It is worth mentioning that modern EBL is capable of producing patterns of different 1D or 2D configurations, and the dimensions of elements can be decreased to approximately 10 nm; thus opening a practical way to combine EBL with cluster beam techniques for the formation of particle arrays, which can be used in nano-electronics and optics. For example, it has been declared that stripes of NPs are promising structures for SERS, wave guiding, and fabrication of strain sensors [140–142].

Another possibility for cluster array formation on polymers is the use of specific topography. This approach was realized in [122], where the silver clusters produced by magnetron sputtering were deposited on triblock copolymer poly (styrene-*b*-butadiene-*b*-styrene) (SBS). The surface of SBS has a typical stripe pattern, which can be seen in the AFM image presented in Fig. 2.19A. After the deposition, the clusters were tended to self-order along the stripes, as shown in the TEM image presented in Fig. 2.19B.

2.4 PROPERTIES OF POLYMER COMPOSITES WITH NANOPARTICLES

The presence of NPs on a polymer significantly changes the characteristics of surface and near-surface layers. It affects the polymer chain mobility and glass

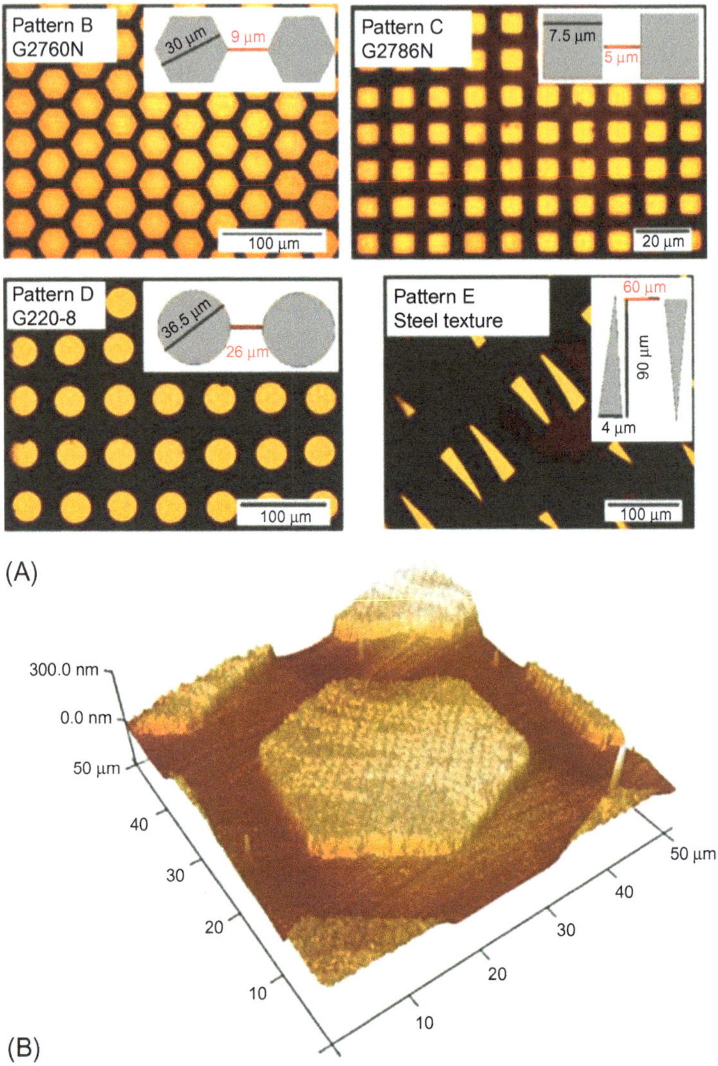

FIG. 2.16 (A) Optical microscope images of gold cluster patterns obtained on PDMS using four different stencil masks (details for the mask dimensions are also given) and (B) 3D AFM image of individual hexagon obtained using pattern B shown in panel (A). *(Reprinted with permission from C. Ghisleri, F. Borghi, L. Ravagnan, A. Podests, C. Melis, L. Colombo, P. Milani, Patterning of gold-polydimethylsiloxane (Au-PDMS) nanocomposites by supersonic cluster beam implantation. J. Phys. D. Appl. Phys. 45 (2014) 015301.)*

transition point [117,143]. NPs covering the polymer surface also change the wettability. Plasma polymers coated with Cu and TiO_x clusters demonstrated very small water contact angles (see Fig. 2.20), or even super-hydrophilic character [70,72]. On the other hand, the experiments on the implantation of gold

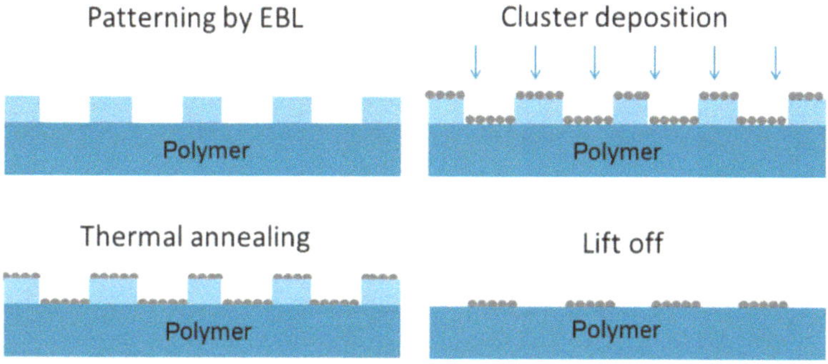

FIG. 2.17 Schematic picture of the steps in formation of ordered arrays of NPs on thin polymer films, including formation of a pattern by electron beam lithography, cluster deposition, and thermal annealing providing partial embedding of NPs into polymer and lift-off.

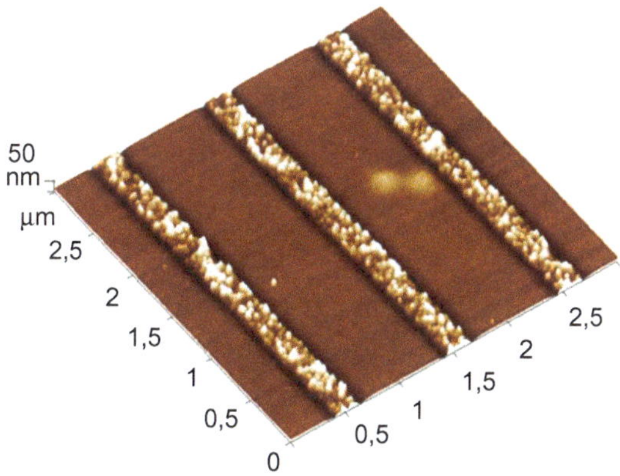

FIG. 2.18 3D AFM image of Cu NP stripes on polymer.

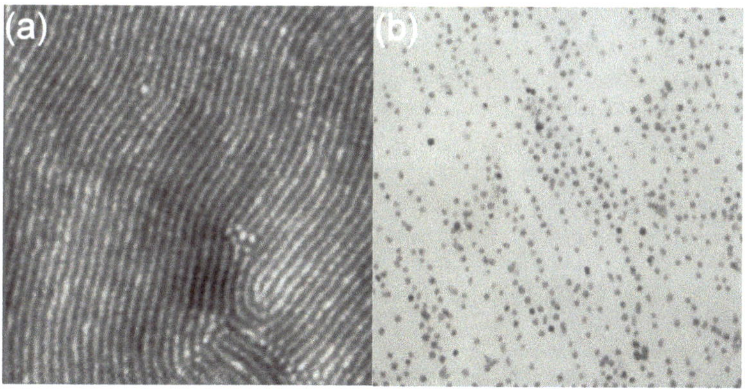

FIG. 2.19 (A) AFM image (1.5 × 1.5 µm) of patterned SBS polymer template and (B) TEM image (680 × 680 nm) of silver clusters deposited on such template (mean cluster size is 10 nm). *(Reprinted with permission from C.H. Xu, X. Chen, Y.J. Liu, B. Xie, M. Han, F.Q. Song, G.H. Wnag, Enhanced thermal stability of monodispersed silver cluster arrays assembled on block copolymer scaffolds. Nanotechnology 21 (2010) 195304.)*

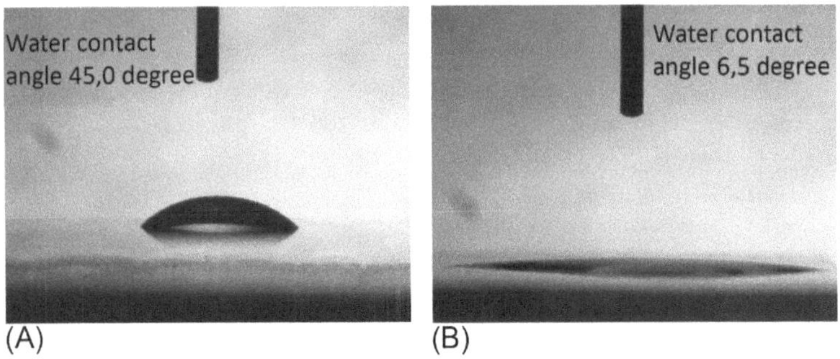

FIG. 2.20 Optical images of water droplets deposited on (A) plasma sputtered nylon and (B) nylon composite with Cu NPs. *(Reprinted with permission from O. Kylian, J. Kratochvil, J. Hanuš, O. Polonskyi, P. Solař, H. Biederman, Fabrication of Cu nanoclusters and their use for production of Cu/plasma polymer nanocomposite thin films. Thin Solid Films, 550 (2014) 46–52.)*

clusters into PDMS showed no change in the contact angle [127]. The surface roughness of the composites with NPs is a very important parameter, by which one can control to some extent the wettability. The roughness is an extremely important parameter for adhesion and growth of biological cells, which will be discussed in more detail below.

Filling polymer films with NPs leads to the evolution of mechanical, electronic, magnetic, and optical properties of the formed composites. The filling factor or ratio of metal in the organic matrix can affect the continuity of the polymer film. High filling factors under cluster implantation can cause reorganization of polymeric chains, termination of links between them, and swelling of the layer, which would lead to corrugation of the surface layer and formation of cracks [127]. These phenomena need to be carefully considered in practical applications; for example, in formation of electrodes or electronic components on polymer surfaces by deposition or implantation of clusters, because the discontinuity or cracking of the surface layer reduces reliability of the fabricated structures.

Recently, it has been shown that the buried layer of metal NPs formed in polymers films by cluster implantation can be transformed into a wrinkled surface structure [74]. Gold clusters were implanted into the near-surface layer of PS, and the film was subjected to anisotropic thermal shrinking at 160°C. This treatment was found to cause anisotropic surface patterns. Cross-sectional TEM images of the samples before and after the shrinking are shown in Fig. 2.21. By changing the implantation dose, it was possible to control the evolution of wrinkling and electrical conductivity of the composite layer, thus showing potential technological applications in catalysis, sensing, formation of biointerfaces, and so forth.

Change of conductance of a polymer layer by increasing the filling factor of metal NPs is another important property, which, unfortunately, is not well

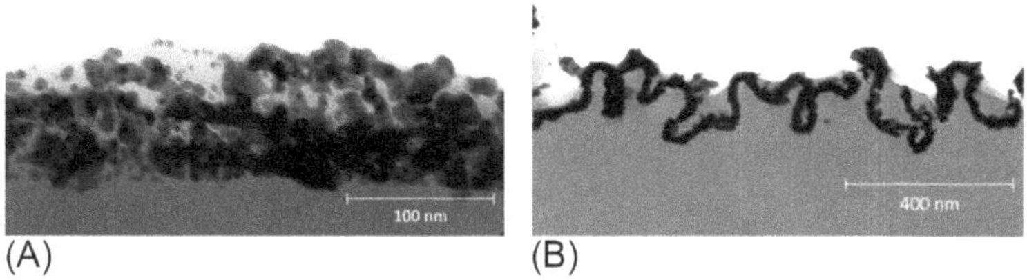

FIG. 2.21 Cross-sectional TEM images of Au NPs/PS composites for (A) as-implanted film and (B) after anisotropic shrinking. *(Reprinted with permission from F. Greco, A. Bellacicca, M. Gemmi, V. Cappello, V. Mattoli, P. Milani, Conducting shrinkable nanocomposite based on Au-nanoparticle implanted plastic sheet: tunable thermally induced surface wrinkling. ACS Appl. Mater. Interfaces 7 (2015) 7060–7065. Copyright (2015) American Chemical Society.)*

studied for the cluster beam synthesis of polymer composites. It is known for the case of conventional ion implantation that the high-fluence embedding of metal leads to the nucleation of NPs, and the conductance increases following the evolution of charge carrier transport mechanisms from VRH of electrons toward formation of percolation paths, and finally to an IMT [32]. One such example is shown in Fig. 2.22, where, with an increase of Fe^+ ion fluence, the temperature

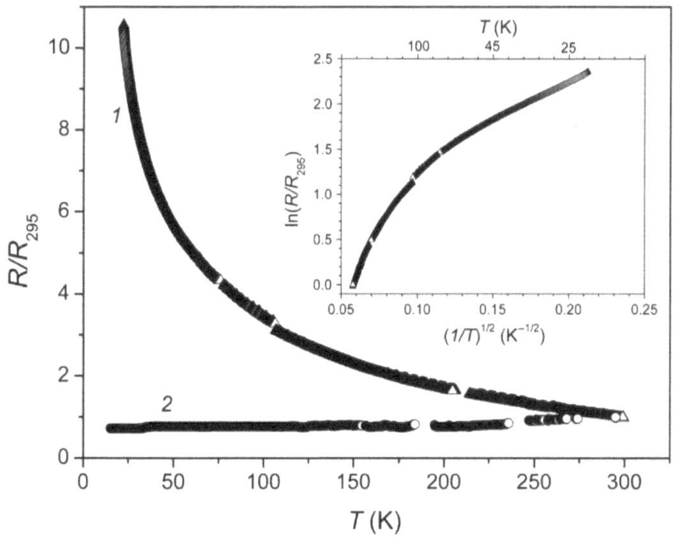

FIG. 2.22 Temperature dependence of relative resistance (with respect to room temperature value R_{295}) for PET implanted by Fe^+ ions with fluences of (1) 0.75 and (2) $1.0 \times 10^{17}\ cm^{-2}$. Dependence (1) is also presented in coordinates $\ln(R/R_{295}) - (1/T)^{1/2}$ in insertion. *(Reprinted from M.G. Lukashevich, V.N. Popok, V.S. Volobuev, A.A. Melnikov, R.I. Khaibullin, V.F. Valeev, A. Wieck, V.B. Odzhaev, Magnetoresistive effect in PET films with iron nanoparticles synthesised by ion implantation. Open Appl. Phys. J. 3 (2010) 1–5.)*

dependence of resistance of the implanted polyethylene terephthalate (PET) changes the slope from "negative" ($\ln(R/R_{295}) \sim T^{-1/2}$), which is typical for the "dielectric" side of the transition, to a "positive" one, evidencing the metallic character of conductance [144]. Formation of Pt porous films by cluster deposition showed a very similar evolution in conductivity (resistivity ρ), in particular, from VRH characterized by the dependence $\ln(\rho/\rho_{300}) \sim T^{-1/2}$ to the metallic conductance with a "positive" slope of the dependence [145].

2.5 APPLICATIONS OF NANOCOMPOSITE POLYMER FILMS

2.5.1 Formation of Electronic Components

As follows from the previous section, implantation of metal NPs into the near-surface polymer layer can be used for the formation of conductive structures; for example, to form elastomeric electrodes. To realize this idea, gold clusters were embedded into the near-surface layer of PDMS and PMMA to the concentrations providing the transition from insulating to conductive behavior of the composite through the percolation of metal NPs [10]. The performance of gold/PDMS films was tested against uniaxial strain cycles, demonstrating that the resistance increases for an order of magnitude by the applied tensile strain, but it recovers when the strain is removed. Repeating the load/unload cycles up to 50,000 times led to a decrease of the difference in resistance between the load and relaxed polymer films. An increase of resistance can be easily explained by the nanocomposite nature of the films: stretching of the films leads to an increase of the distances between NPs, and termination of some percolation paths; while under released strain, the paths are recovered. Repetition of the cycles causes reorganization of NPs in the medium leading to a percolating network less affected by the strain, thus decreasing the resistance difference. Electrical properties of these nanocomposite films are remarkably better compared with those of metal layers evaporated on polymers, which demonstrate degradation (decrease) of conductance under the repeated strains due to cracking, leading to permanent breakage of the metal layer continuity.

Closely spaced chromium NPs deposited between silver interdigital electrodes fabricated on the PET film were shown to be utilized as strain sensors [142]. By tuning the cluster surface coverage, one can control the electrical conductance, and form the network of clusters in which the electron transport is based on the formation of percolation paths similar to the above-mentioned elastomeric electrodes. A strain applied to such film changes the interparticle distance and, thus, the resistance. The sensors have shown a gauge factor of 20 compared with the typical value of 2 for the metal foil-based gauges for small applied strains $\varepsilon \sim 0.00\%–0.36\%$. For $\varepsilon \sim 0.4\%$, which is about the crack limit of semiconductor gauges, the NPs-based devices demonstrated a gauge factor of 100. One more advantage of these sensors is in the easy way to scale down the elements, which can be advantageous for MEMS and NEMS applications.

Nowadays, paper electronics have received a new impulse, mostly due to the possibility of low-cost solutions, the high speed of device fabrication, and an easy way to recycle these components after use [146]. Easy cutting and perforation of paper is an advantage in fabrication of simple passive components. On the other hand, the possibility of folding, unfolding, and creasing for storage in small spaces leads to so-called origami and kirigami electronics, with more complex paper-based devices such as antenas, generators, fuel cells, and pressure sensors [147–149]. Recently, cluster beam deposition has been successfully tried in the fabrication of passive electronic components (resistors and capacitors) on paper [128]. In Fig. 2.23A, one can see a set of resistors (with different resistance values) printed in a one-step process with gold NPs on plain paper using a shadow mask. The process took only 30 min and the devices were found to have sharp borders (see Fig. 2.23B). An advantage of the cluster beam technique is in the wide choice of species (both metals and semiconductors) for deposition. The cluster surface coverage is easily tuned through the deposition time, thus providing the possibility to adjust mean distances between NPs and control electrical conductance through percolative regimes, which is impossible for atom-assembled films produced by evaporation or ink jet printing.

2.5.2 Nanocomposites in Optics and Photovoltaics

As mentioned in the Introduction, NPs of coinage metals deposited on, or embedded in a dielectric matrix are very attractive for nano-optics or optics-related applications, due to the phenomenon of LSPR. The cluster beam technique is a very efficient tool for embedding metal NPs into polymer films, thus

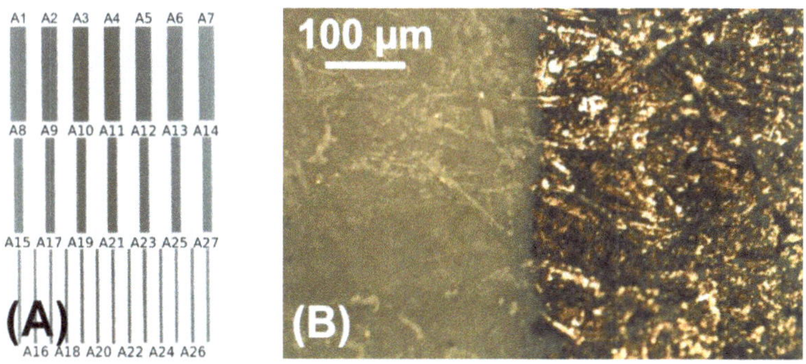

FIG. 2.23 (A) Photograph of a set of resistors printed in a one-step process with gold NPs on plain paper. Height of stripes (resistors) is 30 mm, width varies from 1 (*bottom*) to 5 mm (*top*). (B) Optical microscopy image of an individual resistor border. *(Reprinted with permission from F. Caruso, A. Bellacicca, P. Milani, High-throughput shadow mask printing of passive electrical components on paper by supersonic cluster beam deposition. Appl. Phys. Lett. 108 (2016) 163501.)*

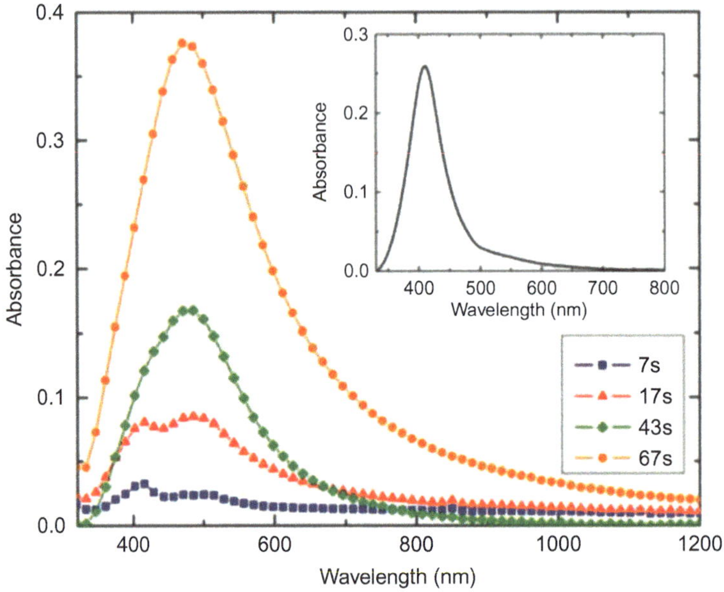

FIG. 2.24 UV–Vis absorption spectra of layered $Ag/SiO_xC_yH_z$ composite films with varying Ag-cluster deposition times from 7 to 67 s. The insert shows an absorption spectrum of 3D nanocomposite with thickness of 90 nm. *(Reprinted with permission from T. Peter, S. Rehders, U. Schürmann, T. Strunskus, V. Zaporojtchenko, F. Faupel, High rate deposition system for metal-cluster/SiOx-CyHz-polymer nanocomposite thin films. J. Nanopart. Res. 15 (2013) 1710.)*

forming media with very attractive plasmonic properties. For example, it was shown in [131] that silver clusters can be co-deposited with a polymer produced by RF plasma, and by varying the cluster deposition time, the intensity of the plasmon band can be tuned in a wide range (see Fig. 2.24) [13]. It is worth mentioning that for short deposition times, the absorption spectra show two distinct plasmon bands. One of them belongs to individual silver clusters, while the other one is assigned to the formation of small agglomerates (for example, dimers), leading to a longitudinal excitation mode along the new, longer axis. Similar extinction spectra with double-band structures were observed for as-deposited silver clusters on PMMA (see Fig. 2.25) [16]. The ratio between the first and second band intensities was found to be a function of cluster coverage. For low coverages, the band for individual clusters dominates. With a coverage increase, there are more particles forming dimers or larger agglomerates; hence, causing a significant rise of intensity of the second band. Even a third band at a higher wavelength corresponding to larger agglomerates can be pronounced [135]. After the thermal annealing facilitating the NPs' immersion into the polymer, there was only one band observed demonstrating the significantly increased intensity and small red shift with respect to the band associated with individual NPs [16]. The intensity was increased, and the band was red shifted, due to the embedding into the medium with a higher dielectric constant (2.6 for PMMA compared with 1.0 for air). The disappearance of the second band was most probably related to preferential immersion of NPs as individual units, that is, the agglomerates became broken. It is also worth mentioning that the LSPR band shown in [16] is much narrower compared with that presented in [13] (compare spectra in Figs. 2.24 and 2.25), which is related to the size-selection of clusters in the former case, and its absence in the latter one, leading to a large dispersion of the extinction.

The experiments on copper cluster deposition (size-selected particles of 15 nm in diameter) on PMMA and PS have also shown the formation of an LSPR band with maximum at $\lambda \approx 590$ nm [138]. However, annealing of the samples with intension to partly embed the deposited NPs led to quenching of the resonance (see Fig. 2.26A). Metallic copper is known to be more reactive compared with silver; therefore, one can expect extensive reactions of copper with environmental species (e.g., oxygen), especially at elevated temperatures, leading to the formation of compounds, and destroying the metallicity and, and thus, the plasmonic properties. One of the suggested solutions to solve this problem and preserve LSPR was to treat the deposited particles in ozone. Ozone can intensively react with the surface copper atoms, forming an oxide shell

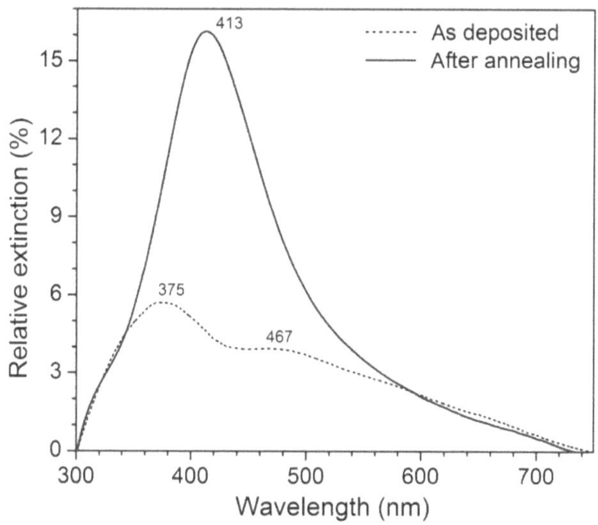

FIG. 2.25 Normalized extinction spectra of PMMA films with as-deposited, size-selected Ag clusters (13.4 nm in diameter), and after thermal annealing at 125°C for 10 min, facilitating the cluster embedding into the polymer. Numbers in the panel show the maxima of intensity for the bands. *(Reprinted with permission from M. Hanif, R.R. Juluri, M. Chirumamilla, V.N. Popok, Poly (methyl methacrylate) composites with size-selected silver nanoparticles fabricated using cluster beam technique. J. Polym. Sci. B Polym. Phys. 54 (2016) 1152–1159.)*

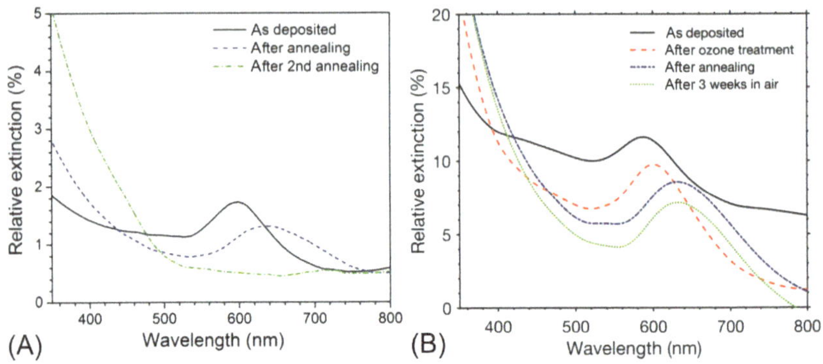

FIG. 2.26 Relative extinction as a function of wavelength for PMMA with copper NPs (A) before and after thermal annealing, (B) before and after ozone treatment, following annealing and long term storage in ambient atmospheric conditions. *(Reprinted with permission from V.N. Popok, M. Hanif, F.A. Ceynowa, P. Foyan, Immersion of low-energy deposited metal clusters into poly(methyl methacrylate). Nucl. Instr. Meth Phys. Res. B 409 (2017) 91–95.)*

protecting the metallic core against further gradual oxidation. As shown in Fig. 2.26B, the particles after such treatment preserve the resonance band on the time scale of a few weeks.

Plasmonic properties of Al NPs are important for a number of applications, because they demonstrate the resonance band in the near UV spectral range. It was shown in [132] that Al clusters of a mean size of 60 nm embedded into 50 nm thick C:H matrix demonstrate a pronounced plasmon band with a maximum at $\lambda \approx 307$ nm. Varying the filling factor of particles in the polymer from 1.6% to 9.3% caused significant rise of the resonance intensity. By changing the size from 60 to 30 nm, the band can be blue shifted to $\lambda \approx 255$ nm. Thus, the cluster beam technique demonstrates great capabilities in obtaining the required optical properties of the nanocomposites.

As mentioned in the part describing the electrical properties of the composites, the plasticity of polymers brings an advantage to fabricate devices in which the conductance is controlled by the applied strain, or the strain can be measured through the change of conductance. A similar approach is found to be possible in relation to optical properties opening a way for systems with stable plasmonic tuning under cyclic strain conditions. Gold clusters were implanted into PDMS forming a near-surface layer filled with NPs [69,127]. Fig. 2.27 shows the evolution of the plasmon band position and intensity for the film filled with 31% volume fraction subjected to a given number of stretching cycles with different elongation represented in percent [23]. It can be seen that the stretching leads to the red shift of the peak position. It was explained by compression of the film, i.e., an increase of the filling factor in the direction of the wave propagation (normal to the surface), due to the stretching of the elastomer in the plane direction.

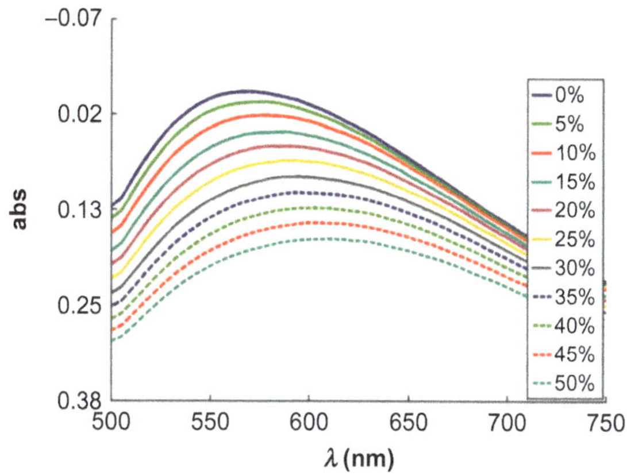

FIG. 2.27 Normalized optical absorbance spectra acquired at different applied strains (shown in %) for PDMS films with gold cluster corresponding to volume filling of 31% in the implanted layer. *(Reprinted with permission from C. Minnai, P. Milani, Metal-polymer nanocomposite with stable plasmonic tuning under cyclic strain conditions. Appl. Phys. Lett. 107 (2015) 073106.)*

An advantage of the cluster beam technique is in the well-controlled cluster volume filling factor, allowing the implantation regimes providing stability in tuning the peak position after a number of cycles. For example, it was observed that for the volume filling superior to 33%, a substantial plasmon peak position dependence on strain was preserved, and the shift was stable, even after 5000 cycles. Thus, one can conclude that the polymer nanocomposites produced by implantation of noble metal clusters are promising media to be used as mechanically tunable optical filters, or as components for stretchable optical devices.

The use of stretchable polymer nanocomposites produced by means of cluster beams can be wider than just utilizing the plasmonic properties. One can proceed toward adaptive optical elements such as lenses, mirrors, gratings, and so forth. An example is paper [73] reports on fabrication of stretchable reflective gratings based on metal-elastomer nanocomposites obtained by implantation of silver clusters into PDMS. An advantage of the nanocomposite films, compared with those obtained by conventional silver evaporation on the polymer surface, is found to be in superior resilience against thousands of stretching cycles, with only minor cracks in the top layer, not affecting the optical quality of the grating. The stretching only increases the distances between the NPs, and when the strain is released, the particles come close to each other again. Of course, thousands of cycles lead to the rearrangement of the NP network in the implanted layer, but this does not break the reflectivity of the entire layer. The stretchability and good resilience of the silver nanocomposite gratings provide a capability to mount them on the curved surfaces, hence, adding optical power to the diffracting elements [73].

Formation of compound clusters and their embedding into organic films is shown to be of interest for photovoltaics. $Cu_{2-x}S$ NPs of about 2.5 nm in size are produced in a cluster source by magnetron sputtering of copper in the presence of reactive hydrogen sulfide (H_2S) gas [150]. Significant efforts have been made toward the development of conditions for the formation of stoichiometric copper sulfide, which are ended in nearly Cu_2S NPs with potentially copper deficient composition. The copper chemical state for the formed compounds is a very important parameter for light-absorbing components. It has been shown that NPs with Cu(I) are better suited for light harvesting, compared with the high-band gap Cu(II) sulfide. The XPS analysis shows that only Cu(I) is present. The clusters are deposited both simultaneously, and in a stepwise manner with pentacene and quaterthiophene as the matrix, allowing control of the thickness of the composite films and filling factor of NPs. These organic films are not observed to impact the NP morphology or chemical structure, thus, making the polymer nanocomposites attractive for hybrid photovoltaic devices.

2.5.3 Polymer Nanocomposites for Biological Applications

Development of nanosensors is a rapidly growing field of research. The increasing interest arises from the unique physical characteristics and properties on the nanoscale, which are not present in bulk materials. Among nanosensors, biorecognition systems are of significant importance for environmental, bioprocess, and food quality controls, as well as for medical and pharmaceutical applications [21,151]. LSPR biosensors were among the first demonstrated, and since then, they have gradually become a very powerful label-free tool. One of the great advantages of label-free detection is that the target molecules are not altered, i.e., they are detected in their natural forms. NPs are typically used as transducers generating optical signals. At the same time, they are similar in size to some organic molecules, such as enzymes and proteins; thus being ideal elements for enhancing detection. However, there are still a number of challenges that need to be addressed to design efficient and reliable detection schemes, to develop appropriate surface immobilization chemistry providing selectivity of sensing, as well as to ensure data registration and analysis. In this context, the formation of polymer films with partly embedded NPs of a controlled size is a promising option for the preparation of transducers. As plastic and flexible polymer materials are easy to form, a wealth of systems with various geometries and dimensions can be easily achieved.

Recently, this approach was realized using silver NPs deposited from cluster beams, and partly embedded into PMMA by thermal annealing [18]. Partial embedding made the particles stable against wet chemical procedures needed for incubation of proteins. On the other hand, size-selected NPs demonstrated a narrow LSPR band with a high quality factor. LSPR properties of silver NPs were exploited for protein detection using a classical antibody-antigen scheme through covalent attachment of antibody molecules to the particles, followed by

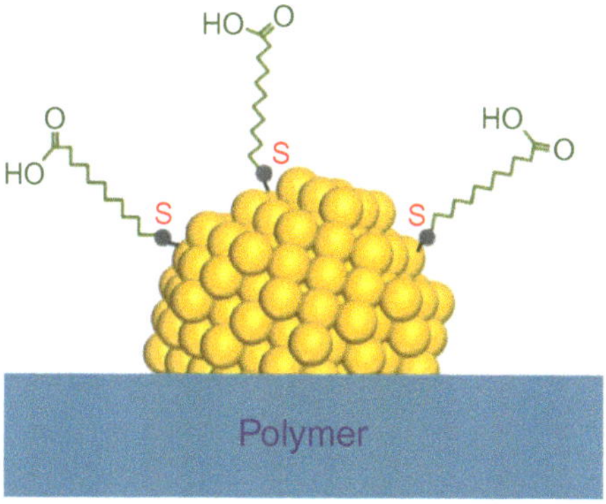

FIG. 2.28 Schematic picture of functionalization of silver surface using MUA. *(Reprinted from M. Hanif, R.R. Juluri, P. Fojan, V.N. Popok, Polymer films with size-selected silver nanoparticles as plasmon resonance-based transducers for protein sensing. Biointerface Res. Appl. Chem. 6 (2016) 1564–1568.)*

incubation with an antigen (chicken egg albumin). Surfaces of NPs were functionalized using 11-mercaptoundecanoic acid (MUA) solution in ethanol providing terminal carboxylic acid groups for the coupling of proteins (see Fig. 2.28). AFM studies revealed that antibodies were predominantly attached to NPs, providing binding sites for the antigens of interest. Specific changes in the LSPR band were observed for each step of protein incubation on NPs. Thus, the transducers demonstrated good sensitivity and selectivity in detection.

Certain metals, when present in excess, fulfil cellular functions that become destructive or lethal to organic cells. Because of these toxic properties, the films and nanostructures of such metals or metal compounds are widely used as antibacterial and antimicrobial agents. Among them are Ag, CuO, ZnO, Fe_3O_4, Al_2O_3, TiO_2 etc. [152–154]. They exhibit bactericidal properties, either through mechanisms based on (i) metal ion selectivity (replacement of original metals leading to cellular dysfunction) or (ii) metal reduction potential (generating or catalysing reactive oxygen species (ROS) damaging cellular proteins, lipids and DNA) [155].

TiO_2 is a well-known and widely used semiconductor material, with a wide band gap, which, under UV illumination, generates electron-hole pairs destructively affecting the cell components through a ROS mechanism, but rendering innocuous products [156,157]. An important starting point was to study adhesion of cells on the titanium NPs. It was shown in [70] that surface roughness of the composite plays an important role in this process. In particular, for the osteoblast-like cells, the growth rate was found to be the highest for the surfaces with nanotopography on the scale of a few tens of nm. It was believed that such nanorough material mimics the architecture of natural tissues, for example, the size of some extracellular matrix molecules, or irregularities on these molecules, and also the size of the extracellular parts of cell adhesion receptors.

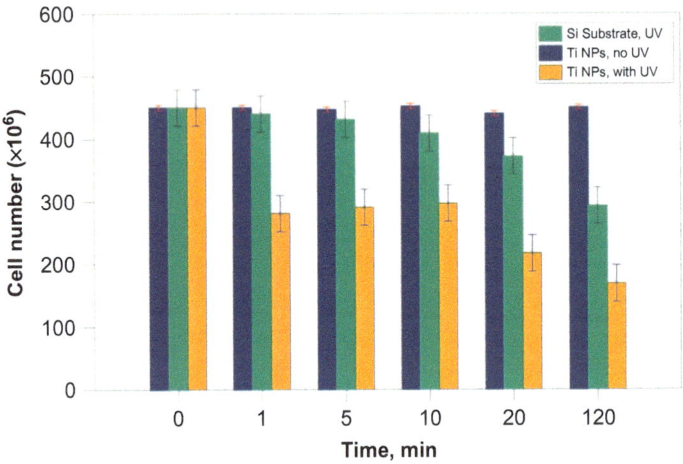

FIG. 2.29 Evolution of cell number with time on the PS films with partly embedded TiO_x NPs illuminated by UV light and with no illumination. Efficiency in killing the bacteria by UV light on Si substrates without particles is also shown.

As a next step, E-coli bacteria were deposited on the PS films with partly embedded TiO_x NPs obtained by the cluster beam deposition [158]. As can be seen in Fig. 2.29, the samples with NPs under the UV illumination demonstrated the highest antibacterial efficiency. The survival ratio (number of alive bacteria with respect to that initially deposited) was reduced to 0.65 within the first 5min, and further decreases to 0.37 after 120min. The samples showed good repeatability of bactericidal efficiency after a number of cycles (removing the bacteria and depositing new E-coli cultures). Thus, TiO_2 nanoparticles embedded into polymer films were demonstrated to be excellent candidates for formation of polymer composites with antibacterial and antimicrobial properties that can be used in food packaging, fabrication of consumables for medicine, and so forth.

The other bactericidal mechanism through the metal ion selectivity was also tested using polymer films with silver and copper NPs deposited from cluster beams [158–160]. In this case, the nanocomposites acted as a reservoir for the out diffusion of metal ions. These ions can affect the microorganisms by impairing the function of their enzymes, damaging nucleic acids or the cell membranes, thus inhibiting growth of bacteria and microbes. The use of metals without UV illumination simplifies the application of the composites. The first experiments have shown good bactericide efficiency of the fabricated metal-polymer composites, and great advantages in regulation of metal ion release by controlling the properties of composites by the plasma polymerization technique utilized in the co-deposition of metal NPs and polymers.

2.6 CONCLUSIONS

This chapter overviews the topic of synthesis of polymer composites with metal and compound metal nanoparticles produced by cluster aggregation techniques. The principles of cluster nucleation and the main technical approaches for the formation of cluster beams are described, showing such advantages as the efficient control of cluster composition, capabilities in formation of single crystal particles, size (mass) selection prior the deposition or implantation on/in a matrix, control of cluster kinetic energy by which the regime of particle-matter interaction can be tuned, possibilities for co-deposition of clusters and organic films, and good control of surface coverage or volume filling factors of nanoparticles on/in polymer. The chapter also discusses the fundamental aspects of metal nanoparticle interaction with polymer surfaces, showing that the driving force for embedding of particles into the polymer bulk is the large difference in surface tension between metals and organic materials. Practical aspects of immersion are discussed, giving a number of examples about the dynamics of the process and its dependence on metal species, as well as on the polymer properties. Combining the knowledge about the physics of nanoparticle-polymer interaction with existing cluster techniques allows several possible approaches to the production of composites with required properties. Namely, one can use (i) supersonic beams with the sufficient kinetic energy for the direct implantation of clusters into thin, near-surface layers, (ii) co-deposition of clusters and plasma polymers, allowing growth of composite matrixes with embedded nanoparticles, and (iii) soft-landing of NPs, followed by thermal annealing, providing the NPs embedding into the near surface layer of polymer in a well-controllable manner. The rest of the chapter describes numerous examples of the formation of polymer composites with partly or fully embedded metal and metal compound nanoparticles, demonstrating a wide spectrum of applications utilizing mechanical, electronic, optical, and other unique properties of the synthesized materials. Analysis of these research publications allows us to draw conclusions about the great potential of the cluster beam approach, combined with advantages of polymers as host materials, on the way toward industrial production of elastic, low-cost, and low-weight electronic, electro-mechanical and optical components, flexible plasmonic media, transducers for biosensing, antibacterial coatings for food production, and components for medical instruments.

REFERENCES

[1] F. Faupel, V. Zaporojtchenko, T. Strunskus, M. Elbahri, Metal-polymer nanocomposites for functional applications, Adv. Eng. Mater. 12 (2010) 1177–1190.

[2] J. Prakash, J.C. Pivin, H.C. Swart, Noble metal nanoparticles embedding into polymeric materials: from fundamentals to applications, Adv. Colloid Interf. Sci. 226 (2015) 187–202.

[3] V. Torrisi, F. Ruffino, Metal-polymer nanocomposites: (co-)evaporation/(co)sputtering approaches and electrical properties, Coatings 5 (2015) 378–424.
[4] M.C. Salvadori, F.S. Teixeira, L.G. Sgubin, M. Cattani, I.G. Brown, Surface modification by metal ion implantation forming metallic nanoparticles in an insulating matrix, Appl. Surf. Sci. 310 (2014) 158–163.
[5] V.N. Popok, M.G. Lukashevich, S.M. Lukashevich, R.I. Khaibullin, V.V. Bazarov, Charge carrier transport in polyimide with Co nanoparticles formed by ion implantation, Surf. Sci. 566–568 (2004) 327–331.
[6] M. Niklaus, H.R. Shea, Electrical conductivity and Young's modulus of flexible nanocomposites made by metal-ion implantation of polydimethylsiloxane: the relationship between nanostructure and macroscopic properties, Acta Mater. 59 (2011) 830–840.
[7] C. Tian-Xiang, Y. Shu-Du, H. Wei, F. Tao, L. Lin, Z. Sheng-Qiang, Charge transport and magnetotransport properties of polyimide irradiated by 80 keV Co ions, Chin. Phys. Lett. 26 (2009) 087201.
[8] M. Tie, A.-A. Dhirani, Conductance of molecularly linked gold nanoparticle films across an insulator-to-metal transition: from hopping to strong Coulomb electron-electron interactions and correlations, Phys. Rev. B 91 (2015) 155131.
[9] E. Barborini, G. Corbelli, G. Bertolli, P. Repetto, M. Leccardi, S. Vinati, P. Milani, The influence of nanoscale morphology on the resistivity of cluster-assembled nanostructured metallic thin films, New J. Phys. 12 (2010) 073001.
[10] G. Corbelli, C. Ghisleri, M. Marelli, P. Milani, L. Ravagnan, Highly deformable nanostructured elastomeric electrodes with improving conductivity upon cyclical stretching, Adv. Mater. 23 (2011) 4504–4508.
[11] G. Di Girolamo, M. Massaro, E. Piscopiello, L. Tapfer, Metal ion implantation in inert polymers for strain gauge applications, Nucl. Instr. Meth. Phys. Res. B 268 (2010) 2878–2882.
[12] G. Maggioni, A. Vomiero, S. Carturan, C. Scian, G. Mattei, M. Bazzan, C. de Julian Fernandez, P. Mazzoldi, A. Quaranta, G. Della Mea, Structure and optical properties of Au-polyimide nanocomposite films prepared by ion implantation, Appl. Phys. Lett. 85 (2004) 5712–5714.
[13] T. Peter, S. Rehders, U. Schürmann, T. Strunskus, V. Zaporojtchenko, F. Faupel, High rate deposition system for metal-cluster/$SiO_xC_yH_z$-polymer nanocomposite thin films, J. Nanopart. Res. 15 (2013) 1710.
[14] A. Heilmann, Polymer Films With Embedded Metal Nanoparticles, Springer, Berlin, 2003.
[15] V.N. Popok, V.I. Nuzhdin, V.F. Valeev, A.L. Stepanov, Copper nanoparticles synthesized in polymers by ion implantation: surface morphology and optical properties of the nanocomposites, J. Mater. Res. 30 (2015) 86–92.
[16] M. Hanif, R.R. Juluri, M. Chirumamilla, V.N. Popok, Poly (methyl methacrylate) composites with size-selected silver nanoparticles fabricated using cluster beam technique, J. Polym. Sci. B Polym. Phys. 54 (2016) 1152–1159.
[17] C. de Julian Fernandez, M.G. Manera, J. Spadavecchia, G. Maggioni, A. Quaranta, G. Mattei, M. Bazzan, E. Cattaruzza, M. Bonafini, E. Negro, A. Vomiero, S. Carturan, C. Scian, G. Della Meac, R. Rella, L. Vasanelli, P. Mazzoldi, Study of the gas optical sensing properties of Au-polyimide nanocomposite films prepared by ion implantation, Sensors Actuators B 111-112 (2005) 225–229.
[18] M. Hanif, R.R. Juluri, P. Fojan, V.N. Popok, Polymer films with size-selected silver nanoparticles as plasmon resonance-based transducers for protein sensing, Biointerface Res. Appl. Chem. 6 (2016) 1564–1568.
[19] W. Gao, G. Chen, W. Xu, C. Yang, S. Xu, Surface-enhanced Raman scattering (SERS) chips made from metal nanoparticle-doped polymer fibres, RSC Adv. 4 (2014) 23838–23845.

[20] S. Fateixa, H.I.S. Nogueira, T. Trindade, Hybrid nanostructures for SERS: materials development and chemical detection, Phys. Chem. Chem. Phys. 17 (2015) 21046–21071.
[21] B. Van Dorst, J. Mehta, K. Bekaert, E. Rouah-Martin, W. De Coen, P. Dubruel, R. Blust, J. Robbens, Recent advances in recognition elements of food and environmental biosensors: a review, Biosens. Bioelectron. 26 (2010) 1178–1194.
[22] E. Petryayeva, U.J. Krull, Localized surface plasmon resonance: nanostructures, bioassays and biosensing—a review, Anal. Chim. Acta 706 (2011) 8–24.
[23] C. Minnai, P. Milani, Metal-polymer nanocomposite with stable plasmonic tuning under cyclic strain conditions, Appl. Phys. Lett. 107 (2015) 073106.
[24] M.K. Hedayati, M. Javaherirahim, B. Mozooni, R. Abdelaziz, A. Tavassolizadeh, V.S.K. Chakravadhanula, V. Zaporojtchenko, T. Strunkus, F. Faupel, M. Elbahri, An omnidirectional transparent conducting-metal-based plasmonic nanocomposite, Adv. Mater. 23 (2011) 5410–5414.
[25] M. Elbahri, M.K. Hedayati, V.S.K. Chakravadhanula, M. Jamali, T. Strunkus, V. Zaporojtchenko, F. Faupel, Design of a perfect black absorber at visible frequencies using plasmonic metamaterials, Adv. Mater. 23 (2011) 1993–1997.
[26] S. Li, M.M. Lin, M.S. Toprak, D.K. Kim, M. Muhammed, Nanocomposites of polymer and inorganic nanoparticles for optical and magnetic applications, Nanotechnol. Rev. 1 (2010) 5214.
[27] C. Lu, C. Guan, Y. Liu, Y. Cheng, B. Yang, PbS/polymer nanocomposite optical materials with high refractive index, Chem. Mater. 17 (2005) 2448–2454.
[28] J. Lee, V.C. Sundar, J.R. Heine, M.G. Bawendi, K.F. Jensen, Full color emission from II–VI semiconductor quantum dot–polymer composites, Adv. Mater. 12 (2000) 1102–1105.
[29] H. Zhang, Z. Cui, Y. Wang, K. Zhang, X. Ji, C. Lu, B. Yang, M. Gao, From water-soluble CdTe nanocrystals to fluorescent nanocrystal–polymer transparent composites using polymerizable surfactants, Adv. Mater. 15 (2003) 777–780.
[30] M.J. Ventura, M. Gu, Engineering spontaneous emission in a quantum-dot-doped polymer nanocomposite with three-dimensional photonic crystals, Adv. Mater. 20 (2008) 1329–1332.
[31] R.I. Khaibullin, B.Z. Rameev, C.Okay, A.L. Stepanov, V.A. Zhikharev, I.B. Khaibullin, L.R. Tagirov, B. Aktas, Ion beam synthesis of magnetic nanoparticles in polymers, in: B. Aktas, L. Tagirov, F. Mikailov (Eds.), Nanostructured Magnetic Materials and Their Applications, NATO Science Series: II Mathematics, Physics and Chemistry, vol. 143, Kluwer, Dordrecht, 2004, pp.33-54.
[32] V.N. Popok, Polymer films with ion-synthesized cobalt and iron nanoparticles: conductance and magnetism, Rev. Adv. Mater. Sci. 36 (2014) 1–12.
[33] A. Munoz-Bonilla, J. Sanchez-Marcos, P. Herrasti, Magnetic nanoparticles-based conducting polymer nanocomposites, in: V. Kumar, S. Kalia, H.C. Swart (Eds.), Conducting Polymer Hybrids, Springer International, 2017, pp. 45–80.
[34] J. Koo, H. Kim, K.-Y. Kim, Y.R. Jang, J.-S. Lee, S.W. Yoon, B.J. Suh, T. Yu, J. Bang, K. Yoon, G. Yuan, S.K. Satija, Controlling the magnetic properties of polymer-iron oxide nanoparticle composite thin films via spatial particle orientation, RSC Adv. 6 (2016) 55842–55847.
[35] C. Okay, B.Z. Rameev, R.I. Khaibullin, M. Okutan, F. Yildiz, V.N. Popok, B. Aktas, Ferromagnetic resonance study of iron implanted PET foils, Phys. Status Solidi 203 (2006) 1525–1532.
[36] J. Kim, B. Van der Bruggen, The use of nanoparticles in polymeric and ceramic membrane structures: review of manufacturing procedures and performance improvement for water treatment, Environ. Pollut. 158 (2010) 2335–2349.

[37] L.Y. Ng, A.W. Mohammad, C.P. Leo, N. Hilal, Polymeric membranes incorporated with metal/metal oxide nanoparticles: a comprehensive review, Desalination 308 (2013) 15–33.
[38] V. Zaporojtchenko, T. Strunskus, J. Erichsen, F. Faupel, Embedding of noble metal nanoclusters into polymers as a potential probe of the surface glass transition, Macromolecules 34 (2001) 1125–1127.
[39] G. Whittell, I. Manners, Metallopolymers: new multifunctional materials, Adv. Mater. 19 (2007) 3439–3468.
[40] L. Tröger, H. Hünnefeld, S. Nunes, M. Oehring, D. Fritsch, Structural characterization of catalytically active metal nanoclusters in poly(amide imide) films with high metal loading, J. Phys. Chem. B 101 (1997) 1279–1291.
[41] S. Li, J. Qin, A. Fornara, M. Toprak, M. Muhammed, D.K. Kim, Synthesis and magnetic properties of bulk transparent PMMA/Fe-oxide nanocomposite, Nanotechnology 20 (2009) 185607.
[42] M. Bhamidipati, L. Fabris, Multiparametric assessment of gold nanoparticle cytotoxicity in cancerous and healthy cells: the role of size, shape, and surface chemistry, Bioconjug. Chem. 28 (2017) 449–460.
[43] K. Akamatsu, S. Deki, Characterization and optical properties of gold nanoparticles dispersed in nylon 11 thin films, J. Mater. Chem. 7 (1997) 1773.
[44] J. Siegel, P. Jurik, Z. Kolska, V. Svorcik, Annealing of silver nanolayers sputtered on polytetrafluoroethylene, Surf. Interface Anal. 45 (2013) 1063–1066.
[45] J. Röder, H.-U. Krebs, Tuning the microstructure of pulsed laser deposited polymer–metal nanocomposites, Appl. Phys. A Mater. Sci. Process. 85 (2006) 15–20.
[46] F. Faupel, V. Zaporojtchenko, H. Greve, U. Schürmann, C. Hanisch, V.S.K. Chakravadhanula, A. Kulkarni, A. Gerber, E. Quandt, R. Podschun, Deposition of nanpocomposites by plasmas, Contrib. Plasma Phys. 47 (2007) 537–544.
[47] H. Biederman (Ed.), Plasma Polymer Films, Imperial College Press, Singapore, 2004.
[48] F. Ruffino, V. Torrisi, G. Marletta, M.G. Grimaldi, Effects of the embedding kinetics on the surface nano-morphology of nano-grained Au and Ag films on PS and PMMA layers annealed above the glass transition temperature, Appl. Phys. A Mater. Sci. Process. 107 (2012) 669–683.
[49] J. Prakash, A. Tripathi, S. Gautam, K.H. Chae, J. Song, V. Rigato, J. Tripathi, K. Asokan, Phenomenological understanding of dewetting and embedding of noble metal nanoparticles in thin films induced by ion irradiation, Mater. Chem. Phys. 147 (2014) 920–924.
[50] A. Meldrum, R.H. Haglund Jr., L.A. Boatner, C.W. White, Nanocomposite materials formed by ion implantation, Adv. Mater. 13 (2001) 1431–1444.
[51] V.N. Popok, Ion implantation of polymers: formation of nanoparticulate materials, Rev. Adv. Mater. Sci. 30 (2012) 1–26.
[52] R. Nathawat, Y.K. Vijay, P. Kumar, P. Kulriya, V. Ganesan, V. Sathe, Physically and chemically modified polycarbonate by metal ion implantation, Adv. Polym. Technol. 27 (2008) 143–151.
[53] V.N. Popok, M. Hanif, A. Mackova, R. Miksova, Structure and plasmonic properties of thin PMMA layers with ion-synthesized Ag nanoparticles, J. Polym. Sci. B Polym. Phys. 53 (2015) 664–672.
[54] R.I. Khaibullin, V.N. Popok, V.V. Bazarov, E.P. Zheglov, B.Z. Rameev, C. Okay, L.R. Tagirov, B. Aktas, Ion synthesis of iron granular films in polyimide, Nucl. Instr. Meth. Phys. Res. B 191 (2002) 810–814.
[55] V.N. Popok, I.I. Azarko, V.B. Odzhaev, A. Toth, R.I. Khaibullin, High fluence ion beam modification of polymer surfaces: EPR and XPS study, Nucl. Instr. Meth. Phys. Res. B 178 (2001) 305–310.

[56] A. Kondyurin, M. Bilek, Ion Beam Treatment of Polymers, Elsevier, Amsterdam, 2008.
[57] V.N. Popok,.I.I. Azarko, R.I. Khaibullin, A.L. Stepanov, V. Hnatowicz, A. Mackova, S.V. Prasalovich, Radiation-induced change of polyimide properties under high-fluence and high ion current density implantation, Appl. Phys. A Mater. Sci. Process. 78 (2004) 1067–1072.
[58] V.N. Popok, Compositional and structural alterations of polymers under low-to-medium-energy ion implantation, in: C.P. Norris (Ed.), Surface Science Research, Nova Science Publishers, New York, 2005, pp. 147–193.
[59] V.B. Odzhaev, I.I. Azarko, I.A. Karpovich, I.P. Kozlov, V.N.P.D.V. Sviridov, V. Hnatowicz, O.N. Jankovskij, V. Rybka, V. Svorcik, The properties of polyethylene and polyamide implanted with B ions to high doses, Mater. Lett. 23 (1995) 163–166.
[60] A. Perez, P. Melinon, V. Dupuis, P. Jensen, B. Prevel, J. Tuaillon, L. Bardotti, C. Martet, M. Treilleux, M. Broyer, M. Pellarin, J.L. Vaille, B. Palpant, J. Lermez, Cluster assembled materials: a novel class of nanostructured solids with original structures and properties, J. Phys. D. Appl. Phys. 30 (1997) 709–721.
[61] P. Milani, S. Iannotta, Cluster Beam Synthesis of Nanostructured Materials, Springer, Berlin, 1999.
[62] K.H. Meiwes-Broer (Ed.), Metal Clusters at Surfaces, Springer, Berlin, 2000.
[63] C. Binns, Nanoclusters deposited on surfaces, Surf. Sci. Rep. 44 (2001) 1–49.
[64] V.N. Popok, S.V. Prasalovich, E.E.B. Campbell, Surface nanostructuring by implantation of cluster ions, Vacuum 76 (2004) 265–272.
[65] V.N. Popok, E.E.B. Campbell, Beams of atomic clusters: effects on impact with solids, Rev. Adv. Mater. Res. 11 (2006) 19–45.
[66] N. Toyoda, I. Yamada, Gas cluster ion beam equipment and applications for surface processing, IEEE Trans. Plasma Sci. 36 (2008) 1471–1488.
[67] V.N. Popok, Cluster ion implantation in graphite and diamond: radiation damage and stopping of cluster constituents, Rev. Adv. Mater. Sci. 38 (2014) 7–16.
[68] V.N. Popok, I. Barke, E.E.B. Campbell, K.-H. Meiwes-Broer, Cluster-surface interaction: from soft landing to implantation, Surf. Sci. Rep. R 66 (2011) 347–377.
[69] L. Ravagnan, G. Divitini, S. Rebasti, M. Marelli, P. Piseri, P. Milani, Poly(methyl methacrylate)–palladium clusters nanocomposite formation by supersonic cluster beam deposition: a method for microstructured metallization of polymer surfaces, J. Phys. D. Appl. Phys. 42 (2009) 082002.
[70] P. Solar, O. Kylian, O. Polonskyi, A. Artemenko, D. Arzhakov, M. Drabik, D. Slavinska, M. Vandrovcova, L. Bacakova, H. Biederman, Nanocomposite coatings of Ti/C:H plasma polymer particles providing a surface with variable nanoroughness, Surf. Coat. Technol. 206 (2012) 4335–4342.
[71] O. Kylian, A. Choukourov, H. Biederman, Nanostructured plasma polymers, Thin Solid Films 548 (2013) 1–17.
[72] O. Kylian, J. Kratochvil, J. Hanuš, O. Polonskyi, P. Solař, H. Biederman, Fabrication of Cu nanoclusters and their use for production of Cu/plasma polymer nanocomposite thin films, Thin Solid Films 550 (2014) 46–52.
[73] C. Ghisleri, M. Siano, L. Ravagnan, M.A.C. Potenza, P. Milani, Nanocomposite-based stretchable optics, Laser Photon. 7 (2013) 1020–1026.
[74] F. Greco, A. Bellacicca, M. Gemmi, V. Cappello, V. Mattoli, P. Milani, Conducting shrinkable nanocomposite based on Au-nanoparticle implanted plastic sheet: tunable thermally induced surface wrinkling, ACS Appl. Mater. Interfaces 7 (2015) 7060–7065.
[75] W.A. de Heer, The physics of simple metal clusters: experimental aspects and simple models, Rev. Mod. Phys. 65 (1993) 611–676.

[76] M.M. Kappes, S. Leutwyler, Molecular beams of clusters, in: G. Scoles (Ed.), Atomic and Molecular Beam Methods, V. 1, Oxford University Press, New York, 1988, pp. 380–415.
[77] B.M. Smirnov, Processes involving clusters and small particles in a buffer gas, Physics-Uspekhi 54 (2011) 691–721.
[78] V. Popok, Energetic cluster-surface collisions, in: K.D. Sattler (Ed.), Handbook of Nanophysics: Clusters and Fullerenes, CRC Press, Boca Raton, 2010, pp. 19-1–19-19.
[79] O.F. Hagena, Cluster ion sources, Rev. Sci. Instrum. 64 (1992) 2374–2379.
[80] H. Haberland, M. Karrais, M. Mall, Y. Thurner, Thin films from energetic cluster impact: a feasibility study, J. Vac. Sci. Technol. A 10 (1992) 3266–3271.
[81] V.N. Popok, S.V. Prasalovich, M. Samuelsson, E.E.B. Campbell, Design and capabilities of a cluster implantation and deposition apparatus: first results on hillock formation under energetic cluster ion bombardment, Rev. Sci. Instrum. 73 (2002) 4283–4287.
[82] K. Wegner, P. Piseri, H.V. Tafreshi, P. Milani, Cluster beam deposition: a tool for nanoscale science and technology, J. Phys. D. Appl. Phys. 39 (2006) R439–R459.
[83] Y. Huttel (Ed.), Gas-Phase Synthesis of Nanoparticles, Wiley-VCH, Weinheim, 2017.
[84] K. Sattler, J. Mühlback, E. Recknagel, Generation of metal clusters containing from 2 to 500 atoms, Phys. Rev. Lett. 45 (1980) 821–824.
[85] I.M. Goldby, B. von Issendorf, L. Kuipers, R.E. Palmer, Gas condensation source for production and deposition of size-selected metal clusters, Rev. Sci. Instrum. 68 (1997) 3327–3334.
[86] H. Abe, W. Schulze, B. Tesche, Optical properties of silver microcrystals prepared by means of the gas aggregation technique, Chem. Phys. 47 (1980) 95–104.
[87] T.P. Martin, T. Bergmann, H. Göhlich, T. Lange, Observation of electronic shells and shells of atoms in large Na clusters, Chem. Phys. Lett. 172 (1990) 209–213.
[88] T.G. Dietz, M.A. Duncan, D.E. Powers, R.E. Smalley, Laser production of supersonic cluster beams, J. Chem. Phys. 74 (1981) 6511–6512.
[89] P. Milani, W. de Heer, Improved pulsed laser vaporization source for production of intense beams of neutral and ionized clusters, Rev. Sci. Instrum. 61 (1990) 1835–1838.
[90] N. Saito, K. Koyama, M. Tanimoto, Cluster generation by laser ablation, Appl. Surf. Sci. 169-170 (2001) 380–386.
[91] S. Vučković, M. Svanqvist, V.N. Popok, Laser ablation source for formation and deposition of size-selected metal clusters, Rev. Sci. Instrum. 79 (2008) 073303.
[92] R.L. Wagner, W.D. Vann, A.W. Castleman, A technique for efficiently generating bimetallic clusters, Rev. Sci. Instrum. 68 (1997) 3010–3013.
[93] S. Nonose, Y. Sone, K. Onodera, S. Sudo, K. Kaya, Structure and reactivity of bimetallic cobalt-vanadium (ConVm) clusters, J. Phys. Chem. 94 (1990) 2744–2746.
[94] W. Bouwen, P. Thoen, F. Vanhoutte, S. Bouckaert, F. Despa, H. Weidele, R.E. Silverans, P. Lievens, Production of bimetallic clusters by a dual-target dual-laser vaporization source, Rev. Sci. Instrum. 71 (2000) 54–58.
[95] G. Ganteför, H.R. Siekmann, H.O. Lutz, K.-H. Meiwes-Broer, Pure metal and metal-doped clusters grown in a pulsed arc cluster ion source, Chem. Phys. Lett. 165 (1990) 293–296.
[96] E. Barborini, P. Piseri, P. Milani, Cluster beam microfabrication of patterns of three-dimensional nanostructured objects, Appl. Phys. Lett. 77 (2000) 1059–1061.
[97] P. Piseri, H.V. Tafreshi, P. Milani, Manipulation of nanoparticles in supersonic beams for the production of nanostructured materials, Curr. Opin. Solid State Mater. Sci. 8 (2004) 195–202.
[98] R.K. Waits, J.L. Vossen, W. Kern (Eds.), Thin Film Processes, Academy Press, London, 1978, pp. 131–173.

[99] S. Pratontep, S.J. Carrol, C. Xirouchaki, M. Streum, R.E. Palmer, Size-selected cluster beam source based on radio frequency magnetron plasma sputtering and gas condensation, Rev. Sci. Instrum. 76 (2005).

[100] H. Hartmann, V.N. Popok, I. Barke, V. von Oeynhausen, K.-H. Meiwes-Broer, Design and capabilities of an experimental setup based on magnetron sputtering for formation and deposition of size-selected metal clusters on ultra-clean surfaces, Rev. Sci. Instrum. 83 (2012) 073304.

[101] O.F. Hagena, Nucleation and growth of clusters in expanding nozzle flow, Surf. Sci. 106 (1981) 101–116.

[102] I. Yamada, J. Matsuo, Z. Insepov, D. Takeuchi, M. Akizuki, N. Toyoda, Surface processing by gas cluster ion beams at the atomic (molecular) level, J. Vac. Sci. Technol. A 14 (1996) 781–785.

[103] A.A. Andreev, V.S. Chernysh, Y.A. Ermakov, A.E. Ieshkin, Design and investigation of gas cluster ion accelerator, Vacuum 91 (2013) 47–53.

[104] V.N. Popok, S.V. Prasalovich, E.E.B. Campbell, Nanohillock formation by impact of small low-energy clusters with-surfaces, Nucl. Instr. Meth. Phys. Res. B 207 (2003) 145–153.

[105] R.E. Palmer, L. Cao, F. Yin, Note: proof of principle of a new type of cluster beam source with potential for scale-up, Rev. Sci. Instrum. 87 (2016) 046103.

[106] Y. Lipatov, A. Feinermann, Surface tension and surface free energy of polymers, Adv. Colloid Interf. Sci. 11 (1979) 195–233.

[107] http://www.accudynetest.com/polymer_surface_data/other_polymers.pdf.

[108] R. Doering, Y. Nishi (Eds.), Handbook of Semiconductor Manufacturing Technology, CRC Press, Boca Raton, 2008.

[109] F. Aqra, A. Ayyad, Surface free energy of alkali and transition metal nanoparticles, Appl. Surf. Sci. 314 (2014) 308–313.

[110] R. Lamber, S. Wetjen, N.I. Jaeger, Size dependence of the lattice parameter of small palladium particles, Phys. Rev. B 51 (1995) 10968–10971.

[111] K.K. Nanda, A. Maisels, F.E. Kruis, H. Fissan, S. Stappert, Higher surface energy of free nanoparticles, Phys. Rev. Lett. 91 (2003) 106102.

[112] K.K. Nanda, A. Maisels, F.E. Kruis, Surface tension and sintering of free gold nanoparticles, J. Phys. Chem. C 112 (2008) 13488–14491.

[113] R.D. Deshmukh, R.J. Composto, Direct observation of nanoparticle embedding into the surface of a polymer melt, Langmuir 23 (2007) 13169–13173.

[114] G.J. Kovacs, P.S. Vincett, Formation and thermodynamic stability of a novel class of useful materials: close-packed monolayers of submicron monodisperse spheres just below a polymer surface, J. Colloid Interface Sci. 90 (1982) 335–351.

[115] G.J. Kovacs, P.S. Vincett, Subsurface particle monolayer and film formation in softenable substrates: techniques and thermodynamic criteria, Thin Solid Films 111 (1984) 65–81.

[116] R. Weber, K.-M. Zimmermann, M. Tolan, J. Stettner, W. Press, O.H. Seeck, J. Erichsen, V. Zaporojtchenko, T. Strunskus, F. Faupel, X-ray reflectivity study on the surface and bulk glass transition of polystyrene, Phys. Rev. E 64 (2001) 061508.

[117] R. Weber, I. Grotkopp, J. Stettner, M. Tolan, W. Press, Embedding of gold nanoclusters on polystyrene surfaces: influence of the surface modification on the glass transition, Macromolecules 36 (2003) 9100–9106.

[118] J.H. Teichroeb, J.A. Forrest, Direct imaging of nanoparticle embedding to probe viscoelasticity of polymer surface, Phys. Rev. Lett. 91 (2003) 016104.

[119] D. Qi, M. Ilton, J.A. Forrest, Measuring surface and bulk relaxation in glassy polymers, Eur. Phys. J. E 34 (2011) 56.

[120] Z. Fakhraai, J.A. Forrest, Measuring the surface dynamics of glassy polymers, Science 319 (2008) 600–604.
[121] K. Paeng, R. Richert, M.D. Ediger, Molecular mobility in supported thin films of polystyrene, poly(methylmethacrylate), and poly(2-vinyl pyridine) probed by dye reorientation, Soft Matter 8 (2012) 819–826.
[122] C.H. Xu, X. Chen, Y.J. Liu, B. Xie, M. Han, F.Q. Song, G.H. Wnag, Enhanced thermal stability of monodispersed silver cluster arrays assembled on block copolymer scaffolds, Nanotechnology 21 (2010) 195304.
[123] V.I. Shulga, P. Sigmund, Penetration of slow gold clusters through silicon, Nucl. Instr. Meth. Phys. Res. B 47 (1990) 236–242.
[124] Y. Yamamura, T. Muramoto, Time-evolution MC simulation of a few 100eV/atom cluster impacts, Rad. Eff. Def. Sol. 130-131 (1994) 225–233.
[125] V.N. Popok, Energetic cluster ion beams: modification of surfaces and shallow layers, Mater. Sci. Eng. Rep. R 72 (2011) 137–157.
[126] R. Cardia, C. Melis, L. Colombo, Neutral-cluster implantation in polymers by computer experiments, J. Appl. Phys. 113 (2013) 224307.
[127] C. Ghisleri, F. Borghi, L. Ravagnan, A. Podests, C. Melis, L. Colombo, P. Milani, Patterning of gold-polydimethylsiloxane (Au-PDMS) nanocomposites by supersonic cluster beam implantation, J. Phys. D. Appl. Phys. 45 (2014) 015301.
[128] F. Caruso, A. Bellacicca, P. Milani, High-throughput shadow mask printing of passive electrical components on paper by supersonic cluster beam deposition, Appl. Phys. Lett. 108 (2016) 163501.
[129] R. Lamber, S. Wetjen, N.I. Jaeger, Size dependence of the lattice parameter of small palladium particles, Phys. Rev. B 51 (1995) 10968–10971.
[130] H. Biederman, Nanocomposites and nanostructures based on plasma polymers, Surf. Coat. Technol. 205 (2011) S10–S14.
[131] T. Peter, M. Wegner, V. Zaporojtchenko, T. Strunskus, S. Bornholdt, H. Kersten, F. Faupel, Metal/polymer nanocomposite thin films prepared by plasma polymerization and high pressure magnetron sputtering, Surf. Coat. Technol. 205 (2011) S38–S41.
[132] O. Polonskyi, O. Kylián, M. Drábik, J. Kousal, P. Solař, A. Artemenko, J. Čechvala, A. Choukourov, D. Slavínská, H. Biederman, Deposition of Al nanoparticles and their nanocomposites using a gas aggregation cluster source, J. Mater. Sci. 49 (2014) 3352–3360.
[133] O. Kylian, A. Shelemin, P. Solar, A. Choukourov, J. Hanus, M. Vaidulych, A. Kuzminova, H. Biederman, Plasma polymers: from thin films to nanocolumnar coatings, Thin Solid Films 630 (2017) 86–91.
[134] T.P. Martin, Shells of atoms, Phys. Rep. 273 (1996) 199–241.
[135] S.M. Novikov, V.N. Popok, A.B. Evlyukhin, M. Hanif, P. Morgen, J. Fiutowski, J. Beermann, H.-G. Rubahn, S.I. Bozhevolnyi, Highly-stable monocrystalline silver clusters for plasmonic applications, Langmuir 33 (2017) 6062–6070.
[136] I. Barke, H. Hartmann, D. Rupp, L. Fluckiger, M. Sauppe, M. Adolph, S. Schorb, C. Bostedt, R. Treusch, C. Peltz, S. Bartling, T. Fennel, K.-H. Meiwes-Broer, T. Moller, The 3D-architecture of individual free silver nanoparticles captured by X-ray scattering, Nat. Commun. 6 (2015) 6187.
[137] J. Zhao, E. Baibuz, J. Vernieres, P. Grammatikopoulos, V. Jansson, M. Nagel, S. Steinhauer, M. Sowwan, A. Kuronen, K. Nordlund, F. Djurabekova, Formation mechanism of Fe nanocubes by magnetron sputtering inert gas condensation, ACS Nano 10 (2016) 4684–4694.
[138] V.N. Popok, M. Hanif, F.A. Ceynowa, P. Foyan, Immersion of low-energy deposited metal clusters into poly(methyl methacrylate), Nucl. Instr. Meth Phys. Res. B 409 (2017) 91–95.

[139] F.A. Ceynowa, M. Chirumamilla, V.N. Popok, Polymer composite films with size-selected metal nanopartciles fabricated by cluster beam technique, in: V.V. Uglov, L.V. Baran, I.I. Azarko (Eds.), Proceedings of the 12th International Conference on Interaction of Radiation with Solids, Minsk, Sep. 19–22, 2017, pp. 301–303.

[140] N. Pazos-Perez, W. Ni, A. Schweikart, R.A. Alvarez-Puebla, A. Fery, L.M. Liz-Marzan, Highly uniform SERS substrates formed by wrinkle-confined drying of gold colloids, Chem. Sci. 1 (2010) 174–178.

[141] W.P. Klein, C.N. Schmidt, B. Rapp, S. Takabayashi, W.B. Knowlton, J. Lee, B. Yurke, W.L. Hughes, E. Graugnard, W. Kuang, Multiscaffold DNA origami nanoparticle waveguides, Nano Lett. 13 (2013) 3850–3856.

[142] M. Zheng, W. Li, M. Xu, N. Xu, P. Chen, M. Han, B. Xie, Strain sensors based on chromium nanoparticle arrays, Nanoscale 6 (2014) 3930–3933.

[143] J. Erichsen, J. Kanzow, U. Schurmann, K. Dolgner, K. Gunter-Schade, T. Strunskus, V. Zaporojtchenko, F. Faupel, Investigation of the surface glass transition temperature by embedding of noble metal nanoclusters into monodisperse polystyrenes, Macromolecules 37 (2004) 1831–1838.

[144] M.G. Lukashevich, V.N. Popok, V.S. Volobuev, A.A. Melnikov, R.I. Khaibullin, V.F. Valeev, A. Wieck, V.B. Odzhaev, Magnetoresistive effect in PET films with iron nanoparticles synthesised by ion implantation, Open Appl. Phys. J. 3 (2010) 1–5.

[145] O. Kylián, J. Prokeš, O. Polonskyi, J. Čechvala, J. Kousal, J. Pešička, J. Hanuš, H. Biederman, Deposition and characterization of Pt nanocluster films by means of gas aggregation cluster source, Thin Solid Films 571 (2014) 13–17.

[146] D. Tobjork, R. Osterbacka, Paper electronics, Adv. Mater. 21 (2011) 1935–1961.

[147] M. Nogi, N. Komoda, K. Otsuka, K. Suganuma, Foldable nanopaper antennas for origami electronics, Nanoscale 5 (2013) 4395–4399.

[148] P.-K. Yang, Z.-H. Lin, K.C. Pradel, L. Lin, X. Li, X. Wen, J.-H. He, Z.L. Wang, Paper-based origami triboelectric nanogenerators and self-powered pressure sensors, ACS Nano 9 (2015) 901–907.

[149] J. Winfield, L.D. Chambers, J. Rossiter, J. Greenman, I. Ieropoulos, Urine-activated origami microbial fuel cells to signal proof of life, J. Mater. Chem. A 3 (2015) 7058–7065.

[150] M.W. Majeski, I.L. Bolotin, L. Hanley, Cluster beam deposition of Cu2 – XS nanoparticles into organic thin films, ACS Appl. Mater. Interfaces 6 (2014) 12901–12908.

[151] M.A. Arugula, A. Simonian, Novel trends in affinity biosensors:current challenges and perspectives, Meas. Sci. Technol. 25 (2014) 032001.

[152] K. Kon, M. Rai, Metallic nanoparticles: mechanism of antibacterial action and influencing factors, J. Comp. Clinic. Pathol. Res. 2 (2013) 160–174.

[153] M.J. Hajipour, K.M. Fromm, A.A. Ashkarran, D.J. de Aberasturi, I.R. de Larramendi, T. Rojo, V. Serpooshan, W.J. Parak, Antibacterial properties of nanoparticles, Trends Biotechnol. 30 (2012) 499–511.

[154] M. Moritz, M. Geszke-Moritz, The newest achievements in synthesis, immobilization and practical applications of antibacterial nanoparticles, Chem. Eng. J. 228 (2013) 596–613.

[155] H. Palza, Antimicrobial polymers with metal nanoparticles, Int. J. Mol. Sci. 16 (2015) 2099–2116.

[156] A. Kubacka, M. Ferrer, M.L. Cerrada, C. Serrano, M. Sanchez-Chaves, M. Fernandez-Garcia, A. de Andres, R.J. Jimenez Rioboo, F. Fernandez-Martın, M. Fernandez-Garcia, Boosting TiO2-anatase antimicrobial activity: polymer-oxide thin films, Appl. Catal. B Environ. 89 (2009) 441–447.

[157] A. Kubacka, M.S. Diaz, D. Rojo, R. Bargiela, S. Ciorgia, I. Zapico, J.P. Albar, C. Barbas, V.A.P. Martins dos Santos, M. Fernandez-Garcia, M. Ferrer, Understanding the antimicrobial mechanism of TiO_2-based nanocomposite films in a pathogenic bacterium, Sci. Rep. 4 (2014) 4134.
[158] V.N. Popok, C.M. Jeppesen, P. Fojan, A. Kuzminova, J. Hanuš, O. Kylián, Comparative study of antibacterial properties of polystyrene films with TiOx and Cu nanoparticles fabricated using cluster beam technique, Beilstein J. Nanotechnol. 9 (2018) 861–869.
[159] A. Kuzminova, J. Beranová, O. Polonskyi, A. Shelemin, O. Kylián, A. Choukourov, D. Slavínská, H. Biederman, Antibacterial nanocomposite coatings produced by means of gas aggregation source of silver nanoparticles, Surf. Coat. Technol. 294 (2016) 225–230.
[160] O. Kylián, J. Kratochvil, M. Petr, A. Kuzminova, D. Slavinska, H. Biederman, Ag/C:F Antibacterial and hydrophobic nanocomposite coatings, Funct. Mater. Lett. 10 (2017) 1750029.

Chapter 3

Thermal Conduction in Polymer Composites

Nitin Mehra, Liwen Mu, Tuo Ji and Jiahua Zhu
Intelligent Composites Laboratory, Department of Chemical and Biomolecular Engineering, The University of Akron, Akron, OH, United States

Chapter Outline

3.1 INTRODUCTION

From prehistoric usage to modern high-tech computing, thermal conductive materials have always been of profound importance. In our everyday lives, heat conduction is a common sight; including the heating of a cooking pan, heaters in homes, melting of ice, heat exchangers in industry, and so on. Such heat conduction is driven by the temperature difference between the heat source and sink, governed by the famous flux law of heat conduction.

Many industries heavily depend on efficient thermal conductive materials for their operations, such as automobiles, aerospace, electronics, and so forth.

Polymer-Based Multifunctional Nanocomposites and Their Applications
https://doi.org/10.1016/B978-0-12-815067-2.00003-2

More recently, the digital revolution's generation of high-speed processors, with their increased lifetime and reliability, has led to significant research in the field of thermal management materials. As the size of electronic devices is shrinking day by day, more heat is generated in smaller and smaller areas. This is mainly due to increased transistor density in the processor-per-unit area. The use of efficient heat dissipating material is vital in this regard. To circumvent various issues and still achieve efficiency, intense research has been conducted in the past two decades to develop efficient thermal dissipating materials with the desired traits. These materials are generally of three kinds—metal matrix composite, ceramic matrix composite, and polymer matrix composite. Among these, polymer-based materials used for thermal conduction are one of the most promising candidates.

Thermally conductive polymer composites have a wide variety of applications across various areas, such as aerospace, automobiles, the oil and gas industry, and electronics packaging. Thermally conductive material based on polymers provides an edge over other matrix composites in terms of ease of fabrication, good mechanical strength, corrosion resistance, light weight, cost-effectiveness, and so forth. Apart from their thermal conduction property, polymer composites are also popular because of their great versatility and ability to be incorporated with multi-functional properties. This chapter presents the detailed underlying principles and mechanisms driving thermal conduction in polymers and polymer composites, along with their applications. First, a theoretical basis is used to explain the heat conducting mechanism in polymers. Then, various strategies and techniques for the development of thermal conductive polymer composites are discussed.

3.2 FUNDAMENTALS OF PHONON TRANSPORT IN SOLID MATERIALS

Phonon, which is analogous to photon [1], is the quantized energy of the lattice vibration of atoms. When atoms are subjected to heat, they vibrate. This vibration leads to the origin of phonon. Acoustic phonons, along with electrons, are responsible for heat conduction in solid materials. Here, thermal conductivity can be written as a summation of $K = K_p + K_e$, where K_p and K_e are the phonon and electron contributions, respectively. Electrons are dominant carriers of heat in metals. For example, in copper, which has very high thermal conductivity (approximately 400 W/m K), almost all heat is carried by electrons, with a phonon contribution of only 1% to 2% of the total [2]. In non-metals such as ceramic, polymers, and so forth, where there is no electron contribution, the phonon drives heat conduction. Therefore, phonons can be called "heat packets" [3]. The role of the phonon is more dominating in transporting heat in materials in the absence of electrons. Phonons, like photons, have a dual wave-particle nature, and obey Bose-Einstein statistics [4,5]. It can have both a wave and a particle nature. For ease of understanding, we have drawn a

FIG. 3.1 Schematic illustration of phonon.

schematic of phonon (Fig. 3.1). The head in Fig. 3.1 with the letter P presents its particle feature, whereas the tail represents its wave-like nature. The origin of phonon from the atom can be better understood by considering the atomic lattice as a spring-mass system [6]. When mass in the spring is subjected to a force, it vibrates. Atoms in lattice can be assumed to follow similar behavior when subjected to heat. This vibration throughout the lattice generates phonons that ultimately carry heat. The thermal conduction ability of the materials, especially non-metals, greatly depends on their ability to efficiently channel phonon across materials. Phonon also experiences various kinds of phonon collisions, generally known as phonon scattering, within the lattice, which can be affected by different factors, such as the presence of impurities, temperature, interfaces, and so forth [6,7]. The concept of the phonon is vital for the study of the fundamental aspect of heat transport at the micro and nanoscale level in materials.

3.3 THERMAL CONDUCTION IN POLYMERS

3.3.1 Why Are Polymers Traditionally Called Thermal Insulators?

Traditionally, polymers have been known as thermal insulators with thermal conductivity in the order of 0.1 W/m K. Various polymers, natural or manmade, are widely used for providing thermal insulation. Polymer-derived materials such as cellulose, polystyrene, urethane, cork, and so forth are employed for thermal insulation applications in various industries. A thermal conductive material requires a well-defined pathway for the transport of phonon. The absence of such pathways leads to poor thermal conduction in materials. Polymer chains in the bulk polymer are present in a random fashion, like spaghetti. This randomness within polymer chains leads to inefficient transport of phonons within polymers. Phonon packets are scattered within the polymer, which is called phonon scattering. There are three different kinds of phonon scattering modes: phonon-phonon scattering, phonon-impurity scattering, and phonon-boundary scattering. Each of these scattering mechanisms renders polymer as thermal insulator in its own way. Phonon-phonon scattering is the collision between two phonons, which can be further classified into umklapp (U) and normal (N) processes. These are inelastic phonon collisions due to the anharmonicity of interatomic potential [8–10]. In N scattering, both energy and momentum are conserved, while in U scattering, only energy is conserved.

U scattering is mainly predominant at higher temperatures. Phonon-impurity scattering is mainly due to the presence of defects, dislocations, vacancy defects, and so forth [11] in the material. Such impurity can substantially affect the thermal conductivity of material. For non-metallic crystals, there is a strong correlation between thermal conductivity and defects [12,13]. For example, lattice oxygen in crystals of AlN and Si_3N_4 act as dominating factors to lower their thermal conductivity, and similarly, for CNT, where defects and vacancy concentration drastically hampers its thermal conduction [13–17]. Phonon-boundary scattering is mainly seen at interfaces or boundaries. When phonon cross-boundaries present within the material, its effective mean free path changes due to scattering [18]. Such phenomena are more significant at lower temperatures, when the effective length of phonon is longest. Boundary scattering also depends upon the characteristic length of the material. Thermal transport in material deviates when the phonon mean-free path is comparable to the characteristic length of the material [19–21]. Phonon can be seen either scattering specularly or diffusively at the interface region. In polymer composites, the filler-matrix interface is one of main reasons for the large phonon scattering, due to an acoustic mismatch [7,22–24]. All of these different factors contribute to the poor thermal conductivity of polymers, rendering them thermal insulators. Circumventing these bottlenecks is one of the key challenges for the development of polymer-based, thermally conductive materials.

3.3.2 Factors Playing a Critical Role in Thermal Conduction in Polymers

As per theoretical predictions, thermal conductivity mainly depends on three factors; namely, specific heat, a phonon mean-free path, and phonon velocity (Eq. 3.1). This expression is based on the kinetic theory of gases, after assuming phonon as a pseudo-particle in nature [11,25].

$$K = \frac{1}{3}\int C_v(\omega)\cdot v(\omega)\cdot L(\omega)d\omega \tag{3.1}$$

where C_v is the heat capacity at a constant volume, v is the velocity of the phonons, L is the phonon mean-free path—the characteristic length before the phonons are scattered. These factors control the microscopic heat transport, and further depend on various other parameters that could be intrinsic to a particular polymer.

"Specific heat" is defined as the ability of the material to store heat. Apart from storage capacity, specific heat can also be used to understand the rate of cooling and the heating ability of the material [26]. In any material, this heat can be stored in the form of lattice vibrations and free conduction electrons. Due to the absence of electrons in most of the polymers, phonons or lattice vibrations play a dominant role. Mathematically, it can be calculated as a partial derivative of the total energy of the phonons [27], with respect to the temperature at a constant volume. The values of specific heat changes with temperature, and can take various forms in different temperature regimes.

Phonon group velocity is the velocity with which the phonon propagates. It can be represented through a dispersion relation, as shown in Eq. (3.2)

$$\lim_{qph\to 0} v = \lim_{qph\to 0} \frac{d\omega}{dq_{ph}} = C_s = \sqrt{\frac{E}{\rho}} \tag{3.2}$$

where ω *is* frequency, q_{ph} is wave vector, E is stiffness and ρ is mass density. The group velocity reaches the speed of sound c_s in the long wavelength region of crystal. Group velocity is affected by the wavelength of the phonon, which is inversely related to the wave vector. In the case of a longer wavelength, phonon velocity is large, and vice-versa. The upper boundary of such velocity is the speed of sound. Stiffness of the material can also be related to its strength of bonding interactions. Phonon velocity generally increases with a high elastic modulus of material. In general, polymers with higher stiffness is assumed to facilitate phonon transport though depending on case to case overall thermal conductivity can be contributed by various other factors. Therefore thermal conductivity in polymers may not be seen in isolation to only one particular factor. The mean-free path is defined as the path travelled by phonon before being scattered due to phonon-scattering. As discussed before, the collision of a phonon with another phonon, impurities, boundaries, and so forth lead to phonon scattering, which results in a significant decrement in thermal conductivity. An effective phonon mean-free path can be expressed as a summation based on Matthiessen's rule

$$\frac{1}{L_{eff}} = \frac{1}{L_{ph}} + \frac{1}{L_b} + \frac{1}{L_{im}} \tag{3.3}$$

where L_{ph}, L_b, and L_{im} are the mean-free path of phonon-phonon, phonon-boundary, and phonon-impurity scattering processes, respectively. Eq. (3.3) provides useful information about the dominating influence of the phonon scattering mechanism.

Phonons require a well-defined pathway for their transport, and in general, bulk polymers are bereft of this characteristic. It is ironic that although carbon-carbon bonds are tremendously strong and can provide efficient pathways for the phonon, bulk polymers are traditionally thermal insulators. In fact, it is reported that the thermal conductivity of a single polymer chain can even reach thermal conductivity of the order of 100 W/m K, even beating values of some of the metals [28]. It's surprising that thermal conductivity of the ultra-drawn polyethylene nano-fiber with a diameter of 50–500 nm can have thermal conductivity as high as 104 W/m K [28], whereas the thermal conductivity of bulk polymer is poor, of the order of 0.1 W/m K. This phenomenon of decrement in thermal conductivity from a single polymer chain to bulk can be better understood through examining Fig. 3.2. As can be seen, thermal conductivity gradually decreases from a single chain to bulk crystal. It signifies that, as the

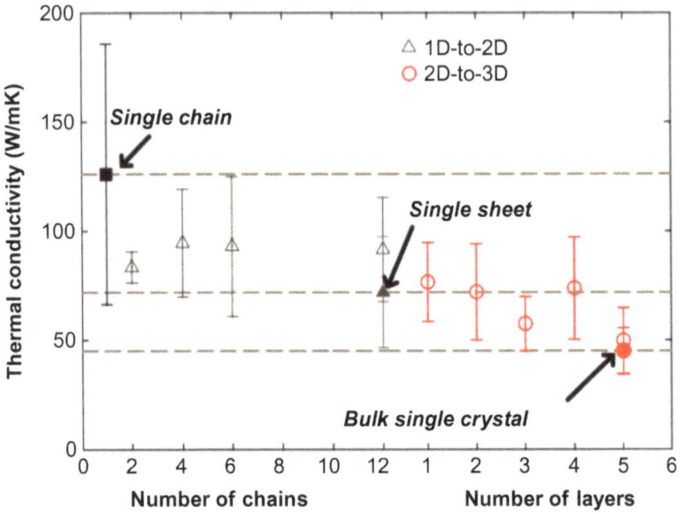

FIG. 3.2 Thermal conductivity results from the Green-Kubo analysis, showing the 1D-2D *left* and 2D-3D *right* transitions. The *left* portion of the horizontal axis is given in a number of molecular chains for the 1D-2D transition, while the *right* portion of the horizontal axis is given in terms of the number of 2D layers being stacked for the 2D-3D transition. Dashed lines indicate the values obtained for the 1D single chain limit, the 2D sheet limit, and the 3D bulk crystal limit, which show the increasing trend with decreased dimensionality [29].

dimension of material changes from 1D to 3D, the phonon transport is significantly obstructed. This phenomenon is more pronounced in the bulk polymers [30,31]. Phonon group velocity for a single polymer chain of polyethylene can reach as high as 16,000 m/s [32,33]. Experimentally, it has also been proven that the transition from two atomic planes to four drastically reduces the thermal conductivity of a few layers of graphene from ~2800 to ~1300 W/m K [33]. Here, U scattering was found to negatively impact phonon propagation.

Branched chains or bulk polymers tend to have lower thermal conductivity than a single-polymer chain. Thermal conductivity witnesses a many-fold decrement from a single-polymer chain to bulk due to drastic phonon scattering present in bulk polymers. A high elastic modulus within carbon atoms is reflected in their thermal conductivity. It's the same reason why carbon-based fillers have some of the highest thermal conductivity values of more than 1000 W/m K [34–37].

Other important factors that can significantly contribute to the thermal conductivity of polymers are intermolecular interactions, crystallinity, molecular weight, and so forth. The arrangement of polymer chains, whether randomly or orderly fashioned, can impact thermal conduction. For this reason, in general, semi-crystalline polymers with ordered chains have higher thermal conductivity than the amorphous polymers [30,31,38–40]. Among these, intermolecular interactions, such as hydrogen bonding, can serve in developing thermal bridges within the polymer chains [3,41]. These bridges drive phonon transport across polymer chains.

3.4 THERMAL CONDUCTION IN POLYMER COMPOSITES

Polymer composites are increasingly employed for the development of novel materials. The versatile nature of polymer composites in terms of tailoring end products for enhancing thermo-mechanical properties make them popular for a wide variety of applications [41–45]. Due to their multifarious nature, they are also an interesting material for thermal conduction. In fact, the past two decades have witnessed immense research in the development of polymer-based composites for thermal conductive applications. These composites, in general, are obtained by the incorporation of thermal conductive fillers in the polymer matrix. Depending on the properties of the filler and matrix, polymer composites can achieve different thermo-mechanical properties. Although the neat polymers have poor thermal conduction, the addition of thermal conductive fillers infuse thermal pathways, which in turn drive thermal phonons. This is how, in general, a thermal insulating polymer becomes thermally conductive. Greater flexibility in the selection of desired properties at the same time being cost effective makes them a widely popular material for thermal conductive applications.

3.4.1 Challenges

Although polymer composites are widely used for thermally conductive purposes, one of the major challenges they have been facing is multiple sites of phonon scattering due to the presence of interfaces. Filler-matrix interfaces present in the polymer composite are the reason for manifold decrements in the effective thermal conductivity of composites. The interface is like a transition area where properties changes locally, and serve as a phonon scattering center. Properties of composites in the region of the interface are different from the properties of bulk [46,47]. Incorporated filler alters the property of the matrix in the vicinity of the filler. This leads to the development of an interfacial layer, which offers interfacial thermal resistance for the flow of phonons. This layer, apart from phonon scattering, also influence glass transition temperature, crystallinity, and the thermal conductivity properties of composites [48]. In nanocomposites, the effect of interfacial resistance is more pronounced due to the large increment of the filler-matrix area.

The requirement of high loading of fillers is another bottleneck for the development of high thermal conductive material. Due to high loading, the polymer loses its intrinsic properties, such as good mechanical properties, ease of fabrication, and so forth. After a certain loading, thermal conductivity increases, but at the cost of mechanical properties. High loading also adds substantially to the weight of the polymer. It's also important to have a cost-effective material. Many of the carbon and ceramic-based fillers such as carbon nanotubes, boron nitride nanotubes, and so forth are expensive, and significantly increase the cost of the composite. To circumvent all these bottlenecks, effective strategies need to be adopted. These are discussed in subsequent sections.

3.4.2 Carbon Filler-Based Polymer Composites

Polymer composites with carbon filler are widely used in industries and serve as promising materials for thermal conductive applications. There are many allotropes of carbon that can be used for thermal conductive purpose. These allotropes' thermal conductivity widely spreads from 0.01 W/m K in amorphous carbon, to 2000 W/m K of graphene (Fig. 3.3). Heat conduction in carbon-based material, in general, is dominated by the phonons; although some materials, such as graphite, behave like metals. Strong covalent bonding with sp^2 hybridization of carbon atoms in the material lead to efficient transport of phonons through lattice vibrations [49]. Various carbon-based fillers are available, such as carbon fibers [50], carbon nanotubes [51–57], carbon nanofibers [58], graphite [59,60], graphite nanoplates [61], graphene [62], fullerenes [63], nano-diamonds [64,65], carbon black [66,67], and so forth. High thermal conductivity is one of the most important features of carbon-based material. Graphene can have tremendously high values, ranging from 3000 to 5000 W/m K [36,68]. CNT can have thermal conductivity as high as 3000 W/m K, or even higher [34,69,70]. In addition to high thermal conductivity, they are lightweight with fairly good dispersion ability. Different theoretical models are available to estimate effective thermal conductivity of composites. For example in CNT, as per Nan's model, for every 0.1 wt% increment of CNT, thermal conductivity increases by 6 times, and similarly, according to Maxwell's mixing theory, for every 1 vol% CNT loading, the value enhances by 55 times [71–73]. However, due to high interfacial resistance by CNT, the theoretical model may not translate into experimental findings [74].

Thermal conductivity of various carbon fillers is presented in Fig. 3.4. CNT, diamonds, and graphene are the ones with the highest values. Fig. 3.4 further

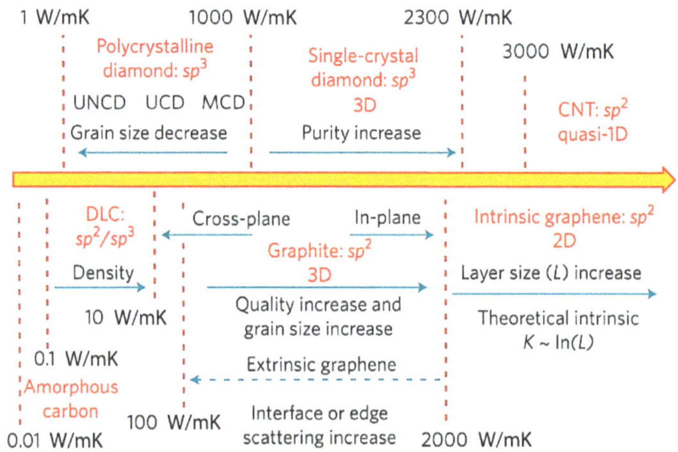

FIG. 3.3 Thermal conductivity of carbon allotropes [2].

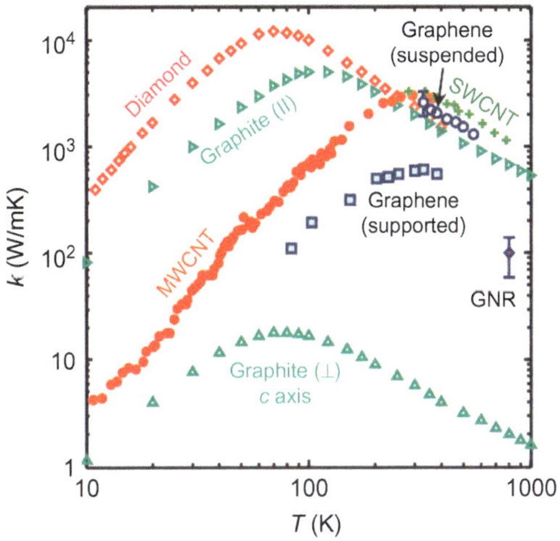

FIG. 3.4 Thermal conductivity of carbon-based filler with change in temperature [26].

presents the change in thermal conductivity with temperature. Thermal conductivity gradually increases as the temperature increases, due to an increase in the specific heat of materials [26] in the region of less than 100 K, and then decreases due to the dominance of phonon scattering.

Some recent, interesting polymer-based composites using carbon fillers are presented here. Depending on the interaction of the filler with the matrix, dispersion ability, fabrication method, thermal conductivity of the individual filler, and the matrix, the effective value of thermal conductivity can acquire a wide range of values. Fig. 3.5 presents a polystyrene composite made of hybrid fillers of nanoplates (GNPs) and multi-walled carbon nanotubes (MWCNT). This is based on a segregated double network for heat conduction. The thermal conductivity of the system increases by almost 1.8 times that of a random dispersed hybrid network. A value greater than 1.0 W/m K was achieved with a loading of 3.5 vol% GNP and 1.5 vol% MWCNT [75]. The fillers lead to the efficient development of thermal pathways by increasing the interconnection through overlapping of GNP and the synergistic effect between GNP and MWCNT. Fig. 3.6 exhibits a high thermally conductive epoxy-CNT composite prepared by infiltration of epoxy in vertically grown CNT arrays. A value as high as 8.2 W/m K was achieved after curing [76]. This composite provides an interesting way to transfer vertically grown CNT arrays to polymer film with improved robustness and resistance to cracking. The coating of carbon fillers with higher thermal conductive materials could be another way to enhance thermal conduction abilities. High conductive material such as copper can be coated on carbon fiber using electroplating followed by rapid thermal annealing. This process results in the development of a continuous thermal network, and at the same time reduces crystal defects in copper that can be borne due to

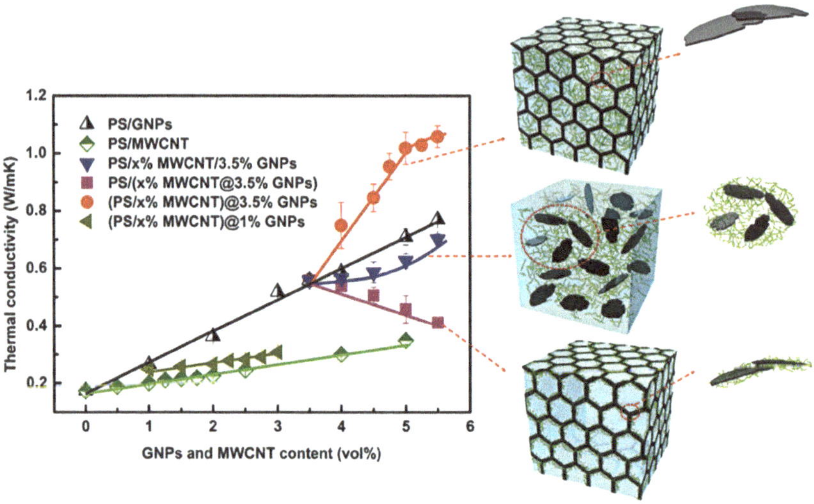

FIG. 3.5 Thermal conduction in epoxy composite using synergistic effect between GNP and MWCNT [75].

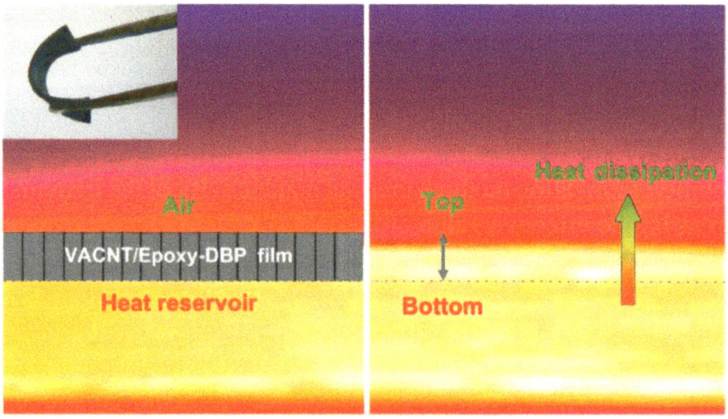

FIG. 3.6 High thermally conductive epoxy-CNT composite prepared by infiltration of epoxy in vertically grown CNT arrays [76].

coating (Fig. 3.7). This epoxy composite using copper-coated carbon fiber resulted in cross-plane thermal conductivity of 3.9 W/mK at a filler loading of 12.0 vol% [76].

3.4.3 Ceramic Filler-Based Polymer Composites

Ceramic fillers are another important class of material that offers excellent thermal conductivity, along with electrical insulation. This combination of interesting properties is desired for the electronic packing industry. Bonding in ceramic materials is usually a combination of ionic and covalent, but sometimes could

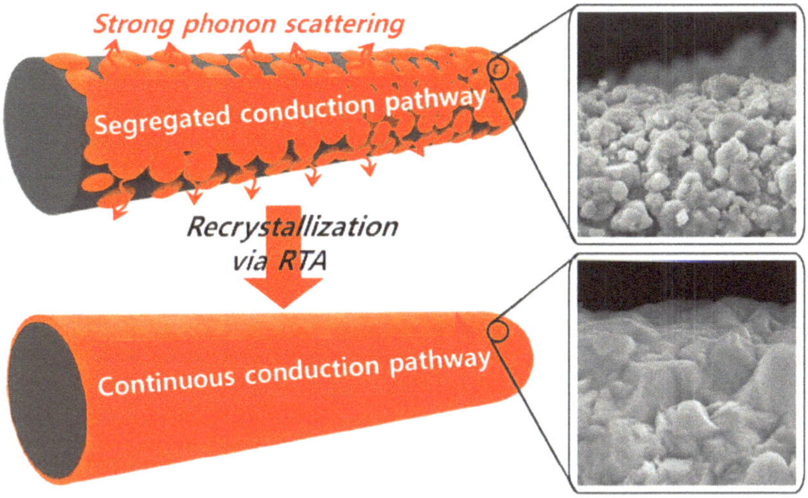

FIG. 3.7 An epoxy composite using copper-coated carbon fiber for thermal conductive applications [77].

also have a Van-der-Waals or metallic component [4]. There are several ceramic fillers available for reinforcement, such as AlN [78–86], Al_2O_3 [87–92], SiC [89,93] [94,95], Si_3N_4 [96,97], $Sr_2Ce_2Ti_5O_{16}$ [98], zirconium silicate ($ZrSiO_4$) [99], wollastonite ($CaSiO_3$) [100], silicon dioxide (SiO_2) [83,92,101], beryllium oxide (BeO), CeO_2 [102], boron nitride (BN) [4], and ZnO [87,103].

Boron nitride is the most extensively used filler for thermal conduction. The relatively high thermal conductivity of boron nitride, along with its electrical insulation, has drawn the attention of various researchers for reinforcement in polymer composites. Silica is another abundantly used material in microelectronic applications, for its high electrical resistivity and low dielectric constant although its thermal conductivity is low [104]. Various other fillers are also being investigated for their ability to develop efficient thermal conductive polymer composites. Polymer composites based on ceramic materials possess high thermal conductivity with electrical insulation, which is ideal for the electronic packaging industry. These composites can also have superior thermal degradation temperatures, depending on the performance of the host matrix.

Several strategies can be developed for efficient ceramic-based polymer composites. Fig. 3.8 presents thermal conductivity enhancement in noncovalently functionalized boron nitride nano-sheets (NF-BNNSs) with chlorosulfonic acid (CSA) and polymethyl methacrylate (PMMA) composite films[105]. PMMA is a poor heat conductor, with a thermal conductivity of 0.22 W/m K. Reinforcement of such NF-BNNS in PMMA leads to increased thermal conductivity of 5.26 W/m K at 50 wt% and 10.22 W/m K at 80 wt% loading. Adsorption of CSA on NF-BNNSs increased the dispersability and interaction with the host matrix, resulting in improved thermal conduction and fabrication. As discussed,

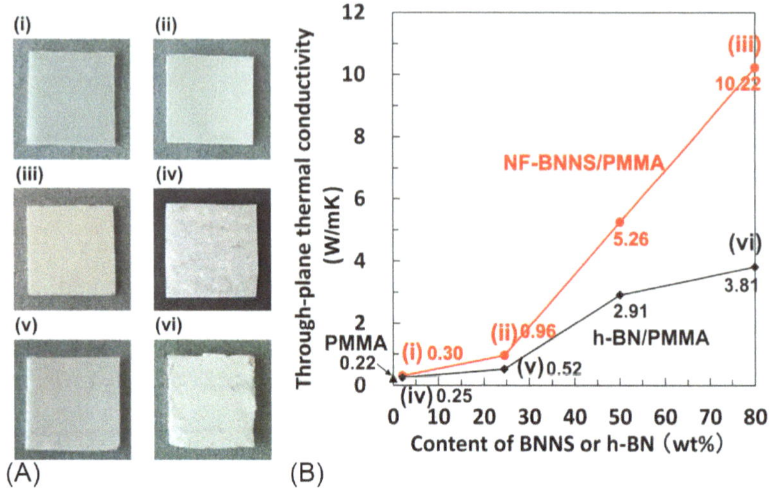

FIG. 3.8 Functionalized boron nitride nanosheets and their use for highly thermally conductive and electrically insulating polymer nanocomposites (A) (i)-(vi) Photographs of films (10 mm × 10 mm × 0.3 mm) of NF-BNNS/PMMA composites having different BNNS loadings. (B) Through-plane thermal conductivities of films presented in (A) [105].

ceramic fillers can also provide electrical insulation in the matrix. When these NF-BNNS fillers were incorporated in polybutylene terephthalates (PBT), volume resistivity in the composite, along with thermal conductivity, were found to improve compared with pristine PBT.

Homogenous dispersion of nano-particles is another challenge for the development of continuous thermal pathways. In the absence of homogenous distribution, agglomeration and phase separation take place, which leads to the formation of boundaries. These boundaries can act as a phonon scattering center, and negatively impact thermal conductivity. In one of the studies, the importance of the homogenous distribution, and control of the particle size have been highlighted through an Si_3N_4/phenolic resins composite (Fig. 3.9) [106]. Here, Si_3N_4 nanoparticles were prepared from a radio frequency (RF) thermal plasma system. The synthesized particles were obtained with a narrow size distribution, and could achieve better dispersions within the matrix. The thermal conductivity of 3.25 W/m K was obtained at 50 vol% filler loading. The dielectric constant of composite gradually decreases with increasing loading, and breakdown strength is enhanced. Additionally, such polymer composites also have excellent thermal stability.

3.4.4 Metallic Filler-Based Polymer Composites

Although metallic fillers come with some of the highest thermal conductivities, such as copper, they can also lead to high weight and fabrication issues. Apart

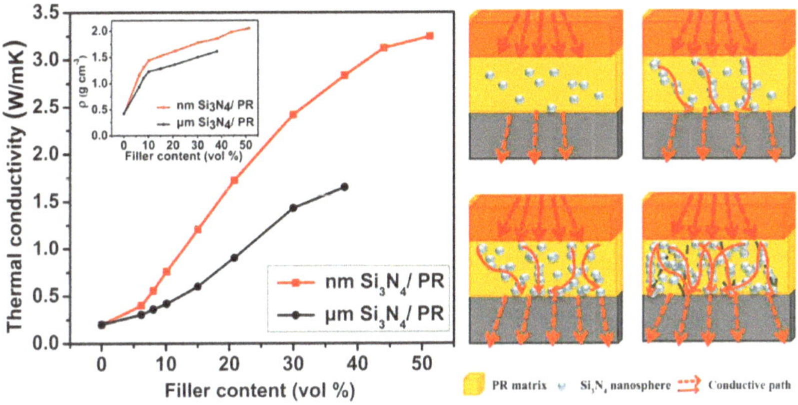

FIG. 3.9 High thermal conductive nanocomposites incorporated with uniform α-Si_3N_4 nanospheres by RF induction thermal plasma [106].

from high thermal conductivity, these composites can also have electrical conduction, unless modified, which may be a disadvantage for the electronic packaging industry. Thermal conductivity of composites is not the only desired property for semiconductor chips, but also low coefficients of thermal expansion (CTE) [107]. Copper, however, has high thermal conductivity, but has also high CTE. The printed circuit boards of many electronic devices are polymer matrix composites with low CTE. When there is a mismatch in the CTE of components and the temperature is varied, the generation of thermal stress may be detrimental to the device. Therefore, copper is incorporated with low CTE fillers, such as carbon fibers or molybdenum. Aircraft electronics and lightweight electronic devices use aluminum, although its thermal conductivity is lower than that of copper. Aluminum fits because of its light weight, which is almost 4 times lower than copper, making it suitable for lightweight applications such as laptop computers and avionics [107]. The thermal conductivity of copper, zinc, bronze, and iron filler reinforcements with high-density polyethylene (HDPE) has been investigated in one of the studies [108]. It has been found that in lower ranges of loading of 0–16 vol%, thermal conductivity follows theoretical predictions; whereas at higher loading, thermal conductivity increases more rapidly than the predicted values. The formation of particle aggregation leads to an exponential increase of thermal conductivity. At 20 vol%, the obtained thermal conductivity was around 0.9 W/m K for zinc and copper, 0.8 W/m K for iron and 1.0 W/m K for bronze. One of the more advanced techniques to incorporate metal into the polymer matrix is through metallic nanowire fillers with heat treatment [109]. When reinforced composite is subjected to heat, micro-wielding of nano-wire takes place. This leads to the formation of a continuous thermal network within the polymer matrix due to coalescence and inter-diffusion [110]. In one such reported work using microwave welding of Ag nano-wire in a PDMS matrix, a 40-fold increment in thermal conductivity was achieved at a filler fraction of ≤ 0.04 [109]. Similarly, a tree-shaped silver network at the nano-scale in epoxy resulted in a 200-fold increment in thermal conductivity [111]. Metallic nanowires such as Ag, Au, and so forth with high aspect ratios are generally reinforced in polymer for welding purposes.

3.5 STRATEGIES TO ENHANCE THERMAL CONDUCTION

Polymer composites, when reinforced with filler, traditionally without any modification, usually result in low thermal and mechanical performance. By keeping in mind that phonon transport takes place more efficiently when there are continuous thermal pathways and less interfacial resistance, different strategies have been employed to further push the limits. Strategies discussed in this section attempt to circumvent challenges present in polymer composites to enhance thermal conduction. The outcome of such strategies could be reduced filler loading while improving thermal conductivity, better thermo-mechanical properties, introducing additional functionality while preserving thermal conduction, and so forth.

3.5.1 Filler Alignment

Ordered pathways of filler for enhanced thermal conductivity can be achieved through various external forces [112–125]. Several methods can be used for this purpose, such as electrical fields [126], tape casting [127], freeze-casting, and so forth [128]. One of the popular techniques for filler alignment is use of the magnetic field [112,118,123,129–133]. It can be used to arrange filler in the desired direction in the polymer composite, depending on the magnetic properties of the filler. Filler has to be magnetically responsive, and this could be achieved by either coating a magnetic particle in the reinforcing particle, or using reinforcing particles that are magnetically responsive [122,133]. Superparamagnetic materials such as iron oxide can be employed as a coating or embedding-reinforcing filler to make them responsive under external magnetic fields. Several fillers such as CNTs [120,121,124,125], graphite nanosheets [113], and graphene, [114,119,123] have been aligned in polymer matrices for such purposes. Fig. 3.10 presents the reinforcement of hexagonal boron nitride platelets in a silicone gel matrix under a magnetic field. SEM images show the alignment of filler from the top and cross view. The thermal conductivity of the composite after magnetic alignment was 44.5% higher with platelets oriented parallel to the heat flow, and 37.9% lower in unaligned composites having 9.14 vol% loading. Various other polymer composites have been prepared with similar approaches.

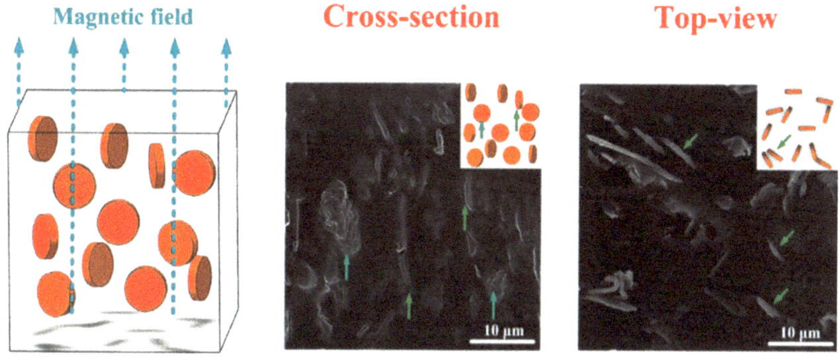

FIG. 3.10 Use of magnetic fields to assemble the hexagonal boron nitride platelets into various microstructures and study of the thermal conductivities of the designed composites [131].

3.5.2 Filler Surface Modification

Another attractive technique that offers more opportunity to manipulate the properties of fillers is surface modification. Many times, the polymer composite suffers from poor dispersability, which leads to agglomeration, and hinders phonon transport. Interfacial thermal resistance, which is one of the critical challenges in polymer composites, also needs to be addressed. The advantage of filler surface modification is that it can address both filler dispersability and interfacial filler-matrix interactions [126]. Modification of the filler can be achieved through various chemistries, depending on its interaction with the filler and the matrix. Usually, different functional groups are non-covalently/covalently attached on the surface of the filler, which, in turn, can enhance both the dispersion and filler-resin interaction. Selection of functional groups specifically depends on their interaction with the matrix. In boron nitride nanotubes (BNNTs)/polyvinyl formal (PVF) and BNNTS/polyvinyl alcohol (PVA) composite films (Fig. 3.11), BNNTs were surface modified with epicatechin gallate molecules, also known as "catechin"[134]. These molecules have numerous aromatic rings and can efficiently interact with PVF with π-π bonding. This helps in the enhanced interaction between filler and matrix, resulting in the increased thermal conductivity by ~160% at 1 wt% in BNNTs/PVF composite and ~270% at 3 wt% loading in BNNTs/PVA. Depending on the chemistry selected for modification, the same filler can be equipped with different functionality. Like in aluminum nitride(AlN)/epoxy composite, six treatments were made on AlN to obtain modified fillers with varied functionalities (Fig. 3.12) [135]. AlN was treated with three kinds of silane, having the terminated groups as amino, epoxy, and mercapto; others were functionalized with polyhedral oligomeric silsesquioxane (POSS), and hyperbranched polymer and graphene oxide (GO). Composites with different functionalized AlN were found to have varied thermal conductivity. The highest was of mercapto-functionalized AlN, which could achieve 6 W/m K at the loading of 65 vol%. This thermal

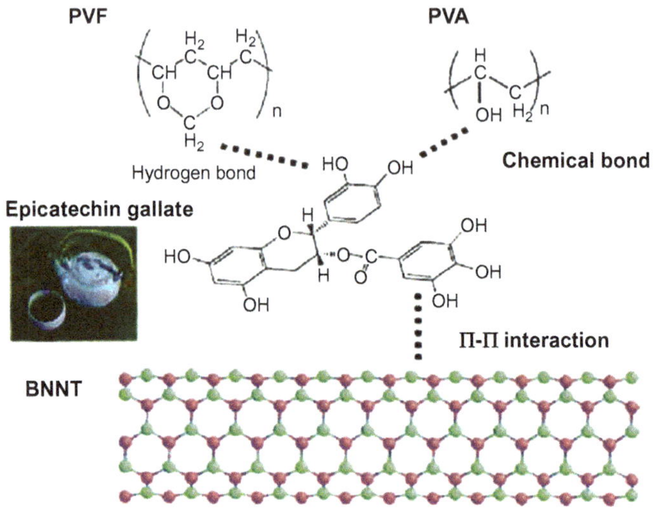

FIG. 3.11 Boron nitride nanotubes (BNNTs) modified by epicatechin gallate molecules for enhanced thermal conductivity in PVA and PVF composites [134].

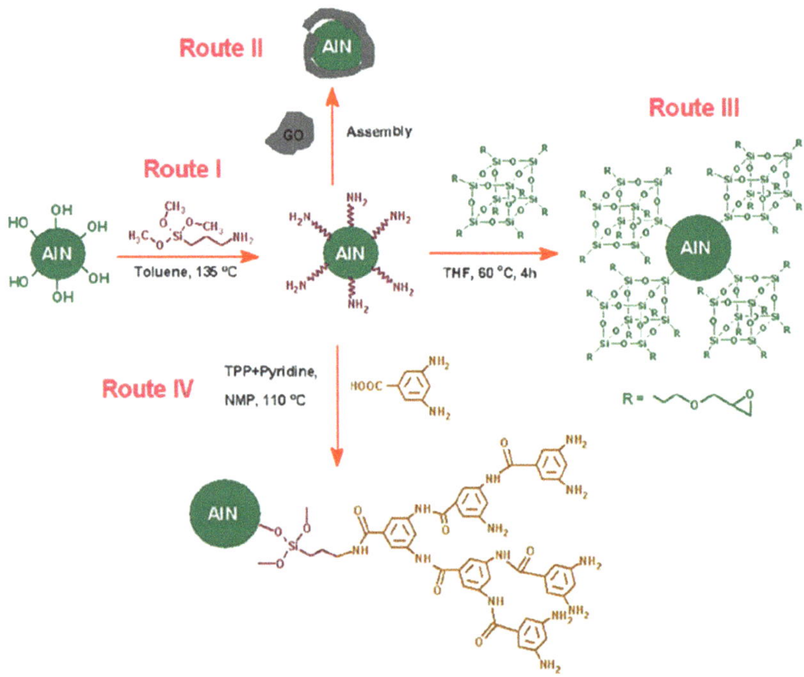

FIG. 3.12 Surface treatment and assembly procedures of AlN particles. Route I: preparation process of amino-functionalized AlN. Route II: assembly of amino-functionalized AlN and GO. Route III: reaction between amino-functionalized AlN and POSS. Route IV: preparation of hyperbranched polymer modified AlN [135].

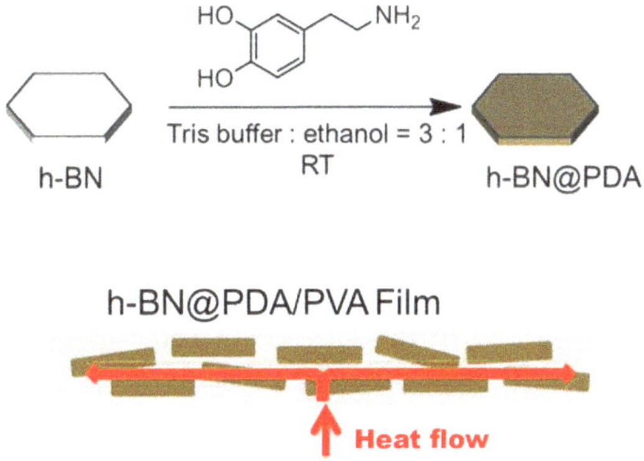

FIG. 3.13 Surface modification of h-BN microplatelets by dopamine chemistry and heat flow in h-BN@PDA/PVA composite [126].

conductivity was around 60% higher than other composites with similar filler loading. In another composite, hexagonal boron nitride microplatelets were modified using polydopamine (PDA) chemistry to enhance the dispersability and interaction in the PVA matrix (Fig. 3.13) [126]. These composites with surface modified fillers were fabricated using a doctor blade, which produced the aligned structure. The resultant in-plane thermal conductivity of composite film due to surface modification and aligned structure was 5.4 W/m K at 10 vol% loading.

3.5.3 Hybrid Fillers

Hybrid filler reinforcement is yet another fascinating technique used to combine properties of one or more fillers in the composite. 'Hybrid' means something made by the amalgamation of two or more different characters or properties. This includes a combination of fillers which are either chemically or physically different, or both. In polymer composites, a symbiotic phenomenon due to the combination of different fillers, which is also known as the synergistic effect, can greatly enhance thermal conductivity. Multiple fillers can be used to integrate the advantages of all fillers in a single composite [136–140]. Selection and pairing of fillers can be critical to achieving thermal conduction with enhanced performance. For example, carbon-based composites usually have high thermal conductivity, but come with the disadvantage of electrical conduction. This makes them unsuitable for electronics industries. To circumvent this, carbon fillers can be mixed with ceramic fillers, such as boron nitride, to reduce electrical conduction while preserving thermal conduction. Various studies have revealed that polymer composites incorporated with hybrid fillers perform better than those with single fillers [141–143]. The pairing of boron nitride-graphite or boron nitride-graphene is found to have a significant synergistic effect in

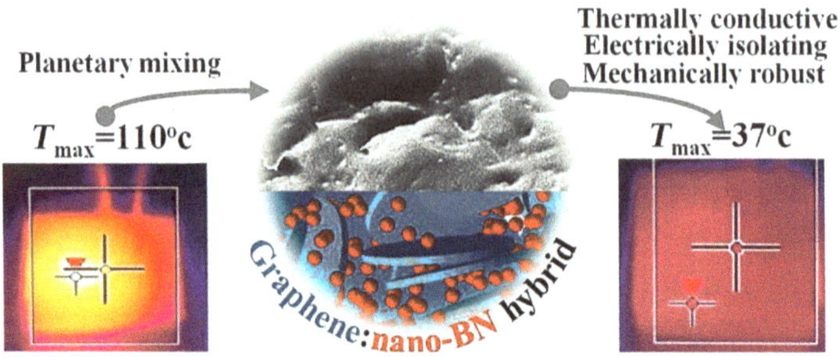

FIG. 3.14 Schematics of epoxy composite reinforced with hybrid filler of graphene and boron nitride. The *blue* platelets and *red* spheres indicate GnP/μm-BN and nm-BN, respectively [42].

driving phonons. When graphene or BN nanoplatelets with lateral dimensions of a micrometer and nanosize thickness were dispersed in epoxy, enhanced thermal conduction of around 4.72 W/m·K was achieved (Fig. 3.14) [42]. Such thermal conductivity was the result of the hybrid combination of graphene nano-platelets and nano-meter boron nitrides at 0.16 vol% and 0.01 vol% in epoxy. Such a composite was found to have low electrical conductivity and enhanced mechanical properties. When used in electronic devices, these composites effectively dissipate heat, reducing the temperature from110°C to 37°C. A combination of fillers that are chemically the same, but with different physical dimensions, can also create a synergistic effect. For example, the combination of a single, bilayer, and multi-layer graphene was found to enhance thermal conductance by 2300% in an epoxy nano-composite (Fig. 3.15) [43]. The combination of

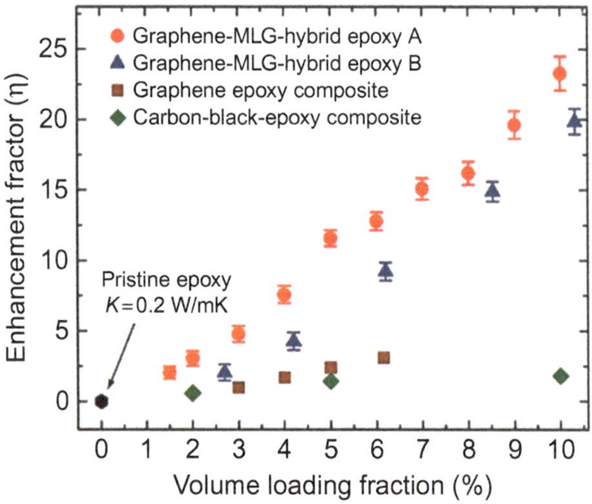

FIG. 3.15 Thermal conductivity enhancement in the graphene-MLG-nanocomposite epoxy TIMs [43].

single and bilayer graphene flakes at a concentration of ~10%–15% with multilayer graphene with a thicker dimension leads to such high enhancement. The thermal conductivity of 5.0 W/m K was achieved at the filler loading of 10 vol% for an optimum nano-composite.

3.5.4 Other Strategies and Materials

3.5.4.1 Optically Transparent, Thermally-Conductive Materials (OPTTCM)

As discussed before, although polymers are traditionally thermally insulating materials, the thermal conduction within a single polymer chain, or within covalent bonds, can be relatively very large. By properly engineering intermolecular interactions within polymers, phonons can be efficiently channelized [41,44]. Hydrogen bonding, or the arrangement of polymer chains to increase crystallinity, can pave the way for thermal conduction within polymer chains. OPTTCM exploits these properties of polymers to achieve enhanced phonon transport, along with optical transparency. The traditional polymer composite made of filler-matrix systems are opaque, and can have fabrication issues or undesired mechanical properties. OPTTCMs are free from fillers such as ceramics/carbonics/metallic, and so can offer a better handle on optical and mechanical properties. Such materials are of great interest in achieving properties such as optical transparency, light weight, mechanical flexibility, and so forth, along with good thermal conduction. Fig. 3.16 exhibits the potential of hydrogen-bonded 'thermal bridges' for phonon propagation [3]. A thermally conductive polymer-blend film was prepared by selecting two compatible polymers. Polyvinyl alcohol (PVA)-Polyethylene glycol (PEG) blend films were prepared using short PEG chains. A thermal conductivity enhancement of 1.6 times the neat PVA was achieved through this system. Such enhancement was inversely related to the molecular weight of PEG, which acts as a thermal bridging chain. The key factor governing phonon movement was homogenous distribution of thermal bridges throughout the polymer chain. Obtained films had a high transparency of more than 80%. Arrangement of the polymer chains in a different fashion to form crystals is another strategy to develop thermal

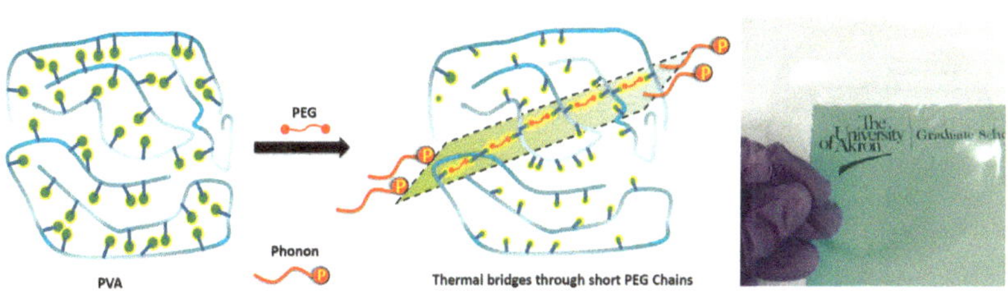

FIG. 3.16 Development of hydrogen bridges through short PEG chains in PVA for optically transparent thermally conductive polymer films [3].

pathways without the addition of any traditional fillers. Fig. 3.17 presents the effect of adding dicarboxylic acids in PVA to form OPTTCM with 75% plus transparency [45]. Such films have a high thermal conductivity of 0.7 W/mK, along with good mechanical properties. The continuous crystal structure in these films was the dominating factor for high thermal conduction. The thermal conduction greatly depends on the quality of intermolecular interactions in polymers films. The microstructure of the coil-coil interaction in a solid polymer can significantly affect thermal conduction. Fig. 3.18 presents the effect of the molecular-level interaction to the macro-scale heat conduction in polymer

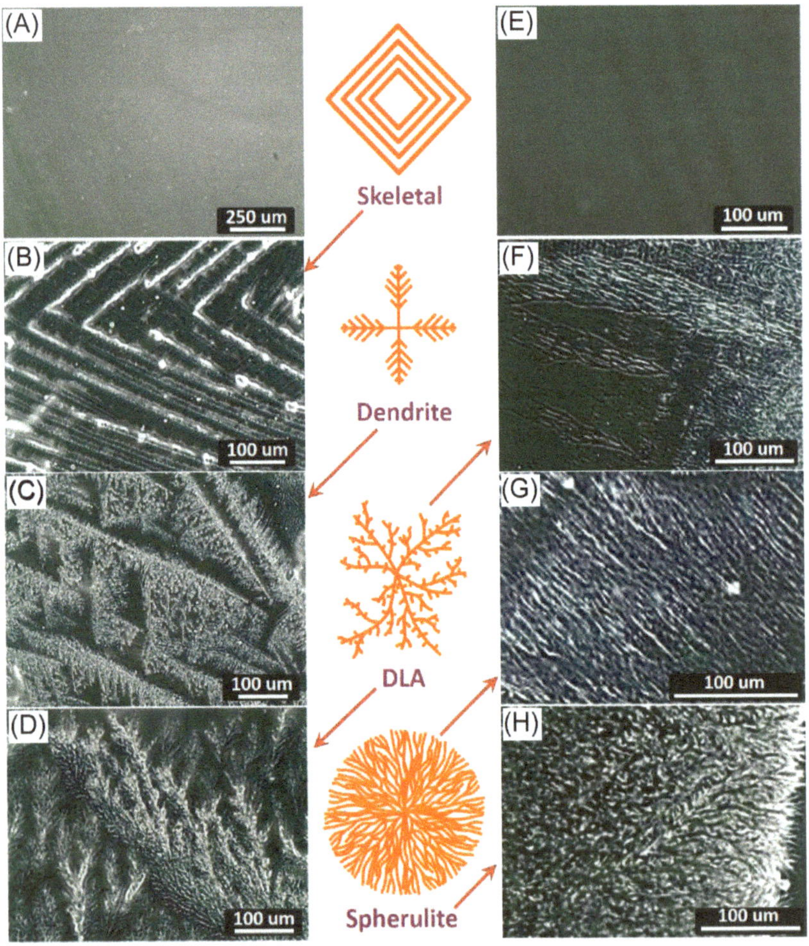

FIG. 3.17 The schematic pictures, skeletal, dendrite, DLA, and spherulite represent the typical crystal structures formed in Dicarboxylic acid-PVA film Optical microscopy images of (A) pure PVA, (B) PVA/OA-10, (C) PVA/OA-20, (D) PVA/OA-30, (E) PVA/SA-10, (F) PVA/SA-20, (G) PVA/SA-30, and (H) PVA/SA-40 [45].

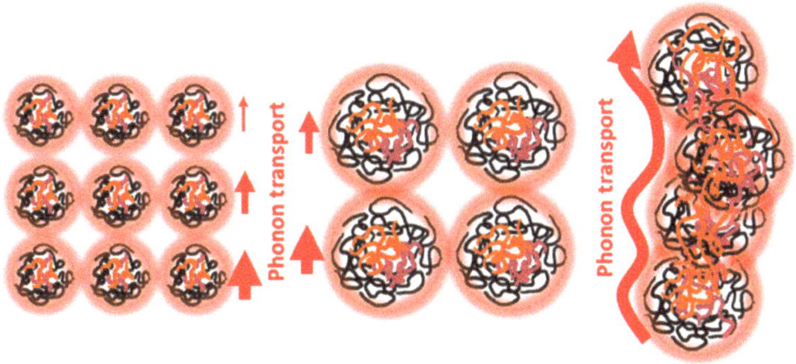

FIG. 3.18 Illustration of intermolecular interaction by hydrogen bonding and photon transfer across polymer blends with small, large, and continuous coil structures [144].

blends of PVA with biopolymers (lignin, gelatin). Development of a continuous coil structure is critical for the flow of phonons. Incorporation of gelatin and lignin in PVA leads to a synergistic effect, which increases thermal conductivity up to 0.71 W/m K at 10 wt% loading of each gelatin and lignin in blend film [144].

3.5.4.2 Thermally Conductive Soft Elastomers

Materials with high thermal conductivity and low mechanical stiffness are very much desired for stretchable electronics, soft robotics, and so forth. Typically, thermally conductive polymer composites reinforced with fillers are rigid, and result in mechanically stiff material that lack stretchable functions required for stretchable electronics and other applications. The soft elastomer material very poorly conducts heat due to a low elastic modulus, which in general is believed to have a positive correlation with thermal conductivity. Even when elastomers are incorporated with high thermal conductive filler, they cannot achieve effective high thermal conductivity, and moreover, their elastic properties substantially degrade. Therefore, advanced techniques are required to combine both thermal conductivity and good elasticity in the same material. In one interesting study, the incorporation of liquid metal (LM) micro-droplets in polymer elastomer led to a thermal conductivity enhancement of 25 times (4.7 W/m K) in a stretch-free condition, and ~50 times (~9.8 W/m K) in a strained condition at 50 vol% loading [145]. LM was comprised of gallium-indium micro-droplets embedded in a silicon elastomer matrix. Such high thermal conductivity enhancement is supposed to be originated from the deformed LM inclusion forming a needle-like micro-structure in the direction of strain, which in turn forms thermal pathways. Fig. 3.19 presents the schematic of liquid metal reinforced in elastomer under stretching and relaxed conditions along with the thermo-mechanical property comparison with other materials.

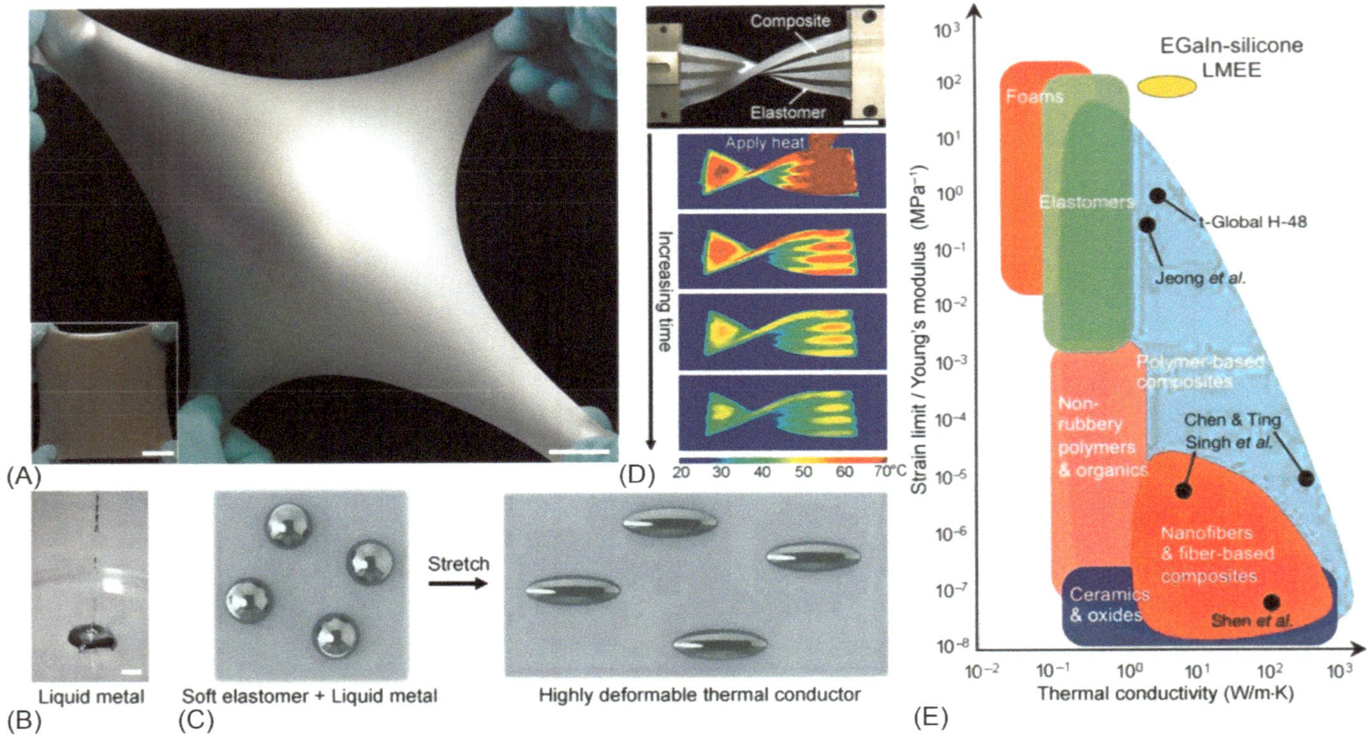

FIG. 3.19 Schematic illustration of the liquid metal embedded elastomer composite where LM micro-droplets are dispersed in an elastomer matrix and, upon deformation, the LM inclusions and elastomer elongate in the direction of stretching (A) Highly deformable Liquid Metal embedded elastomer (LMEE). (scale bars, 25 mm.) (B) EGaIn alloy is liquid at room temperature and shows fluid characteristics as demonstrated by falling droplets. (Scale bar, 10 mm.) (C) Schematic illustration of the LMEE composite where LM microdoplets are dispersed in an elastomer matrix and, upon deformation, the LM inclusions and elastomer elongate in the direction of stretching. (D) Alternating strips of LMEE and unfilled elastomer are heated with a heat gun, and the IR photo time sequence shows the LMEE dissipating heat more rapidly than the elastomer (images correspond to t = 0, 5, 10, and 15 s after the heat source is removed). (Scale bar, 25 mm.) (E) The $\phi = 50\%$ LMEE composites described here occupy a unique region of the material properties space when comparing thermal conductivity with the ratio of strain limit to Young's modulus [145].

3.5.4.3 Thermally Conductive Laminates

Composite laminates are composed of different layers bonded together either by action of heat, pressure, adhesive, and so forth. They can be multilayer polymer films. Such materials are widely used in food and consumer packaging, interior finishing, and thermal coating applications. Thermal transport can vary widely with the thickness of the material. Diffusive and ballistic heat transport in materials is impacted by its thickness and mean free path. When phonon makes the transition from one layer to another, it may lead to phonon scattering, due to interfacial resistance at the boundary. Graphene laminate, as described in Fig. 3.20, is made of a closely packed overlapping structure of graphene, and a few layers of graphene (FLG) flakes of different sizes and shapes deposited on polyethylene terephthalate (PET) substrates [146]. Thermal transport in such graphene laminates (GL) can be quite complex, given the various factors such as random distribution of flakes, sizes, thicknesses, defects, and disorders. Compressed graphene laminate with an average flake size of around 1.8 μm had a thermal conductivity of ~90.0 W/m K, while an uncompressed one has lower values. The size of flakes was found to have a positive correlation with thermal conductivity. Values increase with the size of the flake. Such material can be used for various thermal conductive applications, and one such area can be thermal coating [147].

3.6 APPLICATIONS

With the decreasing size of electronic devices and high computing speed, heat generated per unit area has increased. Efficient heat dissipating material therefore are in great demand in today's e-world. Thermal conductive polymer-based materials have wide applications, from electronics to automobiles, to aerospace. Such materials, based on polymers, have various extra advantages, such as

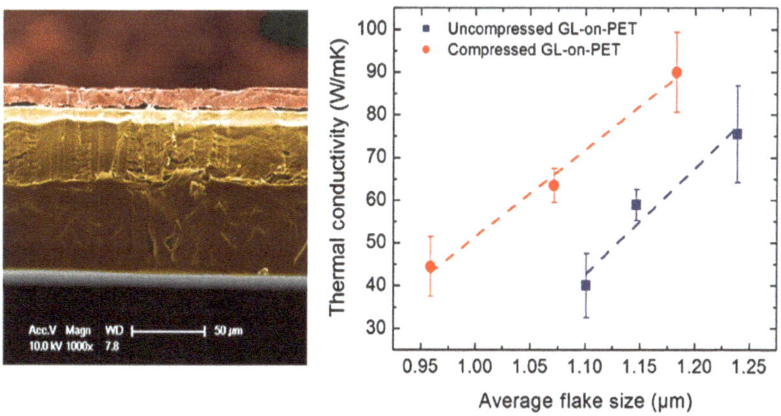

FIG. 3.20 Thermal conductivity of graphene laminate [146].

relatively light weight, ease of fabrication, better mechanical properties, high corrosion resistance, electrical insulative properties, cost effectiveness, and so forth. Everywhere there are heat management issues, thermal conductive polymer composites can play a crucial role. One of the primary areas for such material is the electronic packaging industries. Polymer composites are extensively used as PCB substrates, packaging molds, encapsulations, thermal interface materials (TIM), and so forth [148]. For electronics, polymers have an extra edge in terms of easy miniaturization of electronic components and ease of manufacturing from a simple injection molding without compromising on mechanical properties. Higher manufacturing costs of die-cast of metals such as aluminum, in comparison with polymers, also makes them more preferable than metals. In the computer and semi-conductor industries, TIMs are extensively used in the forms of thermal grease, thermal pads, thermal gels, and thermal taps for heat dissipation purposes. The automobile industry also has many heat management issues. The pressure of making components lightweight with increasing power density due to shrinking size also makes polymer-based materials more promising. For increased reliability and the extended lifetime of electronic control units of automobiles; it is of paramount importance that efficient thermal conductive material should be employed for thermal management [149]. In the telecommunications industry, faster transmission and higher transmission frequency require low dielectric constants and low dissipation factor materials, along with efficient thermal dissipating properties [150]. Development of such materials will result in enhanced performance and reliability in data transfer. Other areas such as oil and gas, aerospace, and so forth require efficient thermal conductive materials, and polymers can pave a promising role in these sectors.

3.7 THERMALLY INSULATIVE MATERIALS

Like high thermal conductivity, thermal insulation is also a desired property in many industrial applications. Such materials are widely used in building construction and piping insulation, apart from the automobile, aerospace, and refrigeration industries [151–153]. Subsea thermal insulation for piping is an emerging market for the safe flow of gas or oil in subsea structures [154]. Without appropriate thermal insulation, formation of hydrate and wax can block the flow of oil, thus resulting in structural damage. Such materials are generally composite layers applied on the pipelines. This can be a combination of polymers, foams, adhesives, and so forth. Good thermal insulation, along with mechanical and corrosion resistance, are required properties for such insulation. Thermal insulating in building structures is also one of the rapidly growing areas. Such materials play key roles in the designing and constructing of energy-efficient buildings. These materials can be inorganic, or organic fiber, or foam [151]. Among the widely used thermal insulation materials in buildings, glass wool and stone wool are the most popular ones. Glass wool is mainly made up of quartz sand, dolomite, resovit, limestone, and so forth. It is further reinforced with oils and adhesives to enhance mechanical properties. Stone wool has similar composition, but has different properties and fiber size. It has a high melting temperature, and hence is suitable for high-temperature applications. For super-insulation purposes, materials such as vacuum insulation panels (VIPs) and aerogels are employed in building construction. Such materials can achieve high insulation while keeping the envelope thickness within required values [155]. VIPs mainly consist of a membrane wall with a core panel mainly made up of silica, fiber, or aerogel materials. Because of the presence of a vacuum, it can achieve better insulation than conventional insulation systems. VIPs can achieve low thermal conductivity values of 0.001 W/m K, which is difficult to achieve in conventional systems with a similar thickness [155]. Another area where low thermal conductivity is desired is thermoelectric materials. Using thermoelectric devices, interconversion between electric energy and thermal energy can be achieved from waste heat sources [156,157]. Semiconductors are one of the promising materials of thermoelectric applications. Lattice thermal conductivity is one of the parameters that can be altered to improve the efficiency of thermoelectric devices. Increasing the defect centers in lattice, which are comparable to the mean free path of phonons, will lead to enhanced phonon scattering. Therefore, introducing nano-structures such as grain boundary, nano-scale precipitate, and point defects will lead to lower thermal conductivity, and enhance thermoelectric performance [158].

3.8 SUMMARY AND OUTLOOK

Thermally conductive polymer-based materials are indeed fascinating because of their versatile nature and cost effectiveness. Systematic knowledge of thermal conductivity of polymers and their composites, from the origin of thermal conductivity in polymers to the selection of materials, and finally, engineering polymers for better thermal conductive composites are presented here. The study of phonons is most fundamental in describing any heat conduction at the microscopic level. Phonons are the heat packets responsible for the transport of heat from one source to sink. At the nanoscale, enhanced thermal conductivity of material hinges on a single pivot point of developing efficient phonon transport pathways. In a material where the phonons, as heat carriers, are able to travel without any resistance like phonon scattering, have higher thermal conductivity. In polymers, three factors that theoretically control the transport of the phonon are group velocity, mean free path, and specific heat. Thermal conductivity in polymers and polymer composites is not only limited to these factors, as these factors are also influenced directly or indirectly by various factors that can be polymers, or its composite intrinsic. In all, one needs to thoroughly understand the system and apply technical know-how to reduce phonon scattering generated by one or more factors discussed herein, and utilize appropriate techniques to improve thermal conductivity of the system. As discussed, phonon scattering is the most important factor that ultimately determines how much heat is able to conduct in a material. Large phonon scattering regions will lead to less thermal conduction, and vice-versa. There is huge untapped potential available in the area of thermal conductivity of polymers. One of the promising and challenging areas that can significantly contribute to the understanding and knowledge of thermal conductivity of polymers and its composite is to understand the nanoscale thermal transport of phonons and polymer chains across the interface.

Various combinations of desired properties can be easily accommodated in polymer based thermal conductive composite materials. Development of polymer composites using metallics, ceramics, and carbon fillers are quite popular. Polymer-filler systems provide great flexibility in achieving desired mechanical and thermal properties. The future is not so far away when by intelligently engineering of matrix-filler properties and intermolecular interactions within polymer chains, highly advanced thermal conductive polymer composites beating the limits of metals can be developed. In conclusion, the principles and strategies discussed here provide tremendous knowledge for the intelligent design of polymeric materials for efficient thermal conduction.

REFERENCES

[1] D.M.N.W. Ashcroft, Solid State Physics, Holt, Rinehart and Winston, New York, 1976.

[2] A.A. Balandin, Thermal properties of graphene and nanostructured carbon materials, Nat. Mater. 10 (8) (2011) 569–581.

[3] N. Mehra, L. Mu, J. Zhu, Developing heat conduction pathways through short polymer chains in a hydrogen bonded polymer system, Compos. Sci. Technol. 148 (2017) 97–105.

[4] C.B. Carter, M.G. Norton, Ceramic Materials: Science and Engineering, Springer Science, New York, 2013.

[5] A. Kumar, Introduction to Solid State Physics, PHI, New Delhi, India, 2015.

[6] W. Kim, Strategies for engineering phonon transport in thermoelectrics, J. Mater. Chem. C 3 (40) (2015) 10336–10348.

[7] Y.S. Xu, D.D.L. Chung, C. Mroz, Thermally conducting aluminum nitride polymer-matrix composites, Compos. Part A 32 (12) (2001) 1749–1757.

[8] G. Chen, Nanoscale Energy Transport and Conversion: A Parallel Treatment of Electrons, Molecules, Phonons, and Photons, Oxford University Press, 2005.

[9] A.M. Chang-Lin Tien, F.M. Gerner, Microscale Energy Transport, Taylor & Francis, USA, 1998.

[10] Y. Son, Study of Anisotropic Thermal Properties and Thermal Transport at The Interface in Nanostructured Materials, Mech. Eng., Rensselaer Polytechnic Institute, Troy, New York, 2008.

[11] D.R.C. Aleksandr Chernatynskiy, S.R. Phillpot, Thermal transport in nanostructured materials, Handbook of Nanoscience, Engineering, and Technology, Pp. 445-572.

[12] K. Hirao, K. Watari, H. Hayashi, M. Kitayama, High thermal conductivity silicon nitride ceramic, MRS Bull. 26 (6) (2001) 451–455.

[13] G.A. Slack, Nonmetallic crystals with high thermal conductivity, J. Phys. Chem. Solids 34 (2) (1973) 321–335.
[14] J.W. Che, T. Cagin, W.A. Goddard, Thermal conductivity of carbon nanotubes, Nanotechnology 11 (2) (2000) 65–69.
[15] G.A. Slack, R.A. Tanzilli, R.O. Pohl, J.W. Vandersande, The intrinsic thermal conductivity of AlN, J. Phys. Chem. Solids 48 (7) (1987) 641–647.
[16] A.V. Virkar, T.B. Jackson, R.A. Cutler, Thermodynamic and kinetic effects of oxygen removal on the thermal conductivity of aluminum nitride, J. Am. Ceram. Soc. 72 (11) (1989) 2031–2042.
[17] T.B. Jackson, A.V. Virkar, K.L. More, R.B. Dinwiddie, R.A. Cutler, High-thermal-conductivity aluminum nitride ceramics: the effect of thermodynamic, kinetic, and microstructural factors, J. Am. Ceram. Soc. 80 (6) (1997) 1421–1435.
[18] G.P. Srivastava, The Physics of Phonons, CRC Press, Taylor & Francis Group, 1990.
[19] M.E. Siemens, Q. Li, R. Yang, K.A. Nelson, E.H. Anderson, M.M. Murnane, H.C. Kapteyn, Quasi-ballistic thermal transport from nanoscale interfaces observed using ultrafast coherent soft X-Ray beams, Nat. Mater. 9 (1) (2010) 26–30.
[20] Y. Hu, L. Zeng, A.J. Minnich, M.S. Dresselhaus, G. Chen, Spectral mapping of thermal conductivity through nanoscale ballistic transport, Nat. Nanotechnol. 10 (8) (2015) 701–706.
[21] L. Zeng, K.C. Collins, Y. Hu, M.N. Luckyanova, A.A. Maznev, S. Huberman, V. Chiloyan, J. Zhou, X. Huang, K.A. Nelson, G. Chen, Measuring phonon mean free path distributions by probing quasiballistic phonon transport in grating nanostructures, Sci. Rep. 5 (2015).
[22] N. Mehra, L. Mu, T. Ji, Y. Li, J. Zhu, Moisture driven thermal conduction in polymer and polymer blends, Compos. Sci. Technol. 151 (2017) 115–123.
[23] Y. Li, N. Mehra, T. Ji, X. Yang, L. Mu, J. Gu, J. Zhu, Stiffness-thermal conduction relationship at composite interface: effect of particle alignment on long-range confinement of polymer chains monitored by scanning thermal microscopy, Nanoscale (2017), https://doi.org/10.1039/C7NR06780A.
[24] S. Pettersson, G.D. Mahan, Theory of the thermal boundary resistance between dissimilar lattices, Phys. Rev. B 42 (12) (1990) 7386–7390.
[25] Z.D. Han, A. Fina, Thermal conductivity of carbon nanotubes and their polymer nanocomposites: a review, Prog. Polym. Sci. 36 (7) (2011) 914–944.
[26] E. Pop, V. Varshney, A.K. Roy, Thermal properties of graphene: fundamentals and applications, MRS Bull. 37 (12) (2012) 1273–1281.
[27] C. Kittel, Introduction to Solid State Physics, John Wiley & Sons, 2004.
[28] S. Shen, A. Henry, J. Tong, R. Zheng, G. Chen, Polyethylene nanofibres with very high thermal conductivities, Nat. Nanotechnol. 5 (2010) 251–255.
[29] A. Henry, G. Chen, S.J. Plimpton, A. Thompson, 1D to 3D transition of phonon heat conduction in polyethylene using molecular dynamics simulations, Phys. Rev. B 82 (14) (2010).
[30] B. Tonpheng, J. Yu, O. Andersson, Thermal conductivity, heat capacity, and cross-linking of polyisoprene/single-wall carbon nanotube composites under high pressure, Macromolecules 42 (23) (2009) 9295–9301.
[31] R. Haggenmueller, C. Guthy, J.R. Lukes, J.E. Fischer, K.I. Winey, Single wall carbon nanotube/polyethylene nanocomposites: thermal and electrical conductivity, Macromolecules 40 (7) (2007) 2417–2421.
[32] A. Henry, G. Chen, Anomalous heat conduction in polyethylene chains: theory and molecular dynamics simulations, Phys. Rev. B 79 (14) (2009).
[33] S. Ghosh, W. Bao, D.L. Nika, S. Subrina, E.P. Pokatilov, C.N. Lau, A.A. Balandin, Dimensional crossover of thermal transport in few-layer graphene, Nat. Mater. 9 (7) (2010) 555–558.

[34] P. Kim, L. Shi, A. Majumdar, P.L. Mceuen, Thermal transport measurements of individual multiwalled nanotubes, Phys. Rev. Lett. 87 (21) (2001).
[35] E. Pop, D. Mann, Q. Wang, K. Goodson, H. Dai, Thermal conductance of an individual single-wall carbon nanotube above room temperature, Nano Lett. 6 (1) (2006) 96–100.
[36] A.A. Balandin, S. Ghosh, W. Bao, I. Calizo, D. Teweldebrhan, F. Miao, C.N. Lau, Superior thermal conductivity of single-layer graphene, Nano Lett. 8 (2008) 902–907.
[37] M. Fujii, X. Zhang, H. Xie, H. Ago, K. Takahashi, T. Ikuta, H. Abe, T. Shimizu, Measuring the thermal conductivity of a single carbon nanotube, Phys. Rev. Lett. 95 (6) (2005).
[38] W.B. Zhang, Z.X. Zhang, J.H. Yang, T. Huang, N. Zhang, X.T. Zheng, Y. Wang, Z.W. Zhou, Largely enhanced thermal conductivity of poly (vinylidene fluoride)/carbon nanotube composites achieved by adding graphene oxide, Carbon 90 (2015) 242–254.
[39] C.L. Choy, W.H. Luk, F.C. Chen, Thermal conductivity of highly oriented polyethylene, Polymer 19 (2) (1978) 155–162.
[40] J. Yu, B. Sundqvist, B. Tonpheng, O. Andersson, Thermal conductivity of highly crystallized polyethylene, Polymer 55 (1) (2014) 195–200.
[41] G.-H. Kim, D. Lee, A. Shanker, L. Shao, M.S. Kwon, D. Gidley, J. Kim, K.P. Pipe, High thermal conductivity in amorphous polymer blends by engineered interchain interactions, Nat. Mater. 14 (3) (2015) 295–300.
[42] M. Shtein, R. Nadiv, M. Buzaglo, O. Regev, Graphene-based hybrid composites for efficient thermal management of electronic devices, ACS Appl. Mater. Interfaces 7 (42) (2015) 23725–23730.
[43] K.M.F. Shahil, A.A. Balandin, Graphene–multilayer graphene nanocomposites as highly efficient thermal interface materials, Nano Lett. 12 (2) (2012) 861–867.
[44] L. Mu, Y. Li, N. Mehra, T. Ji, J. Zhu, Expedited phonon transfer in interfacially constrained polymer chain along self-organized amino acid crystals, ACS Appl. Mater. Interfaces 9 (13) (2017) 12138–12145.
[45] L. Mu, T. Ji, L. Chen, N. Mehra, Y. Shi, J. Zhu, Paving the thermal highway with self-organized nanocrystals in transparent polymer composites, ACS Appl. Mater. Interfaces 8 (42) (2016) 29080–29087.
[46] H.D. Wagner, R.A. Vaia, Nanocomposites: issues at the interface, Mater. Today 7 (11) (2004) 38–42.
[47] W. Gacitua, A. Ballerini, J. Zhang, Polymer nanocomposites: synthetic and natural fillers a review, MaderasCienc. Tecnol. 7 (3) (2005) 159–178.
[48] P.C. Irwin, Y. Cao, A. Bansal, L.S. Schadler, Thermal and mechanical properties of polyimide nanocomposites, Conference on Electrical Insulation and Dielectric Phenomena, Albuquerque, NM, 2003, , pp. 120–123.
[49] H.O. Pierson, Handbook of Carbon, Graphite, Diamonds and Fullerenes: Processing, Properties and Applications, Elsevier, 2010.
[50] J. Yi, G.M. Yuan, X.K. Li, H.D. Du, Z.J. Dong, B.L. Li, J.F. Lin, Preparation and characterization of large diameter pitch based carbon fiber/Abs resin composites with high thermal conductivities, New Carbon Mater. 30 (1) (2015) 63–70.
[51] J. Kong, C.Y. Zhang, X. Cheng, Novel Cu-Cr alloy matrix CNT composites with enhanced thermal conductivity, Appl. Phys. A Mater. Sci. Process. 112 (3) (2013) 631–636.
[52] K. Yang, M. Gu, Y. Guo, X. Pan, G. Mu, Effects of carbon nanotube functionalization on the mechanical and thermal properties of epoxy composites, Carbon 47 (7) (2009) 1723–1737.
[53] W.-T. Hong, N.-H. Tai, Investigations on the thermal conductivity of composites reinforced with carbon nanotubes, Diam. Relat. Mater. 17 (7-10) (2008) 1577–1581.

[54] F.H. Gojny, M.H.G. Wichmann, B. Fiedler, I.A. Kinloch, W. Bauhofer, A.H. Windle, K. Schulte, Evaluation and identification of electrical and thermal conduction mechanisms in carbon nanotube/epoxy composites, Polymer 47 (6) (2006) 2036–2045.

[55] A. Chatterjee, B.L. Deopura, Carbon nanotubes and nanofibre: an overview, Fibers Polym. 3 (4) (2002) 134–139.

[56] M. Moniruzzaman, K.I. Winey, Polymer nanocomposites containing carbon nanotubes, Macromolecules 39 (16) (2006) 5194–5205.

[57] S. Agarwal, M.M.K. Khan, R.K. Gupta, Thermal conductivity of polymer nanocomposites made with carbon nanofibers, Polym. Eng. Sci. 48 (12) (2008) 2474–2481.

[58] W.N. Chang, Nano Fibers: Fabrication, Performance, and Applications, Nova Science, New York, 2009.

[59] Q. Mu, S. Feng, Thermal conductivity of graphite/silicone rubber prepared by solution intercalation, Thermochim. Acta 462 (1-2) (2007) 70–75.

[60] S. Ganguli, A.K. Roy, D.P. Anderson, Improved thermal conductivity for chemically functionalized exfoliated graphite/epoxy composites, Carbon 46 (5) (2008) 806–817.

[61] D.L. Nika, E.P. Pokatilov, A.S. Askerov, A.A. Balandin, Phonon thermal conduction in graphene: role of Umklapp and edge roughness scattering, Phys. Rev. B 79 (15) (2009).

[62] M.J. Allen, V.C. Tung, R.B. Kaner, Honeycomb carbon: a review of graphene, Chem. Rev. 110 (1) (2010) 132–145.

[63] H.W. Kroto, J.R. Heath, S.C. O'brien, R.F. Curl, R.E. Smalley, C60: buckminsterfullerene, Nature 318 (6042) (1985) 162–163.

[64] K.W. Garrett, H.M. Rosenberg, Thermal-conductivity of epoxy-resin-powder composite-materials, J. Phys. D. Appl. Phys. 7 (9) (1974) 1247–1248.

[65] J.H. Yu, R. Qian, P.K. Jiang, Enhanced thermal conductivity for PVDF composites with a hybrid functionalized graphene sheet-nanodiamond filler, Fibers Polym. 14 (8) (2013) 1317–1323.

[66] Y.W. Wong, K.L. Lo, F.G. Shin, Electrical and thermal properties of composite of liquid crystalline polymer filled with carbon black, J. Appl. Polym. Sci. 82 (6) (2001) 1549–1555.

[67] G.G. Tibbetts, M.L. Lake, K.L. Strong, B.P. Rice, A review of the fabrication and properties of vapor-grown carbon nanofiber/polymer composites, Compos. Sci. Technol. 67 (7-8) (2007) 1709–1718.

[68] S. Ghosh, I. Calizo, D. Teweldebrhan, E.P. Pokatilov, D.L. Nika, A.A. Balandin, W. Bao, F. Miao, C.N. Lau, Extremely high thermal conductivity of graphene: prospects for thermal management applications in nanoelectronic circuits, Appl. Phys. Lett. 92 (2008) 15–17.

[69] A.M. Marconnet, N. Yamamoto, M.A. Panzer, B.L. Wardle, K.E. Goodson, Thermal conduction in aligned carbon nanotube–polymer nanocomposites with high packing density, ACS Nano 5 (6) (2011) 4818–4825.

[70] S. Berber, Y. Kwon, D. Tomanek, Unusually high thermal conductivity of carbon nanotubes, Phys. Rev. Lett. 84 (20) (2000) 4613–4616.

[71] J.E. Peters, D.V. Papavassiliou, B.P. Grady, Unique thermal conductivity behavior of single-walled carbon nanotube – polystyrene composites, Macromolecules 41 (20) (2008) 7274–7277.

[72] C.W. Nan, Z. Shi, Y. Lin, A simple model for thermal conductivity of carbon nanotube-based composites, Chem. Phys. Lett. 375 (5–6) (2003) 666–669.

[73] D. Cai, M. Song, Latex technology as a simple route to improve the thermal conductivity of a carbon nanotube/polymer composite, Carbon 46 (15) (2008) 2107–2112.

[74] S.T. Huxtable, D.G. Cahill, S. Shenogin, L. Xue, R. Ozisik, P. Barone, M. Usrey, M.S. Strano, G. Siddons, M. Shim, P. Keblinski, Interfacial heat flow in carbon nanotube suspensions, Nat. Mater. 2 (11) (2003) 731–734.

[75] K. Wu, C. Lei, R. Huang, W. Yang, S. Chai, C. Geng, F. Chen, Q. Fu, Design and preparation of a unique segregated double network with excellent thermal conductive property, ACS Appl. Mater. Interfaces 9 (8) (2017) 7637–7647.

[76] M. Wang, H. Chen, W. Lin, Z. Li, Q. Li, M. Chen, F. Meng, Y. Xing, Y. Yao, C.-P. Wong, Q. Li, Crack-free and scalable transfer of carbon nanotube arrays into flexible and highly thermal conductive composite film, ACS Appl. Mater. Interfaces 6 (1) (2014) 539–544.

[77] S. Yu, B.-I. Park, C. Park, S.M. Hong, T.H. Han, C.M. Koo, RTA-treated carbon fiber/copper core/shell hybrid for thermally conductive composites, ACS Appl. Mater. Interfaces 6 (10) (2014) 7498–7503.

[78] G. Pezzotti, I. Kamada, S. Miki, Thermal conductivity of AlN/polystyrene interpenetrating networks, J. Eur. Ceram. Soc. 20 (8) (2000) 1197–1203.

[79] M. Ohashi, S. Kawakami, Y. Yokogawa, G.-C. Lai, Spherical aluminum nitride fillers for heat-conducting plastic packages, J. Am. Ceram. Soc. 88 (9) (2005) 2615–2618.

[80] S. Konzelmann, C. Hoffmann, R. Merte, D. Peier, Thermal and electrical properties of aluminum nitride filled epoxy-resin compound, dielectrics and electrical insulation, IEEE Trans. Dielectr. Electr. Insul. 15 (2) (2008) 327–333.

[81] E.-S. Lee, S.-M. Lee, D.J. Shanefield, W.R. Cannon, Enhanced thermal conductivity of polymer matrix composite via high solids loading of aluminum nitride in epoxy resin, J. Am. Ceram. Soc. 91 (4) (2008) 1169–1174.

[82] H.Y. Ng, X. Lu, S.K. Lau, Thermal conductivity of boron nitride-filled thermoplastics: effect of filler characteristics and composite processing conditions, Polym. Compos. 26 (6) (2005) 778–790.

[83] C.P. Wong, R.S. Bollampally, Comparative study of thermally conductive fillers for use in liquid encapsulants for electronic packaging, advanced packaging, IEEE Trans. Adv. Packag. 22 (1) (1999) 54–59.

[84] Y. Xu, D.D.L. Chung, Increasing the thermal conductivity of boron nitride and aluminum nitride particle epoxy-matrix composites by particle surface treatments, Compos. Interf. 7 (4) (2000) 243–256.

[85] S.-H. Xie, B.-K. Zhu, J.-B. Li, X.-Z. Wei, Z.-K. Xu, Preparation and properties of polyimide/aluminum nitride composites, Polym. Test. 23 (7) (2004) 797–801.

[86] W. Wang, X. Yang, Y. Fang, J. Ding, J. Yan, Enhanced thermal conductivity and thermal performance of form-stable composite phase change materials by using beta-aluminum nitride, Appl. Energy 86 (7-8) (2009) 1196–1200.

[87] L.C. Sim, S.R. Ramanan, H. Ismail, K.N. Seetharamu, T.J. Goh, Thermal characterization of Al_2O_3 and ZnO reinforced silicone rubber as thermal pads for heat dissipation purposes, Thermochim. Acta 430 (1–2) (2005) 155–165.

[88] X. Lu, G. Xu, P.G. Hofstra, R.C. Bajcar, Moisture–absorption, dielectric relaxation, and thermal conductivity studies of polymer composites, J. Polym. Sci. B Polym. Phys. 36 (13) (1998) 2259–2265.

[89] J. Kim, J. Kim, M. Kim, M. Song, Thermal conductivity and adhesion properties of thermally conductive pressure-sensitive adhesives, Macromol. Res. 14 (5) (2006) 517–523.

[90] Y. Shimazaki, F. Hojo, Y. Takezawa, Highly thermoconductive polymer nanocomposite with a nanoporous alpha-alumina sheet, ACS Appl. Mater. Interfaces 1 (2) (2009) 225–227.

[91] Y. Shimazaki, F. Hojo, Y. Takezawa, Preparation and characterization of thermoconductive polymer nanocomposite with branched alumina nanofiber, Appl. Phys. Lett. 92 (13) (2008).

[92] P. Bujard, G. Kuhnlein, S. Ino, T. Shiobara, Thermal-conductivity of molding compounds for plastic packaging, IEEE Trans. Compon. Packag. Manuf. Technol. Part A 17 (4) (1994) 527–532.

[93] W. Zhou, D. Yu, C. Min, Y. Fu, X. Guo, Thermal, dielectric, and mechanical properties of SiC particles filled linear low-density polyethylene composites, J. Appl. Polym. Sci. 112 (3) (2009) 1695–1703.

[94] T. Zhou, X. Wang, G.U. Mingyuan, X. Liu, Study of the thermal conduction mechanism of nano-SiC/DGEBA/EMI-2,4 composites, Polymer 49 (21) (2008) 4666–4672.

[95] K. Yang, M. Gu, Enhanced thermal conductivity of epoxy nanocomposites filled with hybrid filler system of triethylenetetramine-functionalized multi-walled carbon nanotube/silane-modified nano-sized silicon carbide, Compos. Part A 41 (2) (2010) 215–221.

[96] W. Zhou, C. Wang, T. Ai, K. Wu, F. Zhao, H. Gu, A novel fiber-reinforced polyethylene composite with added silicon nitride particles for enhanced thermal conductivity, Compos. Part A 40 (6–7) (2009) 830–836.

[97] H. He, R. Fu, Y. Han, Y. Shen, D. Wang, High thermal conductive Si_3N_4 particle filled epoxy composites with a novel structure, J. Electron. Packag. 129 (4) (2007) 469–472.

[98] G. Subodh, M.V. Manjusha, J. Philip, M.T. Sebastian, Thermal properties of polytetrafluoroethylene/$Sr_2Ce_2Ti_5O_{16}$ polymer/ceramic composites, J. Appl. Polym. Sci. 108 (3) (2008) 1716–1721.

[99] J. Varghese, D.R. Nair, P. Mohanan, M.T. Sebastian, Dielectric, thermal and mechanical properties of zirconium silicate reinforced high density polyethylene composites for antenna applications, Phys. Chem. Chem. Phys. 17 (22) (2015) 14943–14950.

[100] W. Yu, H. Xie, L. Chen, M. Wang, W. Wang, Synergistic thermal conductivity enhancement of PC/ABS composites containing alumina/magnesia/graphene nanoplatelets, Polym. Compos. 38(10) 2221-2227.

[101] K.W. Garrett, H.M. Rosenberg, The thermal conductivity of epoxy resin/powder composite materials, J. Phys. D. Appl. Phys. 7 (9) (1974) 1247.

[102] P.S. Anjana, V. Deepu, S. Uma, P. Mohanan, J. Philip, M.T. Sebastian, Dielectric, thermal, and mechanical properties of CeO_2 filled HDPE composites for microwave substrate applications, J. Polym. Sci. B Polym. Phys. 48 (9) (2010) 998–1008.

[103] Q. Mu, S. Feng, G. Diao, Thermal conductivity of silicone rubber filled with ZnO, Polym. Compos. 28 (2) (2007) 125–130.

[104] X. Huang, P. Jiang, T. Tanaka, A review of dielectric polymer composites with high thermal conductivity, IEEE Electr. Insul. Mag. 27 (4) (2011) 8–16.

[105] T. Morishita, H. Okamoto, Facile exfoliation and noncovalent superacid functionalization of boron nitride nanosheets and their use for highly thermally conductive and electrically insulating polymer nanocomposites, ACS Appl. Mater. Interfaces 8 (40) (2016) 27064–27073.

[106] G. Hou, B. Cheng, F. Ding, M. Yao, P. Hu, F. Yuan, Synthesis of uniform A-Si_3N_4 nanospheres by RF induction thermal plasma and their application in high thermal conductive nanocomposites, ACS Appl. Mater. Interfaces 7 (4) (2015) 2873–2881.

[107] D.D.L. Chung, Materials for thermal conduction, Appl. Therm. Eng. 21 (16) (2001) 1593–1605.

[108] N.M. Sofian, M. Rusu, R. Neagu, E. Neagu, Metal powder-filled polyethylene composites. V. Thermal properties, J. Thermoplast. Compos. Mater. 14 (1) (2001) 20–33.

[109] I. Seshadri, G.L. Esquenazi, T. Borca-Tasciuc, P. Keblinski, G. Ramanath, Multifold increases in thermal conductivity of polymer nanocomposites through microwave welding of metal nanowire fillers, Adv. Mater. Interfaces 2 (15) (2015), https://doi.org/10.1002/admi.201500186.

[110] K.S. Chou, K.C. Huang, H.H. Lee, Fabrication and sintering effect on the morphologies and conductivity of nano-Ag particle films by the spin coating method, Nanotechnology 16 (6) (2005) 779–784.
[111] K. Pashayi, H.R. Fard, F. Lai, S. Iruvanti, J. Plawsky, T. Borca-Tasciuc, Self-constructed tree-shape high thermal conductivity nanosilver networks in epoxy, Nanoscale 6 (8) (2014) 4292–4296.
[112] K. Gaska, G. Kmita, A. Rybak, R. Sekula, K. Goc, C. Kapusta, Magnetic-aligned, magnetite-filled epoxy composites with enhanced thermal conductivity, J. Mater. Sci. 50 (6) (2015) 2510–2516.
[113] W. Zhao, H. Wang, H. Tang, G. Chen, Facile preparation of epoxy-based composite with oriented graphite nanosheets, Polymer 47 (26) (2006) 8401–8405.
[114] H. Yan, Y. Tang, W. Long, Y. Li, Enhanced thermal conductivity in polymer composites with aligned graphene nanosheets, J. Mater. Sci. 49 (15) (2014) 5256–5264.
[115] B. Weidenfeller, M. Hofer, F. Schilling, Thermal and electrical properties of magnetite filled polymers, Compos. Part A 33 (8) (2002) 1041–1053.
[116] M.R. Sommer, R.M. Erb, A.R. Studart, Injectable materials with magnetically controlled anisotropic porosity, ACS Appl. Mater. Interfaces 4 (10) (2012) 5086–5091.
[117] M.M. Porter, M. Yeh, J. Strawson, T. Goehring, S. Lujan, P. Siripasopsotorn, M.A. Meyers, J. Mckittrick, Magnetic freeze casting inspired by nature, Mater. Sci. Eng. A 556 (2012) 741–750.
[118] Z. Lin, Y. Liu, S. Raghavan, K.-S. Moon, S.K. Sitaraman, C.-P. Wong, Magnetic alignment of hexagonal boron nitride platelets in polymer matrix: toward high performance anisotropic polymer composites for electronic encapsulation, ACS Appl. Mater. Interfaces 5 (15) (2013) 7633–7640.
[119] R. Libanori, R.M. Erb, A.R. Studart, Mechanics of platelet-reinforced composites assembled using mechanical and magnetic stimuli, ACS Appl. Mater. Interfaces 5 (21) (2013) 10794–10805.
[120] T. Kimura, H. Ago, M. Tobita, S. Ohshima, M. Kyotani, M. Yumura, Polymer composites of carbon nanotubes aligned by a magnetic field, Adv. Mater. 14 (19) (2002) 1380–1383.
[121] I.T. Kim, A. Tannenbaum, R. Tannenbaum, Anisotropic conductivity of magnetic carbon nanotubes embedded in epoxy matrices, Carbon 49 (1) (2011) 54–61.
[122] R.M. Erb, J.S. Sander, R. Grisch, A.R. Studart, Self-shaping composites with programmable bioinspired microstructures, Nat. Commun. 4 (2013) 1712.
[123] R.M. Erb, R. Libanori, N. Rothfuchs, A.R. Studart, Composites reinforced in three dimensions by using low magnetic fields, Science 335 (6065) (2012) 199–204.
[124] E. Camponeschi, R. Vance, M. Al-Haik, H. Garmestani, R. Tannenbaum, Properties of carbon nanotube-polymer composites aligned in a magnetic field, Carbon 45 (10) (2007) 2037–2046.
[125] M. Abdalla, D. Dean, M. Theodore, J. Fielding, E. Nyairo, G. Price, Magnetically processed carbon nanotube/epoxy nanocomposites: morphology, thermal, and mechanical properties, Polymer 51 (7) (2010) 1614–1620.
[126] H. Shen, J. Guo, H. Wang, N. Zhao, J. Xu, Bioinspired modification of H-BN for high thermal conductive composite films with aligned structure, ACS Appl. Mater. Interfaces 7 (10) (2015) 5701–5708.
[127] R. Libanori, F.H.L. Münch, D.M. Montenegro, A.R. Studart, Hierarchical reinforcement of polyurethane-based composites with inorganic micro and nanoplatelets, Compos. Sci. Technol. 72 (3) (2012) 435–445.

[128] E. Munch, M.E. Launey, D.H. Alsem, E. Saiz, A.P. Tomsia, R.O. Ritchie, Tough, bio-inspired hybrid materials, Science 322 (5907) (2008) 1516–1520.

[129] L. Ziyin, L. Yan, M. Kyoung-Sik, W. Ching-Ping, Enhanced thermal transport of hexagonal boron nitride filled polymer composite by magnetic field-assisted alignment, IEEE 63rd Electronic Components and Technology Conference 2013, 2013, , pp. 1692–1696.

[130] S. Xu, H. Liu, Q. Li, Q. Mu, H. Wen, Influence of magnetic alignment and layered structure of BN & Fe/EP on thermal conducting performance, J. Mater. Chem. C 4 (4) (2016) 872–878.

[131] C. Yuan, B. Duan, L. Li, B. Xie, M. Huang, X. Luo, Thermal conductivity of polymer-based composites with magnetic aligned hexagonal boron nitride platelets, ACS Appl. Mater. Interfaces 7 (23) (2015) 13000–13006.

[132] K. Kim, J. Kim, Vertical filler alignment of boron nitride/epoxy composite for thermal conductivity enhancement via external magnetic field, Int. J. Therm. Sci. 100 (2016) 29–36.

[133] R.M. Erb, H.S. Son, B. Samanta, V.M. Rotello, B.B. Yellen, Magnetic assembly of colloidal superstructures with multipole symmetry, Nature 457 (7232) (2009) 999–1002.

[134] T. Terao, Y. Bando, M. Mitome, C. Zhi, C. Tang, D. Golberg, Thermal conductivity improvement of polymer films by catechin-modified boron nitride nanotubes, J. Phys. Chem. C 113 (31) (2009) 13605–13609.

[135] X. Huang, T. Iizuka, P. Jiang, Y. Ohki, T. Tanaka, Role of interface on the thermal conductivity of highly filled dielectric epoxy/AlN composites, J. Phys. Chem. C 116 (25) (2012) 13629–13639.

[136] A. Yu, P. Ramesh, X. Sun, E. Bekyarova, M.E. Itkis, R.C. Haddon, Enhanced thermal conductivity in a hybrid graphite nanoplatelet—carbon nanotube filler for epoxy composites, Adv. Mater. 20 (24) (2008) 4740–4744.

[137] E.H. Weber, M.L. Clingerman, J.A. King, Thermally conductive nylon 6,6 and polycarbonate based resins. II. Modeling, J. Appl. Polym. Sci. 88 (1) (2003) 123–130.

[138] E.H. Weber, M.L. Clingerman, J.A. King, Thermally conductive nylon 6,6 and polycarbonate based resins. I. Synergistic effects of carbon fillers, J. Appl. Polym. Sci. 88 (1) (2003) 112–122.

[139] G.-W. Lee, M. Park, J. Kim, J.I. Lee, H.G. Yoon, Enhanced thermal conductivity of polymer composites filled with hybrid filler, Compos. Part A 37 (5) (2006) 727–734.

[140] T.-L. Li, S.L.-C. Hsu, Enhanced thermal conductivity of polyimide films via a hybrid of micro and nano sized boron nitride, J. Phys. Chem. B 114 (20) (2010) 6825–6829.

[141] S.Y. Yang, W.N. Lin, Y.L. Huang, H.W. Tien, J.Y. Wang, C.C.M. Ma, S.M. Li, Y.S. Wang, Synergetic effects of graphene platelets and carbon nanotubes on the mechanical and thermal properties of epoxy composites, Carbon 49 (3) (2011) 793–803.

[142] K.T.S. Kong, M. Mariatti, A.A. Rashid, J.J.C. Busfield, Enhanced conductivity behavior of polydimethylsiloxane (PDMS) hybrid composites containing exfoliated graphite nanoplatelets and carbon nanotubes, Compos. Part B 58 (2014) 457–462.

[143] F. Zhang, Q. Li, Y. Liu, S. Zhang, C. Wu, W. Guo, Improved thermal conductivity of polycarbonate composites filled with hybrid exfoliated graphite/multi-walled carbon nanotube fillers, J. Therm. Anal. Calorim. (2015) 1–7.

[144] L. Mu, J. He, Y. Li, T. Ji, N. Mehra, Y. Shi, J. Zhu, The molecular origin of efficient phonon transfer in modulated polymer blends: effect of hydrogen bonding on polymer coil size and assembled microstructure, J. Phys. Chem. C (2017).

[145] M.D. Bartlett, N. Kazem, M.J. Powell-Palm, X. Huang, W. Sun, J.A. Malen, C. Majidi, High thermal conductivity in soft elastomers with elongated liquid metal inclusions, Proc. Natl. Acad. Sci. 114 (9) (2017) 2143–2148.

[146] H. Malekpour, K.H. Chang, J.C. Chen, C.Y. Lu, D.L. Nika, K.S. Novoselov, A.A. Balandin, Thermal conductivity of graphene laminate, Nano Lett. 14 (9) (2014) 5155–5161.
[147] K.S. Novoselov, V.I. Falko, L. Colombo, P.R. Gellert, M.G. Schwab, K. Kim, A roadmap for graphene, Nature 490 (7419) (2012) 192–200.
[148] X.C. Tong, Electronic packaging materials and their functions in thermal managements, in: X.C. Tong (Ed.), Advanced Materials for Thermal Management of Electronic Packaging, Springer, New York, 2011, pp. 131–167.
[149] S. Mallik, N. Ekere, C. Best, R. Bhatti, Investigation of thermal management materials for automotive electronic control units, Appl. Therm. Eng. 31 (2) (2011) 355–362.
[150] W.P. Ting, F.P. Tseng, K.C. Chiou, in: Advanced materials with low dielectric properties and highly thermal conductivity, 2016 11th International Microsystems, Packaging, Assembly and Circuits Technology Conference (Impact), 2016, pp. 430–433.
[151] A.M. Papadopoulos, State of the art in thermal insulation materials and aims for future developments, Energy Build. 37 (1) (2005) 77–86.
[152] J.M.Z. Fei Wang, Q.G. Tang, Research progress on thermal insulation materials, Adv. Mater. Res. 427 (2012) 157–162.
[153] T. Ojanen, I.P. Seppä, E. Nykänen, Thermal insulation products and applications—future road maps, Energy Procedia 78 (Suppl C) (2015) 309–314.
[154] Q.B. Yong Bai, Subsea Pipelines and Risers, Elsevier, 2005.
[155] A. Capozzoli, S. Fantucci, F. Favoino, M. Perino, Vacuum insulation panels: analysis of the thermal performance of both single panel and multilayer boards, Energies 8 (4) (2015).
[156] G.J. Snyder, E.S. Toberer, Complex thermoelectric materials, Nat. Mater. 7 (2008) 105.
[157] X. Zhang, L.-D. Zhao, Thermoelectric materials: energy conversion between heat and electricity, J. Mater. 1 (2) (2015) 92–105.
[158] J. Li, H. Wu, D. Wu, C. Wang, Z. Zhang, Y. Li, F. Liu, W.-Q. Ao, J. He, Extremely low thermal conductivity in thermoelectric $Ge_{0.55}Pb_{0.45}Te$ solid solutions via Se substitution, Chem. Mater. 28 (17) (2016) 6367–6373.

Chapter 4

Epoxy-Based Multifunctional Nanocomposites

Khalid A. Askar* and Kenan Song†

*Department of Materials Science and Engineering, Masdar Institute of Science and Technology, Abu Dhabi, United Arab Emirates, †Department of Manufacturing Engineering, Arizona State University, Tempe, AZ, United States

Chapter Outline

ABBREVIATIONS

AFM atomic force microscopy
CAB cellulose acetate butyrate
DSC differential scanning calorimetry
EDS Energy-dispersive X-ray spectroscopy
HNTs halloysite nanotubes
SEM scanning electron microscopy
TEM transmission electron microscope
XPS X-ray photoelectron spectroscopy

4.1 Introduction

Polymer coatings have widespread applications in almost every engineering sector, such as aerospace, infrastructure, and the automobile, solar cell, and furniture industries, due to their superior adherence and ability to protect surfaces from corrosion/abrasion [1]. For the majority of these coatings, mechanical durability depends on stiffness, hardness, and toughness, as well as abrasion [2] or corrosion resistance [3]. One method to reinforce the polymer matrix is to include stiff and robust nano-sized segments, such as carbon fillers of carbon nanotubes and graphene [4–10], inorganic particles of clay and metal

Polymer-Based Multifunctional Nanocomposites and Their Applications
https://doi.org/10.1016/B978-0-12-815067-2.00004-4

oxides [11], and bio-fillers of cellulose [12] and wood [13,14]. These fillers can also be categorized as one-dimensional (1D), two-dimensional (2D), and three-dimensional (3D) materials based on their dimensional and geometric features [15]. Among them, 1D nanoparticles have been attractive, mainly due to their anisotropic properties [16]. An example of anisotropic properties would be the dependence of Young's modulus and strength on the direction of the mechanical load. As summarized in Table 4.1, when nanofillers were aligned along the applied force direction in composites, stiffness and strength displayed the highest values (stronger axis). With respect to the orthogonal direction (weaker axis), the composites exhibited worse properties than the polymer matrix. Therefore, control of preferential particle alignment has become one of the most significant issues in composite reinforcement [23].

To utilize a nanotube's anisotropic properties along the axis, studies have been focusing on fabricating both in-plane and out-of-plane oriented assemblies. This research involves dispersing nanotubes in an appropriate solvent and using (i) outside field forces of an electric field [24,25] and magnetic field [26,27] or (ii) shear forces induced by extrusion/injection flow [28] in liquid or the melt polymer matrix (Table 4.1). In the former case, the important technique is to attach either a conductive or super-paramagnetic carrier to the nanofillers. In the latter case, localized, in-plane control of particle alignment has been found to be effective. Other researchers focus on composite fabrication directly from the aligned forest or array, which is synthesized by either the chemical vapor deposition method (CVD) [29] or electrochemical deposition [30]. Carbon nanotubes are popular in these methods due to their flexibility and drawability from the growth substrate [31]. Naturally generated mineral clays, such as montmorillonite, mica, talc, kaolinite, and halloysite, however, are mechanically stiff and hard to synthesize in an economic deposition method due to natural abundance [32]. Under this circumstance, unidirectional out-of-plane reinforcement may not be achieved using the in situ drawing [31,33], infiltration/impregnation [34], or compression molding [35] methods based on CVD growth.

In nanotube-reinforced polymers, dispersion quality and interfacial interactions are two correlated factors influencing reinforcement efficiency, in addition to tube orientation [36]. At a better dispersion quality, exfoliated fillers tend to interact with more of the polymer interfacial area, and transfer stress more efficiently. However, with the increase of concentrations, particles form aggregates in composites, and cause cracks and failure under mechanical loadings. In particle suspensions, the degree of particle interactions defines three concentration regimes as dilute, semidilute, or concentrated [37,38]. These different regimes affect hydrodynamics or rheological properties (i.e., viscosity) in the polymer solution or melt preparations, so precautions should be taken in composite fabrication procedures.

Halloysite nanotubes (HNTs), naturally-occurring clay minerals with a 1D hollow cylindrical structure, are exceptionally stiff and hard for their ceramic

TABLE 4.1 Anisotropic Mechanical Properties Induced by Several Processing Techniques and Various Filler Reinforcement

			Stronger Axis			Weaker Axis		
Year	**Materials**	**Method**	E *(GPa)*	σ *(MPa)*	E_{eff} *(GPa)*	E *(GPa)*	σ *(MPa)*	E_{eff} *(GPa)*
2014 [17], 2013 [18]	Methacrylated gelatin (GelMA) hydrogel	Electric field	$(12.5 \pm 0.1) \times 10^{-6}$	N/A	N/A	$(12.5 \pm 0.1) \times 10^{-6}$	N/A	N/A
	Methacrylated gelatin (GelMA) hydrogel with 0.1 mg/mL CNT		$(59.4 \pm 3.8) \times 10^{-6}$	N/A	28.6[a]	$(20.3 \pm 1.4) \times 10^{-6}$	N/A	4.8[a]
	Methacrylated gelatin (GelMA) hydrogel with 0.3 mg/mL CNT		$(48.5 \pm 2.3) \times 10^{-6}$	N/A	42.9[a]	$(23.4 \pm 1.7) \times 10^{-6}$	N/A	13.1[a]
2012 [19]	Polyurethane based polymer	3D magnetic printing	2.4	30	N/A	2.4	30	N/A
	Polyurethane/ 20 vol% Al_2O_3		6.4	48	100[b]	4.8	24	60[b]
2015 [20]	Urethane acrylate copolymer		0.4	N/A	N/A	0.4	N/A	N/A
	Urethane acrylate copolymer/ 15 vol% Al_2O_3		0.9	N/A	16.5[b]	0.7	N/A	10[b]

Continued

TABLE 4.1 Anisotropic Mechanical Properties Induced by Several Processing Techniques and Various Filler Reinforcement—cont'd

			Stronger Axis			Weaker Axis		
Year	Materials	Method	E (GPa)	σ (MPa)	E_{eff} (GPa)	E (GPa)	σ (MPa)	E_{eff} (GPa)
2013 [21]	PEDOT:PSS	Wet-injected fiber-spinning	3.1	125	N/A	N/A	N/A	N/A
	PEDOT:PSS/ 2 vol% PEG functionalized-SWNTs		12.0	200	417[c]	N/A	N/A	N/A
	PEDOT:PSS/ 12 vol% PEG functionalized-SWNTs		22.8	254	108[c]	N/A	N/A	N/A
2006 [22]	Epoxy/SiC fabrics	CVD-based impregnation	23.1 ± 0.3	62.1 ± 2.1	N/A	3.5	N/A	N/A
	Epoxy/SiC grown MWNT		24.3 ± 0.2	150.1 ± 1.4	N/A[d]	4.5	N/A	N/A

Note: E, *modulus;* σ, *strength;* E_{eff}, *effective reinforcement modulus, which was based on a rough estimation from composite mechanics, namely, analyses of rule-of-mixtures, both orientation factor and dimension efficiency factor were also considered.*

[a] *Multi-wall CNT was reported to possess a length of 40–90 nm, the aspect ratio of 100 and intrinsic modulus around 500 GPa [18].*

[b] *Typical Al_2O_3 particles have a length of 7.5 μm, a diameter of 350 nm and an average aspect ratio of 20 [19]. As a comparison, intrinsic modulus of Al_2O_3 is 410 GPa and strength 70 GPa.*

[c] *Single-wall CNT geometric and dimensional information is unknown, thus length efficiency and orientation efficiency were not taken into consideration.*

[d] *MWNT concentration was not provided; therefore, it was not possible to estimate the intrinsic reinforcement efficiency.*

chemical composition [39]. The hollow lumens of HNTs have been extensively studied for their drug carrier/release properties [40,41] and nanoreactor potential [42]. HNTs have a low surface charge and are easily dispersed in solvents and polymers of medium to high polarity. Significant mechanical and thermal improvements have been demonstrated in starch, chitosan, gelatin, cellulose, pectin, and polyvinyl alcohol [43,44]. However, to achieve their maximum potential, the primary challenge is to eliminate the random distribution of tube orientations. The misalignment of particles will cause inefficiency in stress transfers and lead the properties of HNT-filled nanocomposites to be far below theoretical predictions [45,46]. For this reason, management of particle orientations to achieve the desired macroscopic assembly of HNTs is the focus of this contribution.

In this study, the formation of nanotube alignment structure in epoxy/HNT nanocomposites has been achieved using the simple spray-coating method. The aligned particles were facilitated by aerodynamic flow and controlled with polymer viscosity. An attempt was also made to reveal the relationship between induced clay orientation and mechanical properties in spray-coated, thin films. It was found that the stiffening and hardening efficiencies in composites were increased with higher HNT alignment. The composite mechanics have been examined using the theories of the Cox-Krenchel [47] model. Effective reinforcement of 273 GPa in modulus from HNTs has been observed in more highly oriented particle composites. Controlling particle alignment is meaningful in fabricating one-dimensional, and potentially three-dimensional, reinforced laminate composites with improved functional property manipulations. Incorporating dispersive polymer molecules with low surface energy and tailorable stimulation transition temperatures not only improved structural robustness, but also introduced both self-cleaning and self-healing behaviors in the coatings.

4.2 COMPOSITE PREPARATIONS

Halloysite nanotubes (HNTs) were used to prepare nanocomposites in this research. The as-obtained halloysite nanotubes were characterized using a variety of methods to determine their physical and chemical properties. Fig. 4.1A shows the schematic crystal structure associated with the aluminosilicate composition of halloysite nanotubes: the outer layer is silica, and the inner layer is alumina. As shown in Fig. 4.1B, the halloysite nanotubes used in this work have inner diameters of $\sim$15 nm, outer diameters of $\sim$40 nm, length $\sim$2 microns and an average aspect ratio of around 50. The rolled-up structure is shown in Fig. 4.1C. The average number of layers in these HNTs was calculated to be around 17 [49]. These nanoparticles were chosen due to their high modulus and strength, which can be used to reinforce polymers mechanically. More importantly, the matching of reflective indices between the selected halloysite nanotubes and epoxy can provide composites with transparency of $\sim$90%.

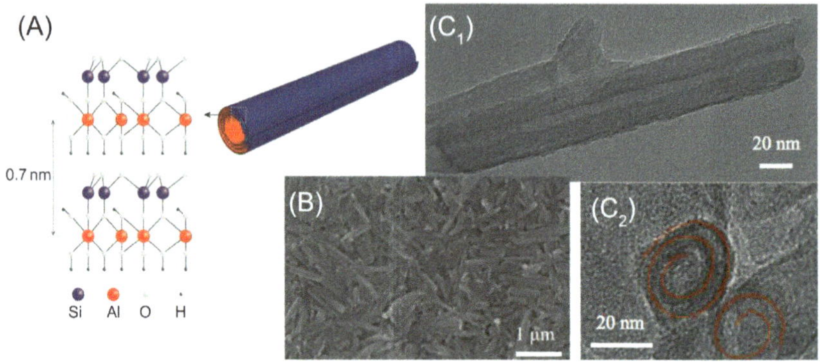

FIG. 4.1 (A) Schematic HNT structure, (B) SEM image of HNTs, TEM images of (C_1) longitudinal view, (C_2) cross-section view of a fractured halloysite nanotube, (D) WAXD of HNT powder with interlayer of 0.7 nm showing a dehydrated structure [48] and (E) TGA of pure epoxy and HNT.

To achieve their maximum potential as reinforcing agents in many applications, it is essential to eliminate the random distributions of tube orientations. The misalignment of particles creates inefficiency in stress transfer and causes the properties of HNT-filled nanocomposites to be far below theoretical predictions [45,46]. Therefore, the development of a simple process for producing nanocomposites in which the nanotubes are vertically aligned is of particular interest. Fabrication methods that produce well-controlled, out-of-plane orientations provide an opportunity to fabricate 3D reinforced nanocomposites [17–22] with highly directional properties. Such nanotube arrangements, for example, can be exploited to generate well-defined nanotemplates or patterns for desirable magnetic, electrical, and barrier properties in a cost-efficient way [50]. HNT-filled epoxy composites were prepared using a simple spray-coating technique, as shown in Fig. 4.2. Upon exiting the spraying nozzle, the particles were accelerated in a gas stream and aligned along the flow. The particle movement was constrained within the viscous polymer fluid as the liquid suspension impacted the surface of the substrate. The concentration of the nanoparticle was found to be effective at 1.0 vol% due to dispersion difficulties at higher loadings. Thus, the viscosity of the polymer suspension was mainly studied to determine its influence on the particle alignment within the polymer matrix.

Two kinds of spraying solutions were prepared based on the properties to be incorporated, namely (i) *structural coatings* for mechanical property optimizations and (ii) *multifunctional coatings* for incorporation of multiple functions (Table 4.2). For the first kind, the fluid phase of each formulation was made starting from 20 parts epoxy by weight and diluting with a specific amount of acetone (i.e., epoxy: acetone weight ratios were 20:0, 20:1, 20:2, and 20:4, which corresponded to epoxy volume percentages of 100, 93, 87, and 77 vol%). HNT was subsequently mixed with the epoxy/acetone solutions.

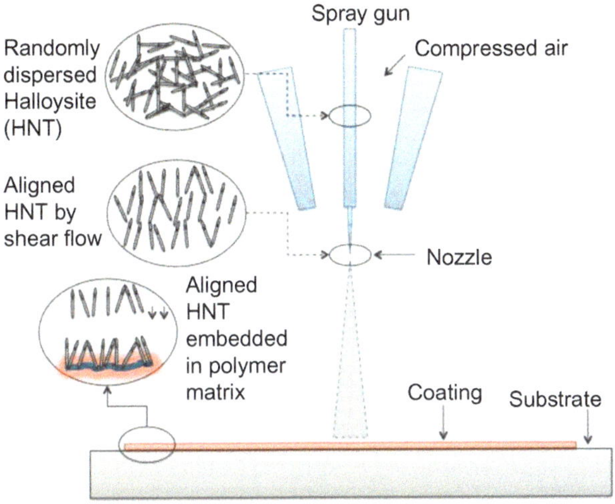

FIG. 4.2 HNT alignment design in spray coating. Batch suspensions of mixed epoxy/acetone/HNT are sprayed out of a spraying gun, and compressed air facilitates HNT alignment in the hydrodynamic flow. The viscosity of epoxy/acetone constrains the movement of HNT and preserves their preferable orientations.

HNT concentration with respect to epoxy in these various suspensions was optimized to be 2 wt% (i.e., corresponding volume percentages of HNTs are approximately 1.0 vol% based on known densities of constituents).

The mixtures were magnetically stirred for 5 min, mechanically mixed for 1 min, and sonicated for around 2 h to eliminate bubbles. The prepared solutions were then sprayed onto glass slides and cured using a Dymax 5000-EC flood lamp (working distance 2–6″) with a total UV energy output of 225 mW/cm^2 at 1.0 in. in direct emission and wavelength, mainly from 350 to 450 nm. The samples were abbreviated as $E_xA_yH_{1.0}$, with x as the volume percentage of epoxy in the liquid phase of acetone and epoxy, x being 100, 93, 87, 77 for the five solutions mentioned herein, and y is the acetone volume percentage, as shown in Table 4.2.

The SEM images of Fig. 4.3A show that the alignment of HNTs along the plane-normal direction in the sprayed composites increases with the increasing viscosity of the liquid suspension. The addition of acetone in spray-coating decreases the viscosity and can improve the ease of processing through high spraying speeds and more continuous ejection of liquid suspensions under the same pressure. However, in this study, the decreased viscosity weakens the constraining of HNTs in as-sprayed films, and the flow-induced orientation that develops at the nozzle exit is not maintained in the composites processed from low viscosity suspensions. As shown in Fig. 4.3A, HNTs in $E_{77}A_{23}$-based composites (23 vol% acetone and viscosity of 0.03 Pa s) are horizontally oriented within the coating plane direction; in contrast, when samples were

TABLE 4.2 Nomenclature of Composites

Samples		Coating Solutions				Final Composites	
		Epoxy:Acetone		*HNT (vol%)*	*CAB (vol%)*	*HNT (vol%)*	*CAB (vol%)*
Epoxy		100:0		0	0	0	0
(i) Structural coatings	$E_xA_yH_{1.0}$	*x:y*	77:23	1.0	0	1.0	0
			87:13				
			93:7				
			100:0				
(ii) Multifunctional coatings	*EC*	93:7		1.0	3.0	1.0	3.0

Note: The composites with only HNT inclusions were tailored in viscosity values by mixing epoxy with four different ratios of acetone (shaded areas). The optimized value of 93:7 was used in epoxy/cellulose acetate butyrate (EC) composites for functional properties.

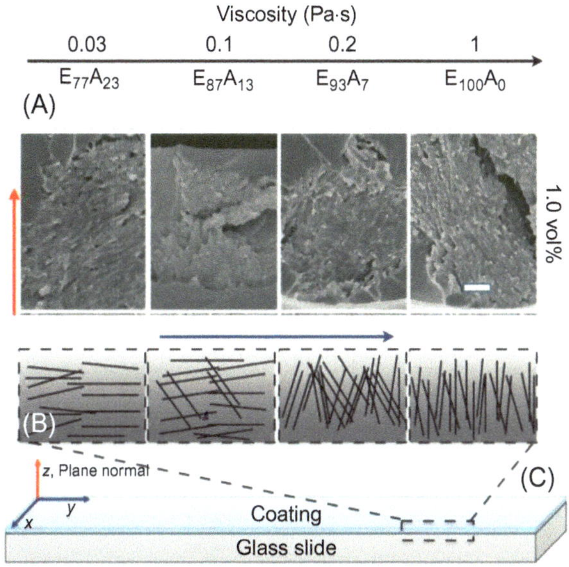

FIG. 4.3 (A) SEM imaged morphologies of cryo-fractured surfaces in coating films containing HNT, (B) schematics of HNT orientations in different viscous solutions and (C) a coating film with labels of the in-plane *x*/*y* axes and plane-normal *z*-axis directions. The scale bars in SEM images are 1 μm.

sprayed from pure epoxy-based $E_{100}A_0$ suspensions (0vol% acetone and viscosity of 1Pas), the alignment of HNTs changed from in-plane orientations to the plane-normal direction (*z*-axis as labeled in Fig. 4.3B). Samples with various HNT concentrations were also tested and showed the same trend of orientations as a function of suspension viscosity.

4.3 MECHANICAL REINFORCEMENTS

Nanoindentation measurements were conducted using a TriboIndentator (Hysitron Inc.) equipped with a Berkovich diamond tip (semi-conical tip with a diameter of 1 μm) to measure the mechanical properties. Indentation tests were operated in a displacement-control mode; the displacement excitation is applied to the sample according to a programmed loading function while the force response is continuously monitored with a resolution of 1 nN. The loading function in this work consisted of a 5-s linear loading and 5s unloading segments with a 10-s force dwelling at the peak load to reduce the influence of creeping effects [51,52]. The maximum displacement used was less than one tenth of the coating thickness (about 200nm) to exclude the influence of the substrate. A total of 36 indents with lateral spacing of 2 μm were taken to obtain average reduced modulus and hardness values on both control and composite samples coated on glass slides.

Nanoindentation hardness is defined as the indentation load divided by the projected contact area of the indentation. It is the mean pressure that material will support under load. From the load-displacement curve, hardness can be obtained at the peak load (P_{max}) as, $H = P_{max}/A$, where A is the projected contact area. For an indenter with a known geometry such as the Berkovich tip used in this study, the projected contact area is a function of contact depth, which is measured by the nanoindenter in situ during indentation [51,52].

The elastic modulus was calculated using the *Oliver-Pharr* data analysis procedure [53], beginning by fitting the unloading curve to a power-law relation,

$$E_r = \frac{\sqrt{\pi}}{2} \frac{S}{\sqrt{A_p}} \tag{4.1}$$

where S is the slope of the tangent to the loading curve at maximal load and A_p is the contact area of the indenter. E_r is related to the constituent properties by the instruments:

$$\frac{1}{E_r} = \frac{1-\upsilon^2}{E} + \frac{1-\upsilon_i^2}{E_i} \tag{4.2}$$

where E and ν are the elastic modulus and Poisson's ratio in the sample, and E_i and ν_i are the same quantities for the diamond indenter. For the diamond, $E_i = 1141$ GPa and $\nu_i = 0.07$ [51,52], and for epoxy used here, $\nu = 0.3$.

The orientation of particles dramatically influences the composite properties [54]. Because the viscosity of the spray processing fluid was shown to affect particle alignment (Fig. 4.3), normalized modulus and hardness values are plotted against viscosity in Fig. 4.4. Non-monotonic behavior is observed for both modulus and hardness. Both parameters increase with viscosity (and orientation) up to about 0.2 Pa s (where the concentration of epoxy is about 93% of

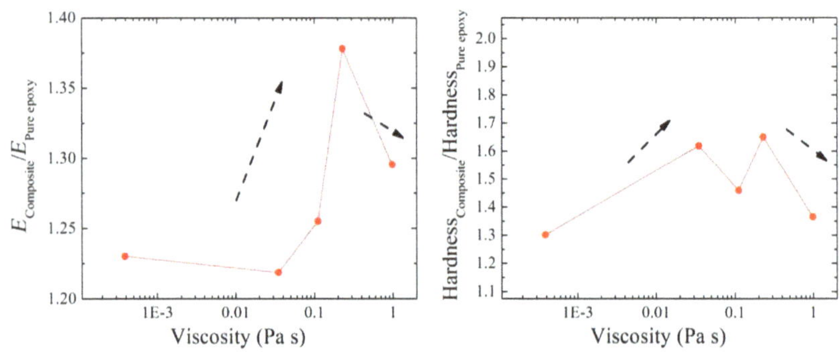

FIG. 4.4 Normalized (A) Young's modulus (E) and (B) hardness values showed the steady increase of mechanical properties with higher viscosity values in spray processing fluids. A drop at highest viscosity is attributed to the appearance of a significant number of aggregates in the samples.

the fluid phase of the processing fluid), but a noticeable decrease in modulus and hardness is observed when the liquid of the spray process is pure epoxy (viscosity = 1 Pa s). We attribute this drop off of properties to the appearance of significant amounts of nanotube aggregates in the highest-viscosity-processed films, an issue discussed in more detail as follows.

To examine the roles of orientation, concentration, and particle dimension in improving modulus and hardness for our set of nanocomposites, we employed the Cox-Krenchel model [47], which is modified from the rule-of-mixture [54], considering volume effects, length efficiency, and orientation factor in reinforcement.

$$E_c = E_m V_m + \eta_l \eta_o E_f V_f \tag{4.3}$$

In Eq. (4.1), E and V represent the modulus and volume fraction for the epoxy matrix (i.e., E_m and V_m) and HNT fillers (i.e., E_f and V_f). Here, the length efficiency factor (η_l) and orientation efficiency factor (η_o) were defined based on the shear lag theory [55] and Krenchel's method [47]. The length efficiency for particles with a specific aspect ratio depends only on volume fraction; these length efficiency values were very similar in all of our composites. Therefore, the analysis of the orientation factor in the following section will expose the most significant influence on the mechanical properties.

Based on our experiments, composites with the same HNT concentration prepared from different viscous formulations showed increases in both modulus and hardness. We attribute these property differences to HNT orientations in the final composite coatings [49]. We have used the misalignment histogram to compute the Krenchel orientation factors. To generate a more quantitative understanding of the HNTs' out-of-plane misalignments in our spray-processed composite films, statistical analyses of the SEM images were conducted using Origin software. The statistical information was plotted and fitted in Fig. 4.5. Assuming Gaussian distribution of HNTs orientations and no bias in out-of-plane direction, the mean angle for in-plane direction should be 90°, and that for out-of-plane direction should be 0°. The fitted values of *full-with at half-maximum (FWHM)* suggest the deviations from average alignment. For $E_{77}A_{23}$, $E_{93}A_7$, and $E_{100}A_0$-based HNT composites, it is possible to fit values to a Gaussian distribution function. The fact that $E_{87}A_{13}$ samples cannot be fitted is due to the purely random particle distributions. The fitted normal function center of $E_{77}A_{23}$ samples was located at ±90 degree, which indicates that the retardation of alignment loss provided by viscosity is weaker than gravitational effects and Brownian motion, as described in Fig. 4.3. Consequently, HNTs are inherently distributed parallel to in-plane x–y axes in this low-viscosity sample. The HNT concentration also influenced particle alignment, as higher concentration was associated with narrower fitted peaks. This observation again demonstrates an efficient route for controllable orientation, from perfectly parallel to vertical, by varying the viscosity.

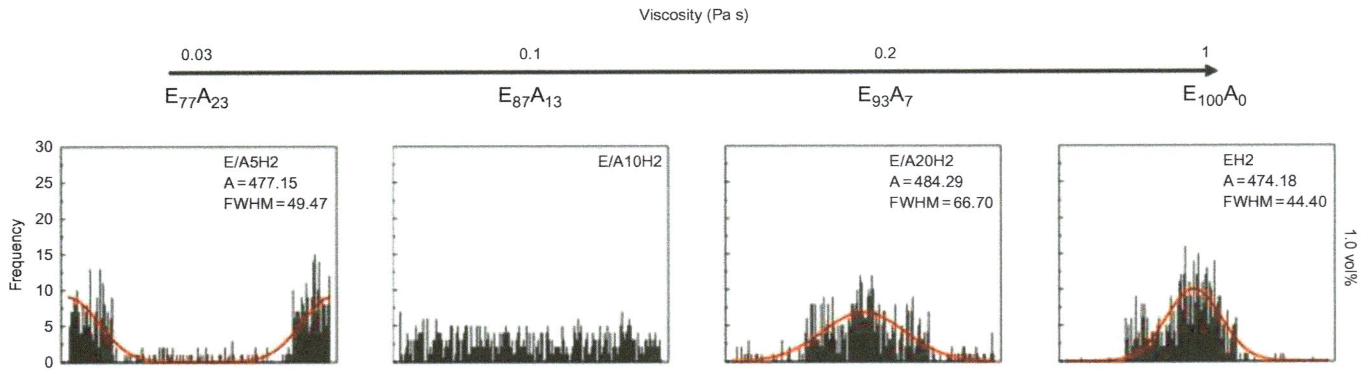

FIG. 4.5 Histogram of HNTs degree of misalignment along the plane-normal direction (*z*-axis in Fig. 4.4) in cured epoxy/HNT composites containing various concentrations of HNT. 500 nanotube orientations were taken on around 15 fracture surfaces for each sample and processed using ImageJ software. Gaussian fitting using Origin 8.5 was conducted based on the fact that the fitted peak center is 0 degree for out-of-plane aligned particles and 90 degree for in-plane aligned particles.

Once all the length efficiency and orientation factor parameters were obtained, the particle intrinsic mechanical properties became accessible based on Eq. (4.1). The effective reinforcements in modulus and hardness have been fitted, and values also listed in Table 4.3. Compared with the literature reports, the effective reinforcement (ηE_{eff}) achieved in this work, up to 182 GPa, is between the reinforcement of carbon nanotubes (up to 417 GPa) [21] and alumina platelets (100 GPa) [19]. It is worth mentioning that the alumina particles are aligned [19] while the carbon nanotubes are not [21]. If length efficiency and orientation factors were taken into consideration, the reinforcement factor for carbon nanotubes would be higher. The HNT particles in our spray coatings have reached rather high reinforcement potential. This improvement is mostly a result of the favorable normal alignment that exists in our coatings, as well as the fact that we have used HNTs with relatively small diameters in our study (around 40 nm). As indicated previously in both experimental measurements [56] and theoretical calculations [39], HNT mechanics are greatly influenced by size effects. For example, HNTs with a diameter of 50 nm have a high modulus of about 600 GPa, while a 160-nm HNT can only reach 20 GPa [56].

4.4 WEAR RESISTANCE

Improved scratch resistance is indicated by *smaller damage regions, more recovery, and higher energy-dissipation values.* The nano-scale wear resistance of the epoxy composites was tested by using the Hysitron TriboIndenter. An estimation of the normalized scratch volume (w_r) can also be determined from the final parts of the scratches. When considering the loss in volume of the material over an average short distance of the scratch lave under an average normal load (F_{Nave}) that acted over this distance,

$$w_r = \Delta V/(F_{Nave}l_{ave}) = C_r l_{ave}/(F_{Nave}l_{ave}) = C_r/F_{Nave} \tag{4.4}$$

where C_r refers to the cross-section of the scratch shortly before its end. The cross-section can be assumed to have the shape of a segment of a circle with a chord length equal to w (residual width) and height equal to the scratch depth h,

$$C_r \doteq h(3h^2+4w^2)/(6w) \doteq h^3/2w + 2wh/3 \tag{4.5}$$

Assuming further that F_{Nave} is nearly equal to F_N at the end of the scratch, the specific scratch wear rate characterized by the normalized scratch volume will reduce to,

$$w_r = C_r/F_N \tag{4.6}$$

The variations of normalized scratch volumes from pure epoxy coatings to composites with different halloysite orientations are shown in Fig. 4.6. The Krenchel orientation factor (η_o) was calculated [49] and is listed along with

TABLE 4.3 HNTs Dimensions, Length Efficiency (η_l), Orientation Factor (η_o), Particle Efficiency Factor (η) and Calculated HNT Modulus From Cox-Krenchel Model for Composites With HNT Concentrations Lower Than 1 vol%

Spraying Solution Batch	Efficiency Parameters			Effective Reinforcement (GPa)		Intrinsic Mechanics (GPa)	
	η_l	η_o	$\eta = \eta_l \eta_o$	ηE_{eff}	ηH_{eff}	E_{eff}	H_{eff}
$E_{87}A_{13}$	0.83	0.37	0.31	131	12	420.97	37.74
$E_{93}A_7$	0.83	0.64	0.54	182	9	336.67	16.67
$E_{100}A_0$	0.83	0.83	0.69	154	12	223.04	16.96

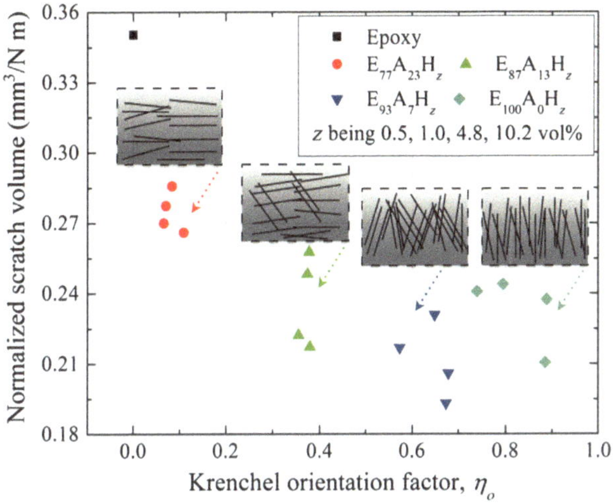

FIG. 4.6 Normalized scratch volume vs. Krenchel orientation factor relationship for pure epoxy and 1 vol% HNT composites. Krenchel orientation factors for halloysite nanotubes (HNTs) are associated with processing compositions, as listed in Table 4.2. Here demonstrated, HNT concentrations of 1.0 vol% plus three other concentrations to show the same trend (i.e., the higher the nanoparticle orientation, the better the wear-resistance).

the sample nomenclature used in this paper in Table 4.2. The results in Fig. 4.6 displayed the nanoparticle orientation-correlated trend, that is, higher nanotube alignment corresponds to better wear resistance. The normalized scratch volumes were also studied using the analysis of variance (ANOVA) method. The differences among groups were significant [49]. Our previous study [49] showed that the mechanical properties do not change when the HNT concentrations go beyond the percolation threshold (1.0 vol%). Similar results were found with tribological behavior; that is, the scratch volumes under the same load did not change much with concentration beyond the mechanical percolation threshold.

The surface morphology of the scratched samples was characterized by atomic force microscope (AFM) 3D imaging and is shown in Fig. 4.7. Fig. 4.7 shows the scratching under a loading of 100 μN for pure epoxy and composite C (η_o=0.678, composition shown in Table 4.4), respectively. Under a normal loading of 100 μN, the pure epoxy coating displayed more distinctive plough marks than those in the HNTs-filled composites, indicating more abrasion wearing (red arrows in Fig. 4.7). The epoxy coating was so fragile that it cracked and broke away from the surface of the specimen, suggesting a brittle cleavage behavior. In contrast, the composite C (η_o=0.678) showed a smoother surface upon scratching. The differences under specific loadings between epoxy and composite C (η_o=0.678) from the profile clearly suggest better resistance to scratching in composites.

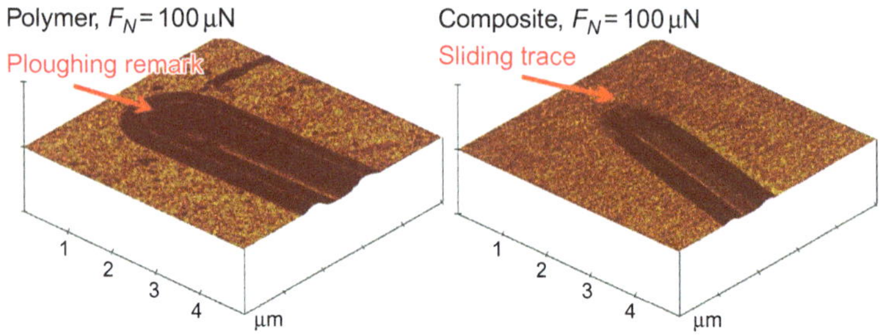

FIG. 4.7 3D profile imaging from atomic force microscopy (AFM) for pure epoxy coating and 1 vol% concentrated composite C ($\eta_o = 0.678$) under 100 μN. The scanning size is 5 × 5 μm.

The indentation depth was set to be 100 nm, less than one-tenth of the coating thickness. The hardness (H) and reduced elastic modulus (E_r) of each coating were obtained from loading-unloading curves by applying the Oliver-Pharr method [53]. As listed in Table 4.4, there is an increase in average H and E_r from epoxy to composites, and the improvement was enhanced as the particles became more vertically oriented. It is well established that the elastic strain and plastic strain are proportional to H/E_r and H^3/E_r^2 values, respectively [57]. Therefore, composite C ($\eta_o = 0.678$) is expected to present the best tribological performance due to the coexistence of the highest H/E_r and H^3/E_r^2 values. This is also consistent with the decreasing plasticity index (ψ, as defined by $(w_{ir})/(w_{ir} + w_r)$, w_{ir} and w_r are irreversible and reversible work done during mechanical tests, see Table 4.4).

The higher self-recovery percentage is an indicator of better wear resistance [58]. Fig. 4.8A illustrates the penetration depth recovery (PDR) versus HNT orientation factor relationship. PDR was defined by the bounced-back-depth after scratching (after a 7-day period, see experimental section) divided by the in situ scratching depth. The self-recovery depends on the soft components (epoxy) and the hard filler particles (HNTs). The temporary shape of the composites is fixed in the soft epoxy that consists of polymer chains, while the permanent shape is determined by the hard segment of nanotubes [60]. Thus the self-recovery percentage of all samples increased significantly with the increase of nanoparticle orientation (Fig. 4.8A). At the normal load of 100 μN, the self-recovery is 60% for pure epoxy and up to 80% for composite C ($\eta_o = 0.678$). These trends are consistent with the observed normalized scratch volume changes (Fig. 4.6) and morphology/topography features (Fig. 4.7). The higher the recovery percentage, the more wear-resistant a sample is.

A proposed recovery mechanism is illustrated in Fig. 4.8B. In Fig. 4.8B, the vertically aligned nanoparticles sustained a majority of the normal loading in the force-controlled nanoscratching test, experiencing much less strain than the horizontally aligned particles. For vertical nanoparticles going through the same strain as epoxy, the modulus is calculated by

TABLE 4.4 Average Mechanical Properties of Coatings in Loading-Unloading Indentation Cycle

Coatings	Composition	Krenchel Orientation Factor, η_o	H (GPa)	E_r (GPa)	H/E_r	H^3/E_r^2 (GPa)	Ψ (%)
Epoxy	Epoxy	0	0.2	4.8	4.2×10^{-2}	3.5×10^{-3}	76.7
Composite C	Epoxy with 1.0vol% HNT	*0.678*	0.6	7.6	7.9×10^{-2}	3.7×10^{-3}	68.0

Note: H *and* E_r *are the hardness and reduced modulus from nanoindentation.* Ψ*, plasticity index, is defined as the ratio between irreversible work and total work done in a loading-unloading cycle.*

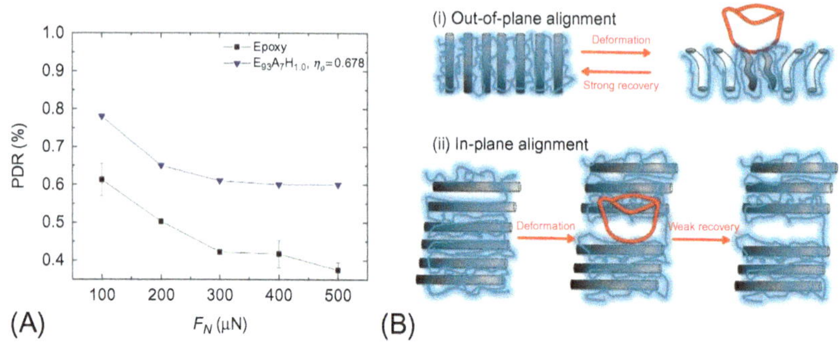

FIG. 4.8 (A) Penetration depth recovery (PDR) vs. Krenchel orientation factors of halloysite. The figures display nanoparticle orientation-related trends (i.e., higher out-of-plane alignment of the HNTs demonstrated better self-recovery after scratching). (B) Scratching in composites with out-of-plane and in-plane nanoparticle alignment (schematic to scale, with a semi-conical tip diameter of 1 μm, HNTs average length of 2 μm and spacing of ~685 nm between nanotube centers). Single halloysite nanotube (length ~400 nm, out diameter ~40 nm and inner diameter ~15 nm) bending has been measured using transmission electron microscopy (TEM) [59], showing an intact structure upon 90° bending and pure elastic deformation up to 5 degree bending.

$E_{composite} = E_{HNT\text{-}axial} \cdot V_{HNT} + E_{Epoxy} \cdot V_{Epoxy}$, where E is the modulus and V is the volume fraction for horizontal nanoparticles experiencing the same stress as epoxy. However, the modulus is computed following $E_{composite} = 1/(1/E_{HNT\text{-}radial} \cdot V_{HNT} + 1/E_{Epoxy} \cdot V_{Epoxy}) \approx E_{Epoxy}$. The HNT modulus is on the order of several GPa in the radial direction ($E_{HNT\text{-}radial}$), two orders of magnitude smaller than that along the axial direction (up to 560 GPa, $E_{HNT\text{-}axial}$) [39]. Therefore, the indentation tip penetrates the epoxy materials when HNTs are aligned horizontally, resulting in cracking and other failures due to the weakness of the polymer and the nanotube along the axis normal to the load [61]. On the other hand, out-of-plane aligned particles experience bending and buckling upon normal loading, and recover more to the original state when the load is released (Fig. 4.8B).

During the tribology test, friction coefficient variations can reflect micro- or nano-structural modifications in the coatings. To exclude the influence of surface roughness, all the sample surfaces were characterized using stylus profilometry, which indicates limited variations among all samples. Fig. 4.9 shows the friction coefficient as a function of particle orientation factors for both pure epoxy (empty marks) and 1 vol% halloysite composite samples (filled symbols). The friction coefficient parameters show an increasing trend as in-plane aligned nanoparticles are included in epoxy. When HNTs are more out-of-plane aligned, the friction coefficient decreases. Such increasing-decreasing trends remain similar in all loading conditions. By comparison, higher loadings allow the indenting tip to penetrate deeper into samples, leading to an increase of friction coefficient. The effect of friction is to add compressive stress to the front edge of the contact and intensify the tensile stress at the back edge with fracture

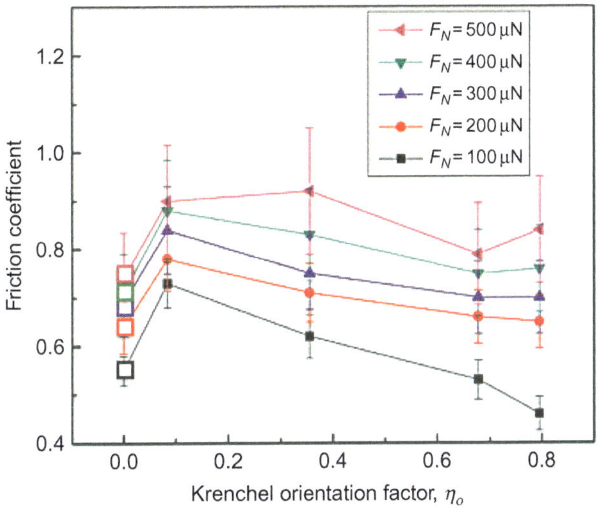

FIG. 4.9 Friction coefficient (μ) vs. Krenchel orientation factor relationship shows the influence of particle alignment on the friction coefficient. In-plane orientations of HNTs increased the friction coefficient, while out-of-plane HNTs alignments caused μ to decrease or remain similar as compared with μ in epoxy. Open marks are pure epoxy, and closed symbols are composites.

events. Thus, a lower value of the friction coefficient correlates with improvements in scratch resistance [62]. Similar research results of vertical nanoparticle alignment improving wear-resistance have been observed in single-wall carbon nanotubes [63], multi-wall carbon nanotubes [64,65], and carbon nanofibers [66]. However, in nanoparticle-reinforced composites, the influence of nanoparticle alignment on the friction coefficient has been studied [67] much less extensively.

4.5 SELF-CLEANING

A water drop placed on a surface exhibits a range of contact angles varying from the advancing (maximal) contact angle, θ_A, to the receding (minimal) contact angle, θ_R. The contact angle hysteresis (CAH) is defined here as CAH $=\theta_A-\theta_R$. Fig. 4.10A shows significant increases of CAH on roughened sample surfaces after Taber Abrasion tests. The healing at 110°C overnight clearly brought the CAH down to values similar to original samples, proving again the healing efficiency of composites when enough CAB_{HM} was added. The water roll-off angles were measured by dropping 30 µL water on tilted coatings (Fig. 4.10A). The addition of CAB_{HM} decreased the roll-off angle of epoxy (87 degree) to values as low as 17 degree (Fig. 4.10B). The healed samples showed similar values. These low water roll-off angles can also benefit the self-cleaning properties of the coatings.

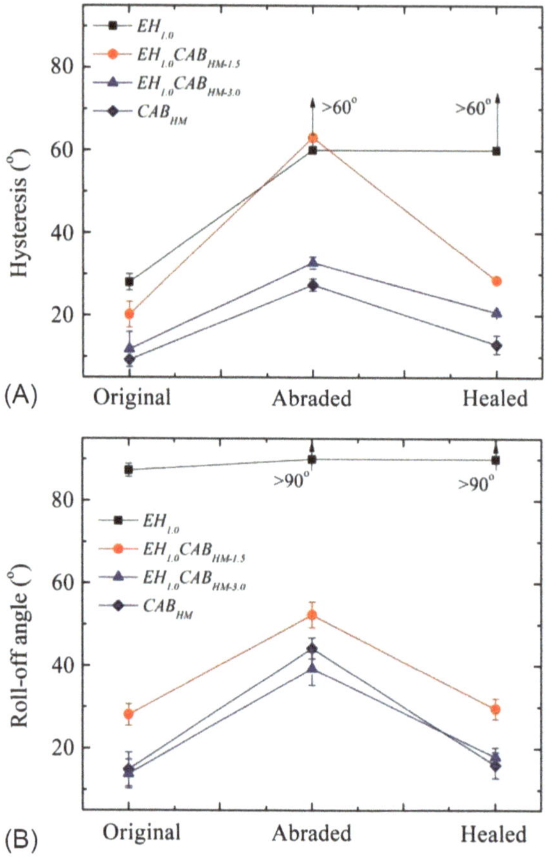

FIG. 4.10 Goniometer measurement of (A) contact angle hysteresis and (B) water roll-off angle on all sample surfaces of $EH_{1.0}CAB_{HM\text{-}y}$, where volume concentration percentage $y=0$, 1.5, 3.0, 6.0, 12.0, 24.0, while HNTs concentration remains 1.0vol% and CAB_{HM}. The surfaces were damaged using Taber Abrasion tests with 1 kg weight on both wheels and abrasion of 40 cycles, and the healing was conducted at 110°C in the oven overnight. For water roll-off angle measurements, distilled water droplet is of 30μL volume. Water roll-off angle was conducted from 0 to 90 degree within 1000 steps until both sides of the water droplet move on tilted sample surfaces. During water roll-off angle measurements, epoxy surfaces after mechanical damages and healing were so rough and hydrophilic that water droplets wouldn't fall at a highest tilted angle of 90 degree. During water cleaning demonstrations, the tilted angle of both samples was 20 degree, with an average sand particle size of microns and sand weight of 200mg on sample surfaces being cleaned by water droplets of 2.5mL.

4.6 SELF-HEALING

Self-healing performance and efficiencies have been evaluated by various testing methods as described herein, and in other methods [68,69], but the present study focused on protective coatings and thin-films, with transparency and sand impact resistance being emphasized. All samples were scratched with the controlled depth of 300nm using a Hysitron NanoIndenter. The healing procedures

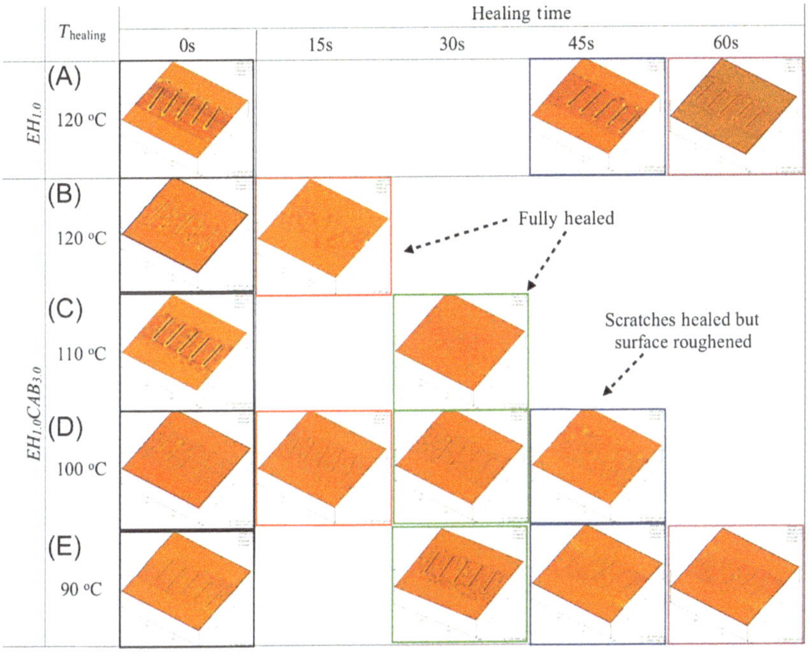

FIG. 4.11 AFM imaging for (A) negligible healing of cracks in $EH_{1.0}$, (B) healing at 120°C showed clear filling of cracks in $EH_{1.0}CAB_{HM\text{-}3.0}$, but at the same time, the surface phase was altered due to over-flow of CAB_{HM}, (C) healing at 110°C, (D) healing at 100°C and (E) healing at 90°C. Required healing periods at each temperature were also studied. AFM 3D images for A–E are 30 by 30 μm for length/width and 1 μm for height.

were captured as a function of healing temperature and healing time (Fig. 4.11). The scratched samples were heated between 90°C and 120°C. The heating at 120°C obviously went beyond the thermal transitions of all samples and resulted in the most efficient recovery of surface smoothness. $EH_{1.0}$ coatings remained constant for the scratch depth, independent of healing time (Fig. 4.11A). $EH_{1.0}CAB_{HM\text{-}3.0}$ composites recovered fully at both 120°C and 110°C, with higher temperatures healing at a shorter time (Fig. 4.11B and C). Healing at 100°C resulted in residual surface roughness and scratches were slightly observable after 45 s healing; 90°C heating for the 60 s also resulted in unhealed scratches on the surfaces (Fig. 4.11D and E). In heating procedures, CAB_{HM} can move within the epoxy network, and makes its way to the surfaces and interfaces.

4.7 Conclusions

This research has demonstrated successful fabrication and characterization of structurally durable and functionally versatile nanocomposites. These composites, upon being filled with specific concentrated nanoparticles, showed improved modulus and hardness. The nanoparticle in these nanocomposites can be tailored to generate better wear-resistance. The introduction of molecules, namely, cellulose acetate butyrate (CAB), provides self-cleaning and self-healing properties due to both the lower surface energy and the responsive movements of these CAB polymer chains upon stimulation by increased temperature.

REFERENCES

[1] A. Mathiazhagan, R. Joseph, Nanotechnology—a new prospective in organic coating-review, Int. J. Chem. Eng. Appl. 2 (4) (2011) 225.

[2] L. Wu, X. Guo, J. Zhang, Abrasive resistant coatings—a review, Lubricants 2 (2) (2014) 66–89.

[3] M. Montemor, Functional and smart coatings for corrosion protection: a review of recent advances, Surf. Coat. Technol. 258 (2014) 17–37.

[4] H. Qian, E.S. Greenhalgh, M.S.P. Shaffer, A. Bismarck, Carbon nanotube-based hierarchical composites: a review, J. Mater. Chem. 20 (23) (2010) 4751–4762.

[5] D.Y. Cai, M. Song, Recent advance in functionalized graphene/polymer nanocomposites, J. Mater. Chem. 20 (37) (2010) 7906–7915.

[6] K. Song, Y. Zhang, M.L. Minus, Polymer interphase self-reinforcement and strengthening mechanisms in low-loaded nanocomposite fibers, Macromol. Chem. Phys. 216 (12) (2015) 1313–1320.

[7] K. Song, Y. Zhang, J. Meng, M.L. Minus, Spectral analysis of lamellae evolution and constraining effects aided by nano-carbons: a coupled experimental and simulation study, Polymer 75 (2015) 187–198.

[8] J. Meng, Y. Zhang, K. Song, M.L. Minus, Forming crystalline polymer-nano interphase structures for high-modulus and high-tensile/strength composite fibers, Macromol. Mater. Eng. 299 (2) (2014) 144–153.

[9] Y. Zhang, K. Song, J. Meng, M.L. Minus, Tailoring polyacrylonitrile interfacial morphological structure by crystallization in the presence of single-wall carbon nanotubes, ACS Appl. Mater. Interfaces 5 (3) (2013) 807–814.

[10] K. Song, Y. Zhang, J. Meng, M.L. Minus, Lubrication of poly (vinyl alcohol) chain orientation by carbon nano-chips in composite tapes, J. Appl. Polym. Sci. 127 (4) (2013) 2977–2982.

[11] A.A. Azeez, K.Y. Rhee, S.J. Park, D. Hui, Epoxy clay nanocomposites' processing, properties and applications: a review, Compos. Part B (2012).

[12] K. Song, X. Ren, L. Zhang, Bamboo fiber-polymer composites: overview of fabrications, mechanical characterizations and applications, in: M. Jawaid, M.S. Salit, O.Y. Alothman (Eds.), Green Biocomposites: Design and Applications, Springer International Publishing, Cham, 2017, pp. 209–246.

[13] H.A. Khalil, A. Bhat, A.I. Yusra, Green composites from sustainable cellulose nanofibrils: a review, Carbohydr. Polym. 87 (2) (2012) 963–979.

[14] K. Song, 2—Micro- and nano-fillers used in the rubber industry, in: Progress in Rubber Nanocomposites, Woodhead Publishing, 2017, , pp. 41–80.

[15] M.S. Dresselhaus, Fifty years in studying carbon-based materials, Phys. Scr. 2012 (T146) (2012).

[16] P.R. Sajanlal, T.S. Sreeprasad, A.K. Samal, T. Pradeep, Anisotropic nanomaterials: structure, growth, assembly, and functions, Nanotechnol. Rev. 2 (5883) (2011) 4–10.

[17] S. Ahadian, J. Ramón-Azcón, M. Estili, X. Liang, S. Ostrovidov, H. Shiku, M. Ramalingam, K. Nakajima, Y. Sakka, H. Bae, T. Matsue, A. Khademhosseini, Hybrid hydrogels containing

vertically aligned carbon nanotubes with anisotropic electrical conductivity for muscle myofiber fabrication, Sci. Rep. 4 (2014) 4271–4282.

[18] J. Ramón-Azcón, S. Ahadian, M. Estili, X. Liang, S. Ostrovidov, H. Kaji, H. Shiku, M. Ramalingam, K. Nakajima, Y. Sakka, Dielectrophoretically aligned carbon nanotubes to control electrical and mechanical properties of hydrogels to fabricate contractile muscle myofibers, Adv. Mater. 25 (29) (2013) 4028–4034.

[19] R.M. Erb, R. Libanori, N. Rothfuchs, A.R. Studart, Composites reinforced in three dimensions by using low magnetic fields, Science 335 (6065) (2012) 199–204.

[20] J.J. Martin, B.E. Fiore, R.M. Erb, Designing bioinspired composite reinforcement architectures via 3D magnetic printing, Nat. Commun. 6 (2015) 8641–8648.

[21] R. Jalili, J.M. Razal, G.G. Wallace, Wet-spinning of PEDOT:PSS/functionalized-SWNTs composite: a facile route toward production of strong and highly conducting multifunctional fibers, Sci. Rep. 3 (2013) 3438–3445.

[22] V.P. Veedu, A. Cao, X. Li, K. Ma, C. Soldano, S. Kar, P.M. Ajayan, M.N. Ghasemi-Nejhad, Multifunctional composites using reinforced laminae with carbon-nanotube forests, Nat. Mater. 5 (6) (2006) 457–462.

[23] Y. Lan, Y. Wang, Z.F. Ren, Physics and applications of aligned carbon nanotubes, Adv. Phys. 60 (4) (2011) 553–678.

[24] C. Martin, J. Sandler, A. Windle, M.-K. Schwarz, W. Bauhofer, K. Schulte, M. Shaffer, Electric field-induced aligned multi-wall carbon nanotube networks in epoxy composites, Polymer 46 (3) (2005) 877–886.

[25] S. Kaida, J. Matsui, T. Sagae, Y. Hoshikawa, T. Kyotani, T. Miyashita, The production of large scale ultrathin aligned CNT films by combining AC electric field with liquid flow, Carbon 59 (2013) 503–511.

[26] E. Camponeschi, R. Vance, M. Al-Haik, H. Garmestani, R. Tannenbaum, Properties of carbon nanotube-polymer composites aligned in a magnetic field, Carbon 45 (10) (2007) 2037–2046.

[27] T. Kimura, H. Ago, M. Tobita, S. Ohshima, M. Kyotani, M. Yumura, Polymer composites of carbon nanotubes aligned by a magnetic field, Adv. Mater. 14 (19) (2002) 1380–1383.

[28] A.B. Sulong, J. Park, Alignment of multi-walled carbon nanotubes in a polyethylene matrix by extrusion shear flow: mechanical properties enhancement, J. Compos. Mater. 45 (8) (2011) 931–941.

[29] M. Zhang, S. Fang, A.A. Zakhidov, S.B. Lee, A.E. Aliev, C.D. Williams, K.R. Atkinson, R.H. Baughman, Strong, transparent, multifunctional, carbon nanotube sheets, Science 309 (5738) (2005) 1215–1219.

[30] T. Chen, Z. Cai, L. Qiu, H. Li, J. Ren, H. Lin, Z. Yang, X. Sun, H. Peng, Synthesis of aligned carbon nanotube composite fibers with high performances by electrochemical deposition, J. Mater. Chem. A 1 (6) (2013) 2211–2216.

[31] K. Jiang, Q. Li, S. Fan, Nanotechnology: spinning continuous carbon nanotube yarns, Nature 419 (6909) (2002) 801.

[32] R. Sivamohan, The problem of recovering very fine particles in mineral processing—a review, Int. J. Miner. Process. 28 (3) (1990) 247–288.

[33] Y.-L. Li, I.A. Kinloch, A.H. Windle, Direct spinning of carbon nanotube fibers from chemical vapor deposition synthesis, Science 304 (5668) (2004) 276–278.

[34] A.M. Marconnet, N. Yamamoto, M.A. Panzer, B.L. Wardle, K.E. Goodson, Thermal conduction in aligned carbon nanotube–polymer nanocomposites with high packing density, ACS Nano 5 (6) (2011) 4818–4825.

[35] N.R. Raravikar, L.S. Schadler, A. Vijayaraghavan, Y. Zhao, B. Wei, P.M. Ajayan, Synthesis and characterization of thickness-aligned carbon nanotube-polymer composite films, Chem. Mater. 17 (5) (2005) 974–983.

[36] M. Liu, Z. Jia, D. Jia, C. Zhou, Recent advance in research on halloysite nanotubes-polymer nanocomposite, Prog. Polym. Sci. 39 (8) (2014) 1498–1525.
[37] M. Doi, S.D. Edwards, Dynamics of concentrated polymer systems. Part 1—Brownian Motion in the Equilibrium State, J. Chem. Soc. Faraday Trans. 2 (74) (1978) 1789–1801.
[38] M. Doi, S. Edwards, Dynamics of concentrated polymer systems. Part 2—molecular motion under flow, J. Chem. Soc. Faraday Trans. 2 (74) (1978) 1802–1817.
[39] L. Guimaraes, A.N. Enyashin, G. Seifert, H.A. Duarte, Structural, electronic, and mechanical properties of single-walled halloysite nanotube models, J. Phys. Chem. C 114 (26) (2010) 11358–11363.
[40] Y.M. Lvov, D.G. Shchukin, H. Mohwald, R.R. Price, Halloysite clay nanotubes for controlled release of protective agents, ACS Nano 2 (5) (2008) 814–820.
[41] S. Levis, P. Deasy, Characterisation of halloysite for use as a microtubular drug delivery system, Int. J. Pharm. 243 (1) (2002) 125–134.
[42] D.G. Shchukin, G.B. Sukhorukov, R.R. Price, Y.M. Lvov, Halloysite nanotubes as biomimetic nanoreactors, Small 1 (5) (2005) 510–513.
[43] T.S. Gaaz, A.B. Sulong, M.N. Akhtar, A.A.H. Kadhum, A.B. Mohamad, A.A. Al-Amiery, Properties and applications of polyvinyl alcohol, halloysite nanotubes and their nanocomposites, Molecules 20 (12) (2015) 22833–22847.
[44] D. Rawtani, Y. Agrawal, Multifarious applications of halloysite nanotubes: a review, Rev. Adv. Mater. Sci. 30 (2012) 282–295.
[45] K. Song, Y. Zhang, J. Meng, E.C. Green, N. Tajaddod, H. Li, M.L. Minus, Structural polymer-based carbon nanotube composite fibers: understanding the processing–structure–performance relationship, Materials 6 (6) (2013) 2543–2577.
[46] X.L. Xie, Y.W. Mai, X.P. Zhou, Dispersion and alignment of carbon nanotubes in polymer matrix: a review, Mater. Sci. Eng. R-Rep. 49 (4) (2005) 89–112.
[47] H. Cox, The elasticity and strength of paper and other fibrous materials, Br. J. Appl. Phys. 3 (3) (1952) 72–75.
[48] F.J. García, S. García Rodríguez, A. Kalytta, A. Reller, Study of natural halloysite from the Dragon Mine, Utah (USA), Z. Anorg. Allg. Chem. 635 (4–5) (2009) 790–795.
[49] K. Song, R. Polak, D. Chen, M.F. Rubner, R.E. Cohen, K.A. Askar, Spray-coated halloysite-epoxy composites: a means to create mechanically robust, vertically aligned nanotube composites, ACS Appl. Mater. Interfaces 8 (31) (2016) 20396–20406.
[50] Z.L. Wang, Zinc oxide nanostructures: growth, properties and applications, J. Phys. Condens. Matter 16 (25) (2004) R829–841.
[51] H. Hu, L. Onyebueke, A. Abatan, Characterizing and modeling mechanical properties of nanocomposites-review and evaluation, J. Miner. Mater. Charact. Eng. 9 (04) (2010) 275–280.
[52] B. Bhushan, X. Li, Nanomechanical characterisation of solid surfaces and thin films, Int. Mater. Rev. 48 (3) (2003) 125–164.
[53] W.C. Oliver, G.M. Pharr, An improved technique for determining hardness and elastic modulus using load and displacement sensing indentation experiments, J. Mater. Res. 7 (06) (1992) 1564–1583.
[54] D. Hull, T.W. Clyne, Introduction to Composite Materials, second ed., Cambridge University Press, Cambridge, 1996.
[55] N.G. McCrum, C.P. Buckley, C.B. Bucknall, Principles of Polymer Engineering, second ed., Oxford University Press, USA, 1997, pp. 276–278.
[56] B. Lecouvet, J. Horion, C. D'Haese, C. Bailly, B. Nysten, Elastic modulus of halloysite nanotubes, Nanotechnology 24 (10) (2013) 105704–105711.

[57] N. Sakharova, J. Fernandes, M. Oliveira, J. Antunes, Influence of ductile interlayers on mechanical behaviour of hard coatings under depth-sensing indentation: a numerical study on TiAlN, J. Mater. Sci. 45 (14) (2010) 3812–3823.

[58] M.A. Haque, T. Kurokawa, G. Kamita, J.P. Gong, Lamellar bilayers as reversible sacrificial bonds to toughen hydrogel: hysteresis, self-recovery, fatigue resistance, and crack blunting, Macromolecules 44 (22) (2011) 8916–8924.

[59] D. Lu, H. Chen, J. Wu, C.M. Chan, Direct measurements of the Young's modulus of a single halloysite nanotube using a transmission electron microscope with a bending stage, J. Nanosci. Nanotechnol. 11 (9) (2011) 7789–7793.

[60] A. Lendlein, S. Kelch, Shape-memory polymers, Angew. Chem. Int. Ed. 41 (12) (2002) 2034–2057.

[61] E.W. Wong, P.E. Sheehan, C.M. Lieber, Nanobeam mechanics: elasticity, strength, and toughness of nanorods and nanotubes, Science 277 (5334) (1997) 1971–1975.

[62] K. Kato, Wear in relation to friction—a review, Wear 241 (2) (2000) 151–157.

[63] K. Miyake, M. Kusunoki, H. Usami, N. Umehara, S. Sasaki, Tribological properties of densely packed vertically aligned carbon nanotube film on SiC formed by surface decomposition, Nano Lett. 7 (11) (2007) 3285–3289.

[64] P. Dickrell, S. Sinnott, D. Hahn, N. Raravikar, L. Schadler, P. Ajayan, W. Sawyer, Frictional anisotropy of oriented carbon nanotube surfaces, Tribol. Lett. 18 (1) (2005) 59–62.

[65] W. Zhang, B. Xu, A. Tanaka, Y. Koga, Frictional behaviour of vertically aligned carbon nanotube films, Carbon 47 (3) (2009) 926–929.

[66] J.P. Tu, L.P. Zhu, K. Hou, S.Y. Guo, Synthesis and frictional properties of array film of amorphous carbon nanofibers on anodic aluminum oxide, Carbon 41 (6) (2003) 1257–1263.

[67] H.-J. Kim, D.-E. Kim, Nano-scale friction: a review, Int. J. Precis. Eng. Manuf. 10 (2) (2009) 141–151.

[68] D. Bekas, K. Tsirka, D. Baltzis, A. Paipetis, Self-healing materials: a review of advances in materials, evaluation, characterization and monitoring techniques, Compos. Part B 87 (2016) 92–119.

[69] H. Ullah, K.A. M Azizli, Z.B. Man, M.B.C. Ismail, M.I. Khan, The potential of microencapsulated self-healing materials for microcracks recovery in self-healing composite systems: a review, Polym. Rev. 56 (3) (2016) 429–485.

Chapter 5

Self-Healing Fiber Composites With a Self-Pressurized Healing System

Yong Zhu*, Xiao Ji Ye*,†, Min Zhi Rong* and Ming Qiu Zhang*
*Key Laboratory for Polymeric Composite and Functional Materials of Ministry of Education, GD HPPC Lab, School of Chemistry, Sun Yat-Sen (Zhongshan) University, Guangzhou, China, †Guangdong Provincial Public Laboratory of Analysis and Testing Technology, China National Analytical Center, Guangzhou, China

Chapter Outline

5.1 INTRODUCTION

Self-healing has been recognized as a next-generation technique for improving reliability and durability of polymers and polymer composites [1,2]. By mimicking naturally occurring substances, nonfatal wounds created during manufacturing or service can be autonomously repaired, leading to fewer returns and less waste.

So far, more and more scientists are involved in research on self-healing materials. Researchers are not only investigating full-scale restoration that is mainly focused on mechanical properties recovery, but also non-structural recovery toward functional properties (e.g., electrical, optical, and physico-chemical properties) [3]. Compared with intrinsic self-healing that operates as a result of reversible inter- and/or intra-macromolecular interaction [4–6], extrinsic self-healing approaches for polymers based on embedded fluidic healing agent have attracted substantial interest [7] because chemical structures of the target materials do not need to be changed, which is different from most intrinsic self-healing systems. Among the available vessels that hold healing agents, micropipelines [8,9] (Fig. 5.1) are able to store/provide larger amounts of healing agent and allow for multiple healing functions when an external

Polymer-Based Multifunctional Nanocomposites and Their Applications
https://doi.org/10.1016/B978-0-12-815067-2.00005-6

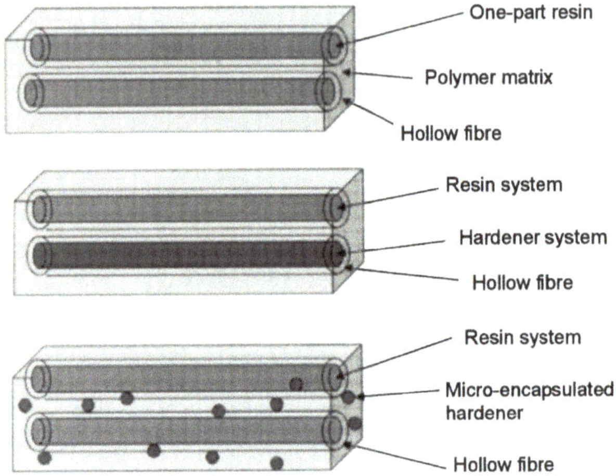

FIG. 5.1 Schematic diagram of repair concept for polymer matrix composites using pre-embedded hollow tubes. *(Source: S.M. Bleay, C.B. Loader, V.J. Hawyes, L. Humberstone, P.T. Curtis. A smart repair system for polymer matrix composites, Compos. Part A 32 (2001) 1767–1776, Copyright 2001. Reproduced with permission from Elsevier.)*

supply is connected, which are the advantages over microcapsules [7]. Accordingly, a variety of micropipeline configurations have been proposed, including hollow glass tubes [10–14], hollow glass fibers [14–19], hollow carbon fibers [20], hollow polymer tubes [21,22], and hollow metallic tubes [11]. Moreover, microvascular networks were also created by incorporating pre-made microchannels [23–27] (Fig. 5.2) or removing pre-loaded microperforms [28–31] from the composites (Fig. 5.3). They have worked well in different composites

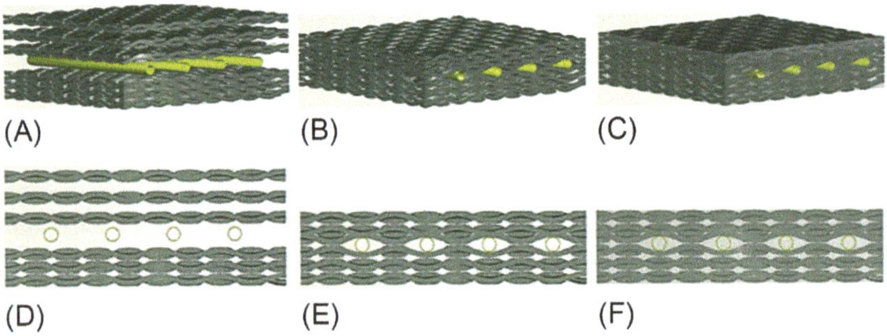

FIG. 5.2 Schematic diagram for manufacturing techniques based on non-removable hollow cores (A and D) placement of hollow cores (glass, polymer or metallic) between dry fabric layers, (B and E) composite processing (hand lay-up, vacuum bagging, resin infusion, RTM or autoclave) and (C and F) cured composite laminate with embedded microchannel. *(Source: M.-U. Saeed, Z.F. Chen, B.B. Li. Manufacturing strategies for microvascular polymeric composites: a review, Compos. Part A 78 (2015) 327–340, Copyright 2015. Reproduced with permission from Elsevier.)*

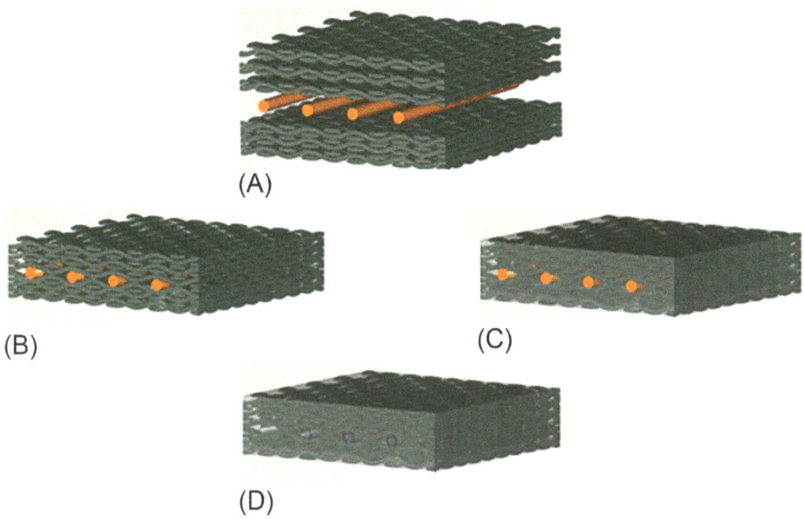

FIG. 5.3 Schematic illustration for microchannel creation by removable solid cores (A) placement of solid cores (polymer or metallic) between dry fabric layers, (B) composite processing (hand lay-up, vacuum bagging, resin infusion, RTM or autoclave), (C) cured composite laminate and (D) exposing microchannels by manual removal of solid mandrels. *(Source: M.-U. Saeed, Z.F. Chen, B.B. Li. Manufacturing strategies for microvascular polymeric composites: a review, Compos. Part A 78 (2015) 327–340, Copyright 2015. Reproduced with permission from Elsevier.)*

(mostly continuous fiber reinforced thermosetting polymer composites) in terms of mechanical properties restoration.

Fiber composites belong to advanced materials, and have wide applications in the aviation and automobile industries. Most damage to fiber reinforced polymer composites results from low velocity, and sometimes high velocity impact [32], and the internal damage is usually more extensive than that seen on the surface. Traditional repairing techniques based on adhesive bonding (such as patch repair) involve covering or replacing the damaged portions with a new material [32–34], and could do nothing about the microcracks that are nearly invisible. As a result, propagation and coalescence of microcracks would bring about catastrophic failure of the materials, and hence, significantly shorten lifetimes of the structures. It reflects the necessity of the introduction of self-healing functionality.

It is worth noting that mobility of the healing agent inside materials released upon damage is a key issue, no matter which type of micropipelines are applied. It determines whether the healing agent can approach the cracked sites, and the envisaged healing reaction can be effectively triggered (for a two-component healant system). In fact, the capillary effect, the main driving factor of healing fluid in self-healing materials with microcracks, becomes rather weak for fiber composites with open fractures [16–18,35–37]. Moreover, the cracks in fiber composites are rather complicated, which may curve and interconnect. Therefore, improvement of flowability of the healing agent becomes a very important issue. Otherwise, overlong healing duration or lower healing efficiency would be observed. In extreme situations, for example, the process of healing may even last 12 months [10]. To solve the problem, assistant measures based on manual intervention, such as evacuation [18], heating [16,35,36], dilution with solvents [18,37] and external pressurization [22,28] were used during authentic healing.

The authors of this chapter describe a novel method that includes mixing a foaming agent with the encapsulated healing agent [38]. As long as it takes effect, internal pressure is greatly increased, and the healing agent would burst out of the reservoir soon after breakage. This self-pressurized system can improve flowability of the healing agent, free of manual intervention. More interestingly, the healing agent filled micropipelines might thus be less dense, favoring maintaining original properties of the composites to a higher extent.

Hereinafter, a proof-of-concept glass fiber/epoxy composite is prepared and verified accordingly, in which polypropylene (PP) tubes respectively filled with epoxy and mercaptan as healing components are incorporated. The healing agent has already shown its ability to conduct crack remending of cured epoxy when it is included in microcapsules [39,40]. The epoxide-mercaptan reaction catalyzed by alkali is a general base catalysis process in which mercaptan first reacts with alkali to give mercaptide ion, and then the ion reacts with epoxide. Alternatively, a nucleophilic catalysis occurs, during which alkali first reacts with epoxide to produce a reactive intermediate, and then the intermediate reacts with mercaptan in a nucleophilic displacement. The PP tubing is selected as a micro-container, not only because of easy realization of large-scale production, but also owing to the following potential expansibility and advantages (Fig. 5.4A). (i) Healing chemicals can be easily filled into larger size tubes first, and then the tube's size can be reduced by thermal stretching. This would avoid the difficulty of introducing a healing agent into the tubes with a fixed small size (glass tubes and fibers, for example), while eventually leading to less deterioration of the composite's mechanical properties, as compared with the larger size tubes. (ii) The healing agent-loaded PP tubes can be converted into compartmented versions [41] by, for example, ultrasonic welding. This would help to realize repeated self-healing. (iii) The PP tubes containing healing agents can be purposely deformed due to their flexible nature, so that they can be knitted together with conventional reinforcing fibers. This may favor fabrication of self-healing structural composites with complicated shapes. (iv) The surface of the PP tubes can be modified by physical or chemical techniques, so that the interfacial interaction in composites could be improved. (v) The chemical inertness of PP ensures that it would not react with the included healing agent.

The current work is the first step of our project concerning self-healing composites based on polymeric micropipelines. The main factors that influence

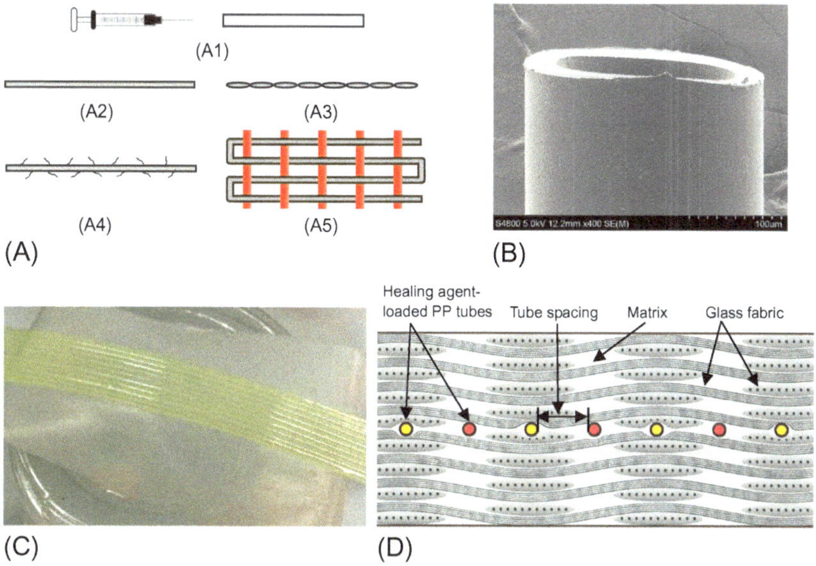

FIG. 5.4 (A) Expansibility and advantages of thermoplastic tubing in constructing microvascular self-healing composites: (A1) easy incorporation of healing fluid with thicker tubes, (A2) thinning, (A3) compartmented, (A4) surface modification, and (A5) knitted with fiber reinforcement. (B) Scanning electronic microscopy image of a PP tube. (C) Prepreg of healing agent-loaded PP tubes. (D) Arrangement of the healing tubes in the composite laminates. The cross-sections of the tubes containing epoxy and hardener are differently colored to distinguish each other. *(Source: Y. Zhu, X.J. Ye, M.Z. Rong, M.Q. Zhang. Self-healing glass fiber/epoxy composites with polypropylene tubes containing self-pressurized epoxy and mercaptan healing agents, Compos. Sci. Technol. 135 (2016) 146–52, Copyright 2015. Reproduced with permission from Elsevier.)*

rehabilitation efficiency of the fiber composites with a self-pressurized healing system, including tube spacing, content of foaming agent, foaming time, and so forth, are carefully discussed.

5.2 COMPOSITES PREPARATION

Because PP is a non-polar semi-crystalline polymer with a certain toughness, PP tubes (outer diameter: 250 μm, inner diameter: 200 μm, Paradigm Optics, Inc., Fig. 5.4B) were first irradiated by UV light (intensity: 95.3 mW/cm^2) at 50°C for 4 h for embrittlement and surface polarization, which ensures that the tubes would be preferentially damaged when a crack propagates in the composites (refer to Fig. 5.5). In the meantime, the glass fibers would not be affected because of their much higher strength than the PP tubes. Afterward, the polymerizable component of the healing agent (i.e., the hand mixing diglycidyl 1,2,3, 6-tetrahydrophthalate (DTHP, Tianjin Institute of Synthetic Materials Industry, China, 100 parts) and foaming agent 2,2′-azobis-(2,4-dimethylvaleronitrile)

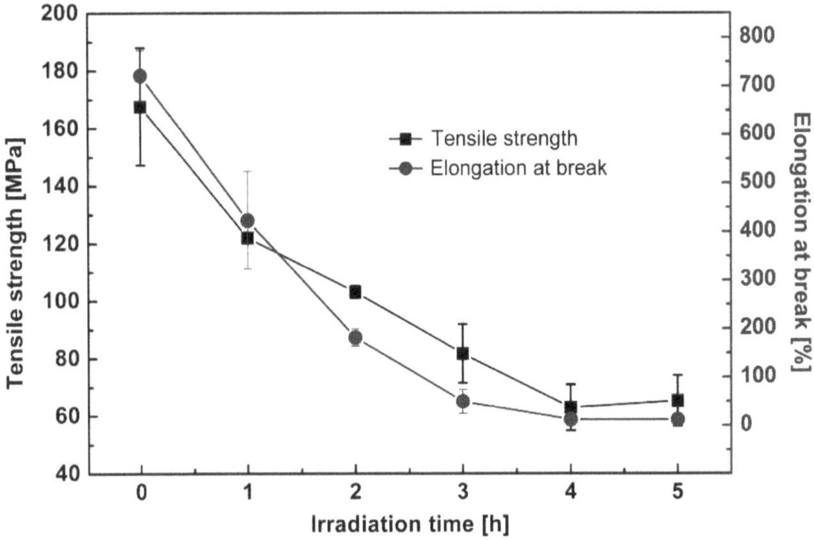

FIG. 5.5 Dependence of tensile strength and elongation at break of PP tubes on UV irradiation time. After 4h of UV irradiation, tensile strength of the PP tube is only about 37.5% of that of the original version, and elongation at break is decreased from 722.0% to 12.0%, indicating that the PP tube is embrittled while remaining suitable strength for the application as container of healing agent.

(ABVN, 3 parts)) and the hardener component (i.e., the hand mixing pentaerythritol tetrakis(3-mercaptopropionate) (PETMP, Fluka Chemie AG, 100 parts), ABVN (3 parts) and *N*-benzyl dimethylamine (NBDMA, Shanghai Medical Group Reagent Co., Ltd., China, 0.4 parts)) were injected into the UV treated PP tubes by a syringe, respectively. Then, the tube ends were sealed with a Pattex Glue Flex Gel adhesive (Henkel AG). Finally, prepregs of healing agent loaded PP tubes (Fig. 5.4C) were made through impregnation in the mixture of Diglycidyl ether of bisphenol A (EPON 828, Shell Chemical Inc., 100 parts) and its hardener (21 parts, consisting of acrylonitrile (Guangdong Guanghua Sci-Tech Co., Ltd., China) and tetramethylene pentamine (TEPA, Sinopharm Chemical Reagent Co., Ltd., China) at a molar ratio = 0.7) by hand lay-up. In each prepreg, the tube with epoxy and that with the curing agent were alternately arranged to promote the future mixing and reaction of healing agents after release.

To prepare the composite laminates, the woven glass fabric (120 × 150 mm) was also impregnated in the preceding mixture of EPON 828, TEPA, and acrylonitrile. Next, eight plies of the glass prepregs with a single ply of the PP tube prepreg in the middle (Fig. 5.4D) were degassed for 15 minutes, compressed to keep the laminates thickness of 2.5 mm, and cured at 25°C for 24 h and 40°C for 24 h. A volume fraction of the glass fiber in the composites was estimated to be

41%. To generate higher internal pressure in the PP tubes, the composite laminates were heated to 70°C for 4 or 8h allowing gasification of the blowing agent. Specimens (120 × 25 × 2.5 mm) were cut from the laminates, along with the longitudinal direction of the PP tubes for further mechanical tests. Each specimen contains 6vol% healing agent (tube spacing = 750 μm).

The control laminates that do not contain any PP tube prepreg were manufactured under the same conditions.

5.3 BASIC CHARACTERIZATION

Because the foaming agent is incorporated with the healing agent to produce high internal pressure in the PP tubes, it is worth noting whether the reactivity of the healing agent would be changed after addition of the foaming agent. In this context, curing kinetics of the healing agent (epoxy/mercaptan pair) containing the foaming agent should be measured. For this purpose, mixtures of epoxy (DTHP)/foaming agent (ABVN, 100 parts/3 parts) and hardener (PETMP)/ABVN (100 parts/3 parts) were pretreated, by heating at 70°C for 8h to simulate the foaming process. Afterward, the pretreated DTHP/ABVN and PETMP/ABVN were quickly mixed with NBDMA (0.4 parts) and then tested by DSC scans under different heating rates (Fig. 5.6). For comparison, non-isothermal curing kinetics of the mixture of DTHP (100 parts)/PETMP (100 parts)/NBDMA (0.4 parts), which did not contain any foaming agent, were also measured (Fig. 5.7). As listed in Table 5.1, the systems with and without the

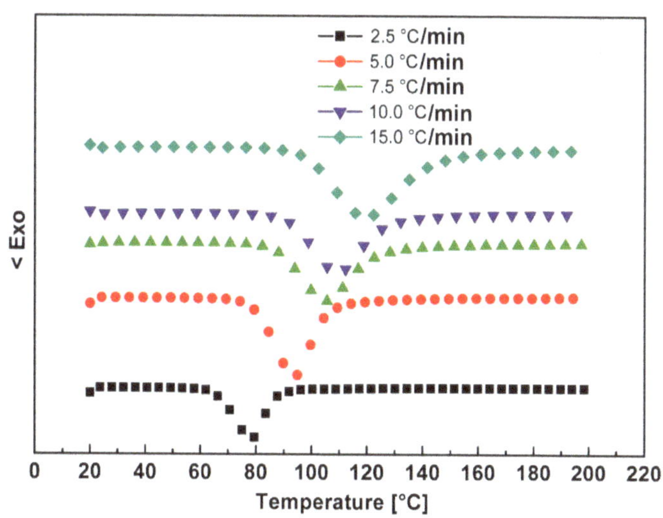

FIG. 5.6 Non-isothermal DSC scanning curves of the mixture of 100.0 parts DTHP (with 3.0 parts ABVN), 100.0 parts PETMP (with 3.0 ABVN) and 0.4 parts NBDMA collected at various heating rates (2.5, 5, 7.5, 10 and 15°C/min) in N_2. The DTHP with ABVN and PETMP with ABVN were preheated at 70°C for 8.0h before mixing.

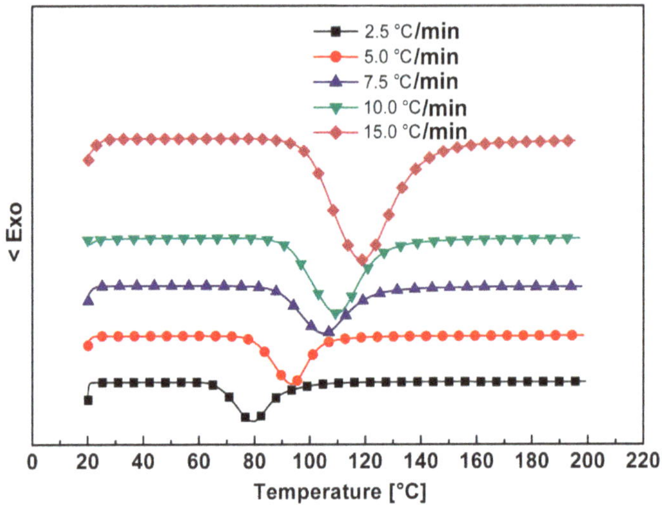

FIG. 5.7 Non-isothermal DSC scanning curves of mixture of 100.0 parts DTHP, 100.0 parts PETMP and 0.4 parts NBDMA at various heating rates.

TABLE 5.1 Non-Isothermal Curing Kinetics of the Healing Agent With and Without the Foaming Agent[a]

Foaming Agent	φ (°C/min)	T_i (°C)	T_p (°C)	T_f (°C)	ΔH (J/g)	Characteristic Parameters
Excluded	2.5	66.1	79.6	93.6	439.4	$\ln A = 5.08$
	5.0	79.6	93.8	105.5	433.0	$E_a = 46.6$ kJ/mol
	7.5	88.1	104.6	122.9	433.8	$n = 0.88$
	10	91.6	109.1	126.0	438.9	
	15	98.3	118.0	141.8	433.0	
Included	2.5	65.7	78.5	87.5	445.4	$\ln A = 4.57$
	5.0	78.9	93.9	106.4	436.2	$E_a = 45.1$ kJ/mol
	7.5	86.7	104.9	124.3	437.1	$n = 0.89$
	10	91.1	109.0	127.3	431.3	
	15	97.5	119.1	144.8	416.9	

[a]φ, *heating rate;* T_i, *onset temperature of the exothermic peak;* T_p, *peak exothermic temperature;* T_f, *ending temperature of the exothermic peak;* ΔH, *heat of reaction;* A, *pre-exponential factor;* E_a, *activation energy;* n, *reaction order.*

Source: Y. Zhu, X.J. Ye, M.Z. Rong, M.Q. Zhang. Self-healing glass fiber/epoxy composites with polypropylene tubes containing self-pressurized epoxy and mercaptan healing agents, Compos. Sci. Technol. 135 (2016) 146–52, Copyright 2015. Reproduced with permission from Elsevier.

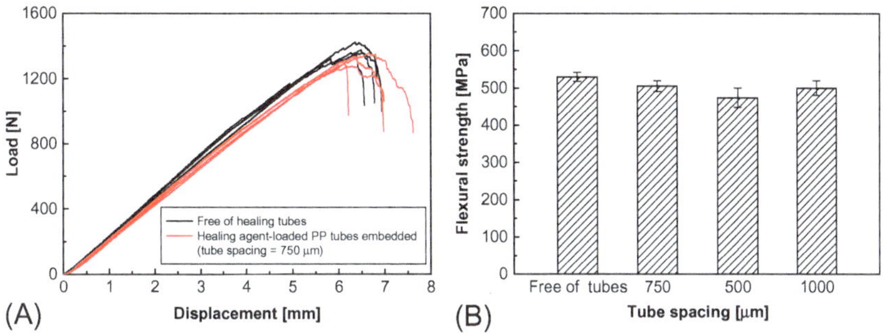

FIG. 5.8 (A) Load-displacement curves and (B) flexural strengths of the composite laminates with and without the healing agent-loaded PP tubes measured by four-point bend tests. *(Source: Y. Zhu, X.J. Ye, M.Z. Rong, M.Q. Zhang. Self-healing glass fiber/epoxy composites with polypropylene tubes containing self-pressurized epoxy and mercaptan healing agents, Compos. Sci. Technol. 135 (2016) 146–52, Copyright 2015. Reproduced with permission from Elsevier.)*

foaming agent exhibit a similar curing activation energy and reaction order. Evidently, the foaming agent, foaming process, and remainder of foaming agent do not affect curing of the healing agent because of the absence of chemical interaction between the foaming agent and healant.

On the basis of the preceding study, the influence of incorporation of healing agent-loaded PP tubes on mechanical properties of the fiber composites is further examined. As non-polar heterogeneous substances in the laminates, the tubes would most likely weaken the interfacial interaction, despite the fact that UV pretreatment might somewhat increase surface polarity. It is interesting to see from Fig. 5.8A that the composite containing the PP tubes has a similar deformation behavior as the one without the tubes. Moreover, flexural strengths of the composites are only slightly reduced after insertion of the tubes (Fig. 5.8B). For example, the highest decrease with a rate of decay of 10.6% is observed in the case of tube spacing of 500 μm. This may be due to the fact that the overall content of the PP tubes is relatively low for the present range of tube spacing of interest, as compared with previous works [16,18,35,37].

According to the data in Fig. 5.8, the composite with healing agent-loaded PP tubes spaced 750 μm apart is used for the following characterization, unless otherwise specified, because it has the greatest retention of strength after all.

5.4 SELF-HEALING PERFORMANCE

To assess healability of the system, the composite laminates were damaged by indentation, healed, and evaluated by a four-point bend flexural test according to the protocol of Bond and co-workers [18,35,37] as follows. The specimen with a back face supported by a steel ring (inner diameter: 20 mm, outer diameter: 35 mm, depth: 10 mm) was indented to the displacement of 1.8 mm by a steel hemispherical end (diameter: 5 mm) at a crosshead speed of 3 mm/min.

After indentation, the specimen was allowed to heal the internal cracks at room temperature for a period of time without additional lateral pressure. The four-point bend flexural test was conducted at a crosshead rate of 4.0 mm/min based on ASTM D6272-02. Five specimens were tested for each recipe to get the average value. Healing efficiency was determined by the strength ratio of the healed specimen to the virgin specimen. Typical load-displacement curves of the virgin, damaged, and healed laminates containing the PP tubes refer to Fig. 5.9. All the mechanical tests were carried out by a WD-5A universal mechanical testing machine. It is noted that the epoxy healing system not only heals the cracks, but also provides the composite with higher displacement, which should be ascribed to the higher ductility of the cured healing material [39].

As previously mentioned, the composite laminates have to be damaged by indentation prior to healing. The load-displacement curves in Fig. 5.10A indicate that with a rise in the indentation depth, the indentation load gradually increases, and then decreases. There is no sudden drop of load on the upward slope. It means that delamination and matrix cracking, rather than glass fiber breakage, predominate the failure process at this stage. Although the fracture of PP tubes is not highlighted in this case, it can be reasonably deduced that they must have ruptured together with the surrounding matrix as a result of UV light pretreatment. In this context, the maximum indentation displacement at 1.8 mm is appropriate. Otherwise, plenty of glass fibers would be broken, which cannot be reconnected by the present healing agent, and the results have to diverge from the theme of this study.

In the meantime, model experiments were conducted, showing the increase of pressure and its effect on the outflow of the healing agent due to the activated foaming agent. As shown in Fig. 5.11A, PP tubes, which are filled with the

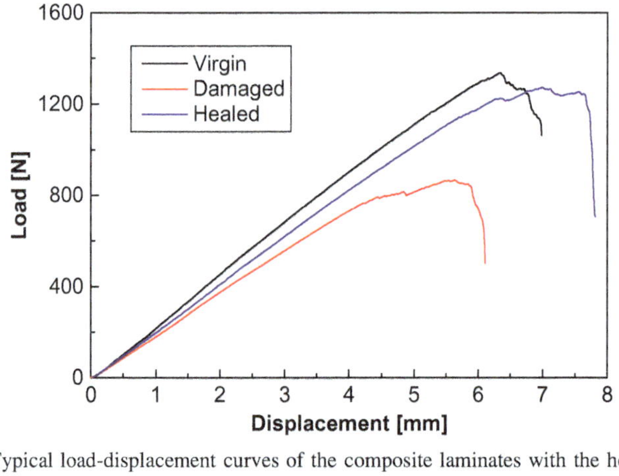

FIG. 5.9 Typical load-displacement curves of the composite laminates with the healing agent-loaded PP tubes measured by four-point bend tests (tube spacing: 750 μm; foaming agent content: 3 parts; foaming condition: 70°C, 8 h; healing condition: 25°C, 7 days).

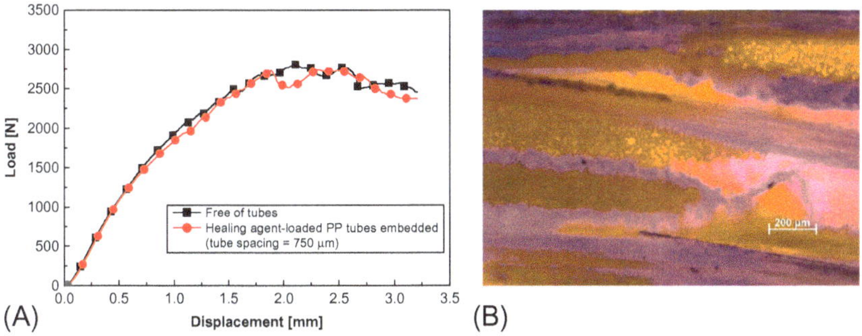

FIG. 5.10 (A) Typical load-displacement curves recorded during indentation on the composite laminates with and without the healing agent-loaded PP tubes. (B) Fluorescent microscopy photo of cross-section of the composite laminates with the healing agent-loaded PP tubes. The composite was damaged by indentation. To visualize the microstructure, Rhodamine B (0.2 parts) was introduced in the matrix of the specimen, while another fluorescent penetrant, Ardrox 985 (0.2 parts), was added into hardener component of the healing agent. The two chemicals emitted *orange* and purplish *blue* glows under illumination of the lights of 525 and 365 nm, respectively. Tube spacing: 750 μm; foaming agent content: 3 parts; foaming condition: 70°C, 8 h. *(Source: Y. Zhu, X.J. Ye, M.Z. Rong, M.Q. Zhang. Self-healing glass fiber/epoxy composites with polypropylene tubes containing self-pressurized epoxy and mercaptan healing agents, Compos. Sci. Technol. 135 (2016) 146–52, Copyright 2015. Reproduced with permission from Elsevier.)*

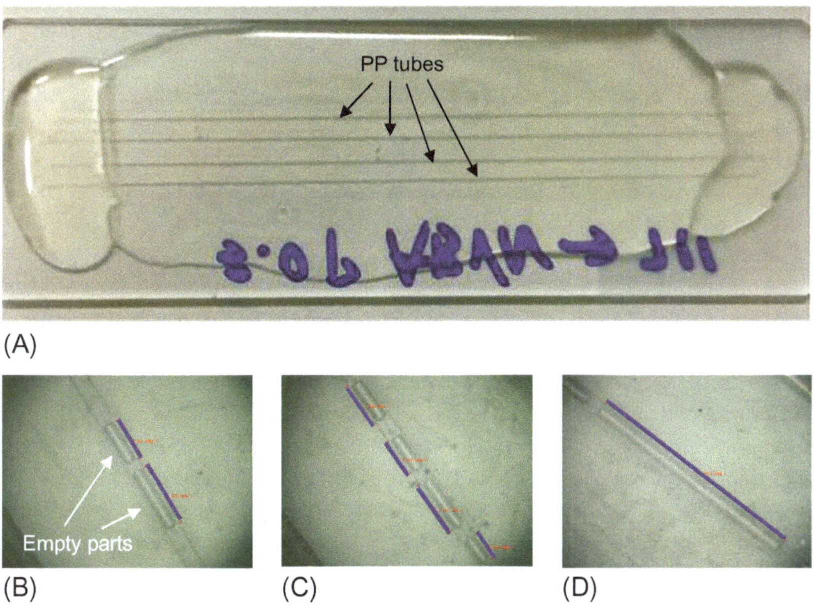

FIG. 5.11 (A) Sealed PP tubes containing the mixture of epoxy monomer DTHP (100 parts) and foaming agent (ABVN) are embedded in cured epoxy (the mixture of EPON 828 (100 parts) and its hardener (21 parts, consisting of acrylonitrile and TEPA at a molar ratio = 0.7)) on a piece of glass slide. Average thickness of the matrix epoxy: 1.2 mm. (B–C) The model composites like (A) with different ABVN contents ((B) 1.5 parts, (C) 3 parts, and (D) 4.5 parts) are heated to 70°C for 8 h allowing gasification of the blowing agent and then sawed in an end region. Most healing agent flows out of the cuts, while some remain in the tubes (the empty parts are labeled). All photos were taken by the digital microscope of Keyence VHX-1000c.

TABLE 5.2 Relative Outflow[a] of Epoxy Healing Agent From PP Tubes

Foaming Condition	Outflow[a] (%)		
	ABVN (1.5 Parts)	*ABVN (3 Parts)*	*ABVN (4.5 Parts)*
70°C, 8h	36.2±17.3	66.2±22.1	77.7±29.4
25°C, 10 days	1.5±0.9	13.8±3.0	18.5±2.0

[a]*Outflow of the fluidic healing agent is represented by the ratio of the total lengths of the empty parts to the length of the entire tube (see Fig. 5.11B and C).*

mixture of epoxy monomer (DTHP) and foaming agent (ABVN) and sealed in advance, are embedded in cured epoxy that has the same formulation as the authentic composite matrix used in this work (i.e., the mixture of EPON 828, acrylonitrile and TEPA). Afterward, the model composite is heated to 70°C for 8h, allowing gasification of the blowing agent, and then sawed in an end region. As a result, most healing agent flows out of the cuts, while some remain in the tubes (Fig. 5.11B and C). Clearly, the higher content of the foaming agent ABVN corresponds to the longer empty part of the tubes, meaning much more healing agent is released. A quantitative evaluation of the experiments based on Fig. 5.11 is shown in Table 5.2, which further evidences the preceding analysis. When the model composite is simply stored at room temperature for ten days without activating the foaming agent at 70°C, the healing agent can also be released after cutting, probably due to gradual decomposition of ABVN. Nevertheless, the outflow is significantly reduced. It is thus known from these results that the foaming agent is able to increase the internal pressure of the PP tubes.

Similarly, the PP tubes filled with DTHP epoxy monomer and the hardener (mixture of PETMP (100 parts) and NBDMA (0.4 parts)) are alternately arranged in parallel and wrapped by matrix epoxy on a piece of glass plate. The foaming agent ABVN and fluorescent indicators are included in advance. After treatment at 70°C for 8h, the model prepregs are cut and observed by fluorescent microscopy. As exhibited in Fig. 5.12A and B, both the DTHP epoxy monomer (with orange glows) and the mercaptan curing agent (with purplish blue glows) are liberated, and their distances of traveling are 500~750μm, at least. Comparatively, much more mercaptan curing agent flows out (more empty parts are left in the tube) because of its lower viscosity. It is interesting to see from Fig. 5.12C that very little healing agent flows out of the cuts in the absence of the foaming agent, as a result of lack of internal pressure. The results manifest that the pressurized healing agents are expelled at longer distances.

In the case of authentic composite, the fluorescent microscopy photo in Fig. 5.10B evidences that the PP tubes can be indeed broken after the indentation, coinciding with the results of the preceding model experiments.

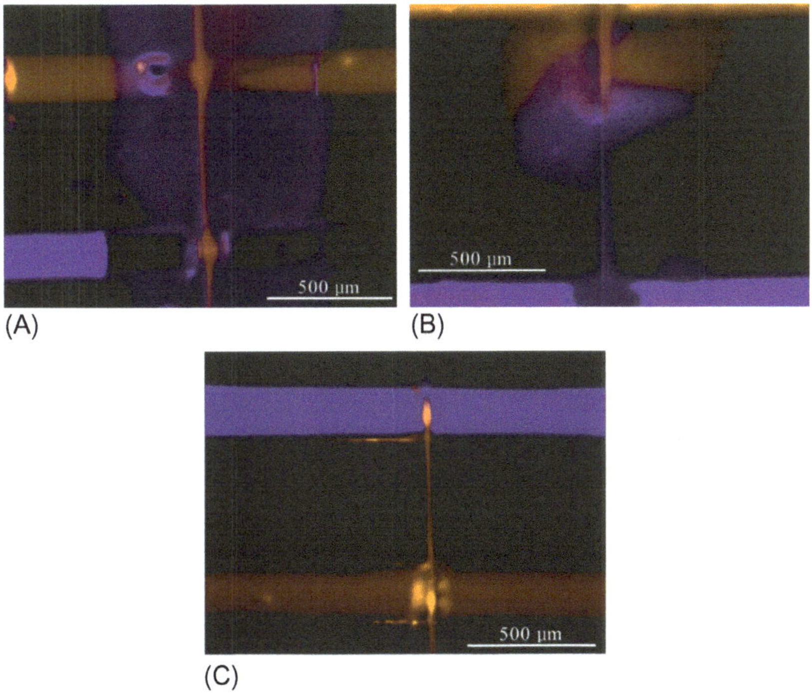

FIG. 5.12 Typical fluorescence microscopic images (taken by Axio Oberver Z1 fluorescence microscope) of cut model prepregs on a piece of glass slide under the lights of 525 and 365 nm, respectively. The prepregs consist of PP tubes, which are firstly filled with DTHP epoxy monomer and the hardener (mixture of PETMP (100 parts) and NBDMA (0.4 parts)), and then alternately arranged in parallel and wrapped by matrix epoxy. The foaming agent ABVN ((A) and (B) 1.5 parts, (C) 0 part) and fluorescent indicators (Rhodamine B (0.2 parts) for the epoxy monomer, and Ardrox 985 (0.2 parts) for the curing agent) are included in advance. After treatment at 70°C for 8 h, the model prepregs are cut and observed by fluorescent microscopy. Tube space: (A, C) 500 μm, and (B) 750 μm. The tubes with *orange* glows contain epoxy monomer, while those with purplish *blue* glows contain mercaptan curing agent.

The healing agent was liberated to the cracks in the resin-rich region, and bundled fibers/matrix interface. The cured version of the healing agent should help to recover the deteriorated mechanical properties of the composite laminates. The deduction is supported by the data exhibited in Fig. 5.13, which clearly demonstrates that the healing agent must have been bled out so that the damaged specimens containing the healing tubes have higher flexural strength after healing.

A careful survey of Fig. 5.13A reveals that the healing efficiency of the laminates with PP tubes approaches the maximum at a spacing of 750 μm. The lower value for the tube spacing of 500 μm might result from the harsher damage of the composite under indentation represented by a larger amount of broken glass fibers that cannot be reconnected (Fig. 5.14). In the meantime, the

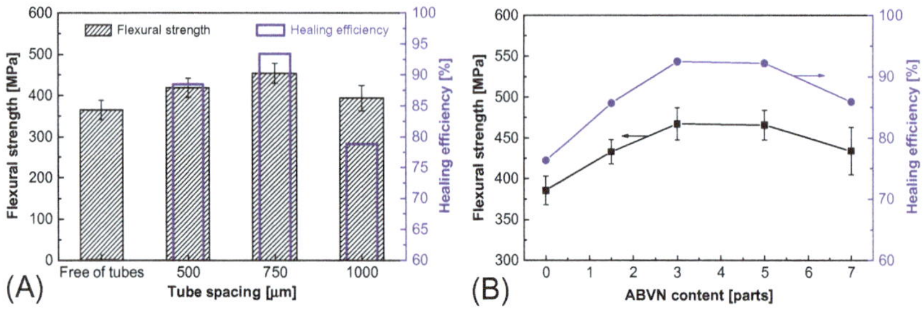

FIG. 5.13 Effect of (A) tube spacing and (B) foaming agent content on flexural strength and healing efficiency of the damaged and healed composite laminates with the embedded healing agent-loaded PP tubes. Flexural strength of the damaged laminates without PP tubes is shown in (A) as reference. Tube spacing for (B): 750 μm. Foaming agent content: 3 parts; foaming condition: 70°C, 8 h; healing condition: 25°C, 7 days. *(Source: Y. Zhu, X.J. Ye, M.Z. Rong, M.Q. Zhang. Self-healing glass fiber/epoxy composites with polypropylene tubes containing self-pressurized epoxy and mercaptan healing agents, Compos. Sci. Technol. 135 (2016) 146–52, Copyright 2015. Reproduced with permission from Elsevier.)*

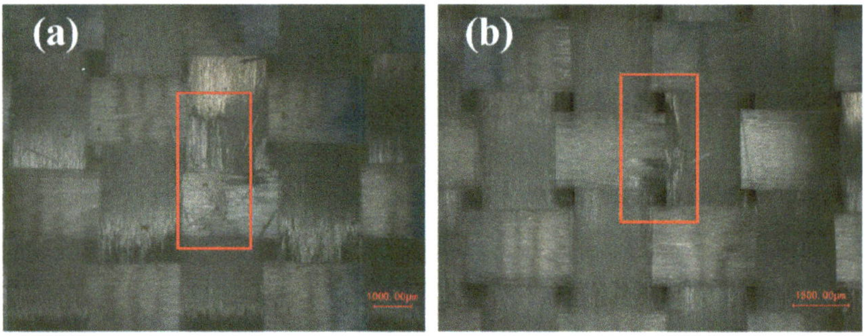

FIG. 5.14 Photos of the back surface of the composite laminates with the healing agent-loaded PP tubes after indentation damaging and subsequent calcination for removing the matrix. Tube spacing: (A) 500 μm, (B) 750 μm. The zones in the red frames show the broken glass fibers.

healing agent released from the tubes spaced 1000 μm might be insufficient for healing, which also leads to a poorer healing effect. The competition of these factors yields the highest healing efficiency in the case of tube spacing of 750 μm.

Because the self-pressurized healing system has proved to work in the composite laminates, the effect of the foaming agent (ABVN) concentration, which is directly related to the gas volume created inside the tubes, and hence, the internal pressure, is investigated (Fig. 5.13B). Initially, the healing efficiency increases with increasing the amount of ABVN. It means that the higher internal pressure has forced diffusion of the healing agent to wider regions. More and more cracked sites are re-bonded. When the AVBN content exceeds 3 parts, however, healing efficiency starts to decrease instead of further increasing. It is because the internal pressure was high enough to expel quite a bit of healing

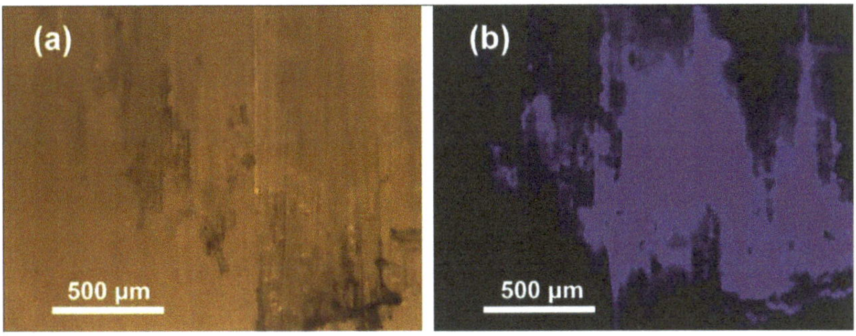

FIG. 5.15 Fluorescent microscopy photos of back face of the composite laminates with the healing agent-loaded PP tubes (tube spacing: 750 μm; foaming agent content: 3 parts; foaming condition: 70°C, 8 h). The composite was damaged by indentation. The photos were taken under the lights of (A) 525 nm and (B) 365 nm, respectively. Information of the fluorescent penetrant refers to Fig. 5.10B. *(Source: Y. Zhu, X.J. Ye, M.Z. Rong, M.Q. Zhang. Self-healing glass fiber/epoxy composites with polypropylene tubes containing self-pressurized epoxy and mercaptan healing agents, Compos. Sci. Technol. 135 (2016) 146–52, Copyright 2015. Reproduced with permission from Elsevier.)*

agent out of the bulk specimen via the surface damages (Fig. 5.15), and the amount of the healing agent remaining inside the composites has to be reduced. Besides, the cracks must have interconnected with each other from the middle to the surface of the composite after indentation damaging. With increasing the dosage of the foaming agent, internal pressure in the PP tubes increases, which expels more healing agent out of the composite. Therefore, the dosage of foaming agent or internal pressure should be optimized in favor of maximum utilization of the healing agent.

Fig. 5.16A depicts the dependence of healing efficiency on healing time. It is found that the healing speed is much slower than that of the epoxy composite filled with microencapsulated healing agent (which has the same composition as the present version) [39,40]. The Raman spectra in Fig. 5.16B also show that the epoxide and thiol groups of the released healing agent are almost completely consumed after 7 days, which coincides with the result of the mechanical test in Fig. 5.16A. In fact, most of the damaged zones have been repaired within 36 h (Fig. 5.17). Only a very small amount of cracks need a prolonged healing period. According to these observations, we believe that as soon as the released epoxy monomer and hardener meet on the crack plane, the curing reaction takes place, which increases the viscosity of the healing agent, and hinders the movement of the late-comers. Accordingly, the subsequent healing agent can only travel slowly by penetrating the remaining tiny gaps. As a result, the reaction among these unreacted chemicals has to be considerably postponed. For the microencapsulated healing agent, which possesses a shorter inter-sphere distance, polymerization of the released healing agent is localized, and there is no similar problem.

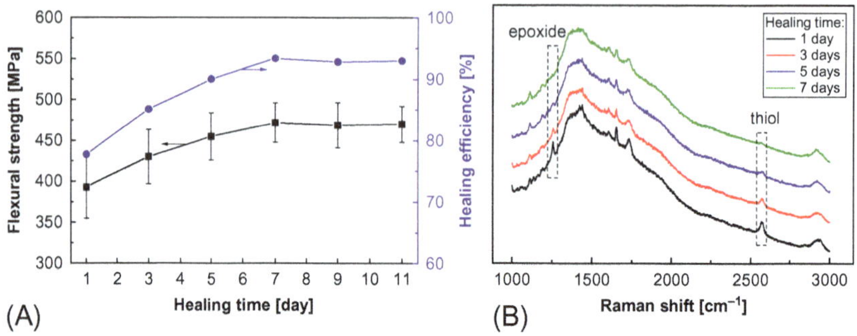

FIG. 5.16 (A) Effect of healing time on flexural strength and healing efficiency of the healed composite laminates with the embedded healing agent-loaded PP tubes (tube spacing: 750 μm; foaming agent content: 3 parts; foaming condition: 70°C, 8 h; healing condition: 25°C). (B) In situ confocal Raman microscopy spectra collected from the fractured surface of the above composites. *(Source: Y. Zhu, X.J. Ye, M.Z. Rong, M.Q. Zhang. Self-healing glass fiber/epoxy composites with polypropylene tubes containing self-pressurized epoxy and mercaptan healing agents, Compos. Sci. Technol. 135 (2016) 146–52, Copyright 2015. Reproduced with permission from Elsevier.)*

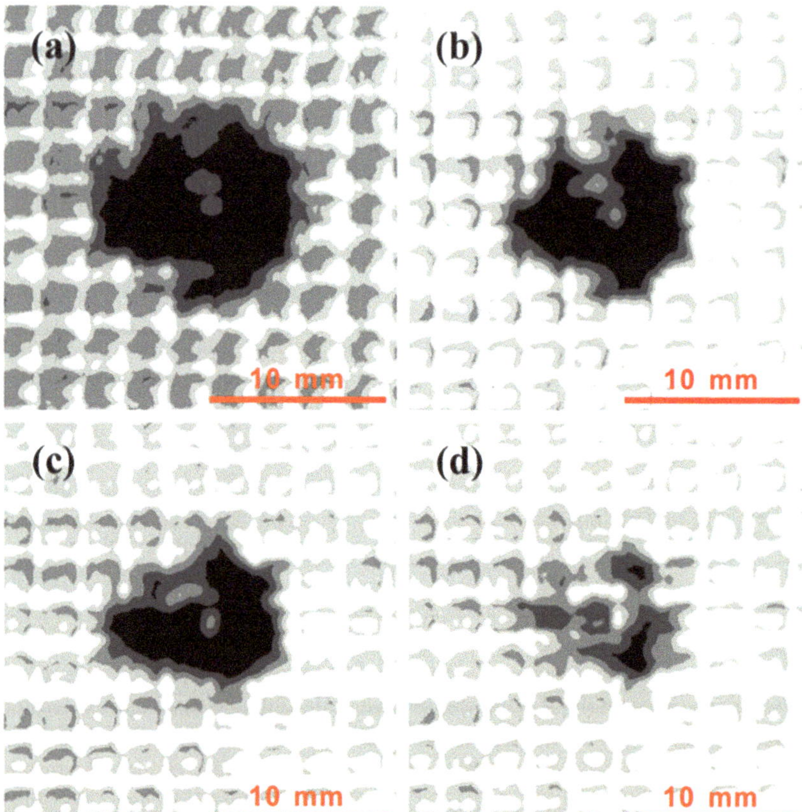

FIG. 5.17 Ultrasonic T-scan images of the indendation damaged composite laminates with the embedded healing agent-loaded PP tubes (tube spacing: 750 μm; foaming agent content: 3 parts; foaming condition: 70°C, 8 h; healing temperature: 25°C). Time of healing: (A) 2.0 h, (B) 12.0 h, (C) 24.0 h, and (D) 36.0 h. Damage area: (A) 74.4 mm^2, (B) 53.8 mm^2, (C) 39.1 mm^2, and (D) 5.4 mm^2. *(Source: Y. Zhu, X.J. Ye, M.Z. Rong, M.Q. Zhang. Self-healing glass fiber/epoxy composites with polypropylene tubes containing self-pressurized epoxy and mercaptan healing agents, Compos. Sci. Technol. 135 (2016) 146–52, Copyright 2015. Reproduced with permission from Elsevier.)*

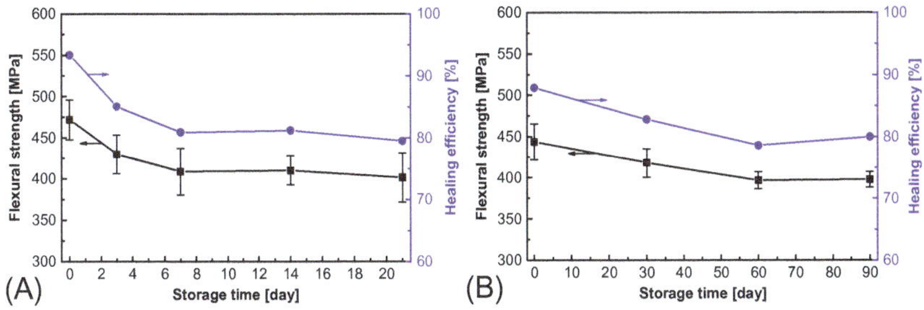

FIG. 5.18 Effect of room temperature storage time on flexural strength and healing efficiency of the healed composite laminates with the embedded healing agent-loaded PP tubes (foaming agent content: 3 parts). Healing condition: 25°C, 7 days. Foaming condition: (A) 70°C, 8h; (B) 70°C, 4h.

Last, the effect of room temperature storage time on the healing capability is evaluated. It is related to the durability of the composites. As shown by Fig. 5.18, the healing efficiency decreases with the storage time because of gradual leakage of the pressurized gas from the embedded tubes. The depressurization reduces the coverage area of the released healing agent and the amount of healed cracks. The analysis is supported by the comparison between Fig. 5.18A and B. That is, the decline of healing efficiency becomes milder when the foaming time is decreased from 8h to 4h. Because shorter foaming time corresponds to less decomposition of the foaming agent and lower internal pressure of the tubes, the gas leakage-induced pressure relief is less severe. Accordingly, the healing efficiency plotted in Fig. 5.18B takes a much longer time to approach the equilibrium than that in Fig. 5.18A. The results reflect a dilemma. Longer foaming time is beneficial for higher healing efficiency, but results in poorer durability. The solution lies in improvement of the impermeability of the PP tubes.

5.5 CONCLUSIONS

This chapter shows the feasibility of a proof-of-concept, self-pressurized healing system for self-healing glass fiber/epoxy composites. The core issue lies in the introduction of the foaming agent into healing the agent-loaded thermoplastic PP tubes. Decomposition of the foaming agent generates gas inside the sealed tubes and pressurizes the included healing fluid beforehand. Upon damage of the composites, the released healing agent is driven by the high internal pressure, and spreads out over larger cracked planes, as compared with the unpressurized version. As a result, higher healing efficiency is detected.

The factors that influence healing performance of the composites are interconnected. Polymerization speed of the healing agent had better be lower than its diffusion rate, allowing complete mixing of the components during long-distance delivery [42,43]. This is true, especially in the case of extensive damage. Additionally, impermeability of the PP tubes should be improved (e.g., by applying a gastight coating) for reducing gas leakage. Although the processing and materials parameters have not yet been optimized, the results of this preliminary exploration have shown that our idea works, which might have broad applicability to other microvascular self-healing composites.

Recently, micro- and nano-sized healing fibers were successively prepared by electrospinning [44–46]. This actually points out one of the future development directions of the present work because of their continuous production nature. That is, fabrication of the polymeric hollow fibers and infiltration of the fluidic healing agent are completed within one step. When a foaming agent is incorporated into the formulation of the healing agent, throughput of the self-pressurized healing system must be greatly increased, which is critical for large-scale application. Because PP might be no longer appropriate for electrospinning, however, new wall materials and outside surface treatment techniques have to be explored accordingly.

Another possible upgrading of the present design may be the combination with healing microcapsules. Under the guidance of the idea that cracks should be eliminated soon after their initiation, a series of ultrafast extrinsic self-healing systems based on the cure of epoxy-Lewis acid or epoxy-Bronsted acid were developed in our laboratory [42,43,47–50]. Mechanical strength of the epoxy filled with the healing agent-loaded microcapsules can be recovered within seconds at room temperature without manual intervention. If internal pressure inside the microcapsules could be raised in advance by using the preceding method, the delivery speed of the healing liquid would be further increased so as to match its ultrafast polymerization rate.

ACKNOWLEDGMENT

The authors are thankful for the support of the Natural Science Foundation of China (Grants: 51333008 and 51673219).

REFERENCES

[1] M.Q. Zhang, M.Z. Rong, Self-Healing Polymers and Polymer Composites, Wiley, 2011.
[2] W.H. Binder, Self-Healing Polymers: From Principles to Applications, Wiley-VCH, 2013.
[3] M. Zhu, M.Z. Rong, M.Q. Zhang, Self-healing polymeric materials towards non-structural recovery of functional properties, Polym. Int. 63 (2014) 1741–1749.
[4] S.D. Bergman, F. Wudl, Mendable polymers, J. Mater. Chem. 18 (2008) 41–62.
[5] Y. Yang, X. Ding, M.W. Urban, Chemical and physical aspects of self-healing materials, Prog. Polym. Sci. 49–50 (2015) 34–59.
[6] M.Q. Zhang, M.Z. Rong, Intrinsic self-healing of covalent polymers through bond reconnection towards strength restoration, Polym. Chem. 4 (18) (2013) 4878–4884.
[7] D.Y. Zhu, M.Z. Rong, M.Q. Zhang, Self-healing polymeric materials based on microencapsulated healing agents: from design to preparation, Prog. Polym. Sci. 49–50 (2015) 175–220.
[8] S.M. Bleay, C.B. Loader, V.J. Hawyes, L. Humberstone, P.T. Curtis, A smart repair system for polymer matrix composites, Compos. Part A 32 (2001) 1767–1776.

[9] M.-U. Saeed, Z.F. Chen, B.B. Li, Manufacturing strategies for microvascular polymeric composites: a review, Compos. Part A 78 (2015) 327–340.
[10] C. Dry, Procedures developed for self-repair of polymer matrix composite materials, Compos. Struct. 35 (3) (1996) 263–269.
[11] M. Motuku, U.K. Vaidya, G.M. Janowski, Parametric studies on self-repairing approaches for resin infused composites subjected to low velocity impact, Smart Mater. Struct. 8 (1999) 623–638.
[12] A. Kousourakis, A.P. Mouritz, The effect of self-healing hollow fibres on the mechanical properties of polymer composites, Smart Mater. Struct. 19 (8) (2010).
[13] C.J. Norris, J.A.P. White, G. McCombe, P. Chatterjee, I.P. Bond, R.S. Trask, Autonomous stimulus triggered self-healing in smart structural composites, Smart Mater. Struct. 21 (9) (2012).
[14] H.-B. Yue, J.P. Fernández-Blázquez, D.F. Beneito, J.J. Vilatela, Real time monitoring of click chemistry self-healing in polymer composites, J. Mater. Chem. A 2 (11) (2014) 3881–3887.
[15] I.P. Bond, Self-repairing hollow fibre composites, Reinf. Plast. 48 (8) (2004) 16.
[16] R.S. Trask, I.P. Bond, Biomimetic self-healing of advanced composite structures using hollow glass fibres, Smart Mater. Struct. 15 (3) (2006) 704–710.
[17] M. Hucker, I. Bond, A. Foreman, J. Hudd, Optimisation of hollow glass fibres and their composites, Adv. Compos. Lett. 8 (4) (1999) 181–189.
[18] J.W.C. Pang, I.P. Bond, A hollow fiber reinforced polymer composite encompassing self-healing and enhanced damage visibility, Compos. Sci. Technol. 65 (11–12) (2005) 1791–1799.
[19] S. Kling, T. Czigány, Damage detection and self-repair in hollow glass fiber fabric-reinforced epoxy composites via fiber filling, Compos. Sci. Technol. 99 (2014) 82–88.
[20] S. Kling, T. Czigány, Analysis of applicability of the hollow carbon fibres for self-repairing composites, Mater. Sci. Forum 729 (2013) 246–251.
[21] H.R. Williams, R.S. Trask, I.P. Bond, Self-healing composite sandwich structures, Smart Mater. Struct. 16 (4) (2007) 1198–1207.
[22] H.R. Williams, R.S. Trask, I.P. Bond, Self-healing sandwich panels: restoration of compressive strength after impact, Compos. Sci. Technol. 68 (15-16) (2008) 3171–3177.
[23] D. Therriault, R.F. Shepherd, S.R. White, J.A. Lewis, Fugitive inks for direct-write assembly of three-dimensional microvascular networks, Adv. Mater. 17 (4) (2005) 395–399.
[24] K.S. Toohey, N.R. Sottos, J.A. Lewis, J.S. Moore, S.R. White, Self-healing materials with microvascular networks, Nat. Mater. 6 (8) (2007) 581–585.
[25] C.J. Hansen, S.R. White, N.R. Sottos, J.A. Lewis, Accelerated self-healing via ternary interpenetrating microvascular networks, Adv. Funct. Mater. 21 (22) (2011) 4320–4326.
[26] J.-H. Huang, J. Kim, N. Agrawal, A.P. Sudarsan, J.E. Maxim, A. Jayaraman, V.M. Ugaz, Rapid fabrication of bio-inspired 3D microfluidic vascular networks, Adv. Mater. 21 (35) (2009) 3567–3571.
[27] A.P. Esser-Kahn, P.R. Thakre, H. Dong, J.F. Patrick, V.K. Vlasko-Vlasov, N.R. Sottos, J.S. Moore, S.R. White, Three-dimensional microvascular fiber-reinforced composites, Adv. Mater. 23 (32) (2011) 3654–3658.
[28] A.R. Hamilton, N.R. Sottos, S.R. White, Pressurized vascular systems for self-healing materials, J. R. Soc. Interface 9 (70) (2012) 1020–1028.
[29] A.S. Wu, A.M. Coppola, M.J. Sinnott, T.-W. Chou, E.T. Thostenson, J.-H. Byun, B.-S. Kim, Sensing of damage and healing in three-dimensional braided composites with vascular channels, Compos. Sci. Technol. 72 (13) (2012) 1618–1626.
[30] R.S. Trask, I.P. Bond, Bioinspired engineering study of Plantae vascules for self-healing composite structures, J. R. Soc. Interface 7 (47) (2010) 921–931.

[31] C.-Y. Huang, R.S. Trask, I.P. Bond, Characterization and analysis of carbon fibrereinforced polymer composite laminates with embedded circular vasculature, J. R. Soc. Interface 7 (49) (2010) 1229–1241.
[32] J. Paul, R. Jones, Repair of impact damaged composites, Eng. Fract. Mech. 41 (1992) 127–141.
[33] C. Soutis, D.-M. Duan, P. Goutas, Compressive behaviour of CFRP laminates repaired with adhesively bonded external patches, Compos. Struct. 45 (1999) 289–301.
[34] H. Zhang, J. Motipalli, Y.C. Lam, A. Baker, Experimental and finite element analyses on the post-buckling behaviour of repaired composite panels, Compos. Part A 29 (1998) 1463–1471.
[35] G. Williams, R.S. Trask, I.P. Bond, A self-healing carbon fibre reinforced polymer for aerospace applications, Compos. Part A 38 (6) (2007) 1525–1532.
[36] R.S. Trask, G.J. Williams, I.P. Bond, Bioinspired self-healing of advanced composite structures using hollow glass fibres, J. R. Soc. Interface 4 (13) (2007) 363–371.
[37] J.W.C. Pang, I.P. Bond, 'Bleeding composites'-damage detection and self-repair using a biomimetic approach, Compos. Part A 36 (2) (2005) 183–188.
[38] Y. Zhu, X.J. Ye, M.Z. Rong, M.Q. Zhang, Self-healing glass fiber/epoxy composites with polypropylene tubes containing self-pressurized epoxy and mercaptan healing agents, Compos. Sci. Technol. 135 (2016) 146–152.
[39] Y.C. Yuan, M.Q. Rong MZ Zhang, J. Chen, G.C. Yang, X.M. Li, Self-healing polymeric materials using epoxy/mercaptan as the healant, Macromolecules 41 (14) (2008) 5197–5202.
[40] Y.C. Yuan, X.J. Ye, M.Q. Rong MZ Zhang, G.C. Yang, J.Q. Zhao, Self-healing epoxy composite with heat-resistant healant, ACS Appl. Mater. Interfaces 3 (11) (2011) 4487–4495.
[41] M. Prajer, X. Wu, S.J. Garcia, S. van der Zwaag, Direct and indirect observation of multiple local healing events in successively loaded fibre reinforced polymer model composites using healing agent-filled compartmented fibres, Compos. Sci. Technol. 106 (2015) 127–133.
[42] X.J. Ye, J.L. Zhang, Y. Zhu, M.Q. Rong MZ Zhang, Y.X. Song, H.X. Zhang, Ultrafast self-healing of polymer toward strength restoration, ACS Appl. Mater. Interfaces 6 (5) (2014) 3661–3670.
[43] X.J. Ye, Y.X. Song, Y. Zhu, G.C. Yang, M.Q. Rong MZ Zhang, Self-healing epoxy with ultrafast and heat-resistant healing system processable at elevated temperature, Compos. Sci. Technol. 104 (2014) 40–46.
[44] Z. Shang, D. Wang, J. Guo, Q.S. Zhang, N. Zhao, J. Xu, L. Chen, Preparation of micrometer-scale self-repaired fibers made by electrospinning and their self-healing characterization, Acta Polym. Sin. 6 (2015) 727–732.
[45] R.E. Neisiany, S.N. Khorasani, J.K.Y. Lee, S. Ramakrishna, Encapsulation of epoxy and amine curing agent in PAN nanofibers by coaxial electrospinning for self-healing purposes, RSC Adv. 6 (74) (2016) 70056–70063.
[46] S. Sinha-Ray, D.D. Pelot, Z.P. Zhou, A. Rahman, X.F. Wu, A.L. Yarin, Encapsulation of self-healing materials by coelectrospinning, emulsion electrospinning, solution blowing and intercalation, J. Mater. Chem. 22 (18) (2012) 9138–9146.
[47] G. Szebenyi, T. Czigany, B. Vermes, X.J. Ye, M.Z. Rong, M.Q. Zhang, Acoustic emission study of the TDCB test of microcapsules filled self-healing polymer, Polym. Test. 54 (2016) 134–138.
[48] Y.C. Song, X.J. Ye, M.Z. Rong, M.Q. Zhang, Self-healing epoxy with a fast and stable extrinsic healing system based on BF_3-amine complex, RSC Adv. 6 (103) (2016) 100796–100803.
[49] X.J. Ye, Y. Zhu, Y.C. Yuan, Y.C. Song, G.C. Yang, M.Z. Rong, M.Q. Zhang, Fatigue life extension of epoxy materials using ultrafast epoxy-SbF_5 healing system introduced by manual infiltration, Express Polym Lett 9 (3) (2015) 177–184.
[50] M.Q. Zhang, Polymer self-heals in seconds, Express Polym Lett 9 (2) (2015) 84.

Chapter 6

Multifunctional Nanocomposite Sensors for Environmental Monitoring

Yang Lu*, Linh Doan†, Adarsh Bafana‡, Guoqiang Yu§, Clayton Jeffryes‡, Tracy Benson†, Suying Wei¶ and Evan K. Wujcik*

*Materials Engineering And Nanosensor [MEAN] Laboratory, Department of Chemical and Biological Engineering, The University of Alabama, Tuscaloosa, AL, United States, †Operando Spectroscopy Laboratory, Dan F. Smith Department of Chemical Engineering, Lamar University, Beaumont, TX, United States, ‡Nanobiomaterials and Bioprocessing Laboratory, Dan F. Smith Department of Chemical Engineering, Lamar University, Beaumont, TX, United States, §Dan F. Smith Department of Chemical Engineering, Lamar University, Beaumont, TX, United States, ¶Department of Chemistry and Biochemistry, Lamar University, Beaumont, TX, United States

Chapter Outline

6.1 BACKGROUND

Due to rapidly developing international globalization and industrialization, countless toxic chemicals have been released into the environment in terms of airborne, water, and solid pollutants. During the past century, waste products have risen 10-fold. And this amount will double again by 2025 according to projections [1,2]. Chemical pollutants, such as heavy metals, aromatic compounds, pesticides, microorganisms, and so forth can cumulatively cause severe damage to human health and the environment [3]. Thus, it is of great importance to have

Polymer-Based Multifunctional Nanocomposites and Their Applications
https://doi.org/10.1016/B978-0-12-815067-2.00006-8

a sensitive, selective, portable, reliable, and rapid response to this issue, and cost-effective sensors for monitoring pollutants in the environment. Nanocomposites are a multiphase solid material in which at least one phase has one, two, or three dimensions less than 100 nm [4]. Commonly seen nanocomposite matrices include ceramics [5], polymers [6], and metals [7]. Due to the presence of nanomaterials, the mechanical [8], electrical [9], optical [10], electrochemical [11], and catalytic properties [12] will greatly differ from bulk materials. Nanomaterials with exceptional electrochemical properties, high surface area, high adsorption, and effective catalytic activities are mostly used for fabricating nanocomposite sensors. The aforementioned properties that nanocomposites possess offer great opportunities and possibilities for fabricating portable nanosensors for successful detection of environmental pollutants present in the air, soil, and water.

During the past two decades, considerable progress has been made on nanocomposite-based sensors for environmental applications. Taking the electrochemical sensor as an example, owing to the development of nanotechnology, the significant amplification of electrochemical signals for enhancing sensor sensitivity and selectivity has been achieved based on advanced nanocomposite electrodes through selective bindings with analytes, effective catalytic activities, fast electrode kinetics, and control over the microenvironment of electrodes [13]. This chapter summarizes and evaluates the recent developments of nanocomposite-based sensors on environmental monitoring in three major sections (i) air, (ii) soil, and (iii) water. The aim of the chapter is to provide insight into potential applications and benefits of nanocomposites for environmental monitoring, and inspire broader interests in this field.

6.2 AIR MONITORING

Energy use and production by burning fossil fuels or biomass will release inorganic gases and organic vapors into the air, which are the main source of air pollution. Most of these gases are detrimental to human health and planet Earth [14]. Major pollutants include carbon dioxide (CO_2), sulfur oxides (SO_x), nitrogen oxides (NO_x), carbon monoxide (CO), volatile organic compounds (VOC), ammonia (NH_3), chlorofluorocarbons (CFCs), particulates, and so forth. For resistive gas sensors, semiconducting nano-metal oxides, such as SnO_2, In_2O_3, ZnO, TiO_2, WO_3, Fe_2O_3, and NiO are of great interest, because of the benefits of cost effectiveness, high sensitivity, fast response/recovery time, a simple electronic interface, and the ability to detect a large number of gases [3,15]. Metal oxide-based gas sensors are usually workable between 200°C and 500°C. Working under high temperatures and long-term instability are the two main disadvantages, while polymer-based gas sensors can operate under room temperature [16]. Nanocomposites containing both metal oxides and conductive polymers can overcome these problems, and are promising for developing high performance gas sensors. The principles for most nanocomposite-based gas sensors for gas detection are based on changes in electrical properties (e.g., resistance) due to charge transfers or a change in adsorption with the gas molecules on an active surface of the sensors [17].

In this section, the analyte gases will be introduced by their three categories: inorganic gases, organic gases, and carcinogenic gases.

6.2.1 Monitoring Inorganic Gases in Air

Nitrogen dioxide (NO_2) is a toxic gas produced mainly from the high temperature combustion of chemical plants and motor vehicles. NO_2 can also react with atmosphere-forming ground-level ozone (O_3), a constituent of smog [15]. Navale et al. reported a room temperature gas sensor based on PPy/α-Fe_2O_3 hybrid nanocomposites doped with camphor sulfonic acid (CSA) [18]. The nanocomposites were prepared using a solid state synthesis method and deposited on a glass substrate. The CSA doped PPy/α-Fe_2O_3 hybrid nanocomposites exhibited a chemiresistive response of 64% with a rapid response time of 148 s. By exposing it to oxidizing (NO_2, Cl_2), and reducing (NH_3, H_2S, CH_3OH, and C_2H_5OH) gases, the nanocomposite sensor showed excellent selectivity (Fig. 6.1).

Sulfur oxide (SO_2) is another toxic gas produced from volcanic eruption or combustion of sulfur containing coal and petroleum. SO_2 is also prone to form H_2SO_4 in the presence of a catalyst such as NO_2, which causes acid rain. Tyagi et al. have developed an SO_2 gas sensor using an SnO_2 thin film integrated with a 10nm thin NiO dotted cluster that performs as the catalyst [19]. For

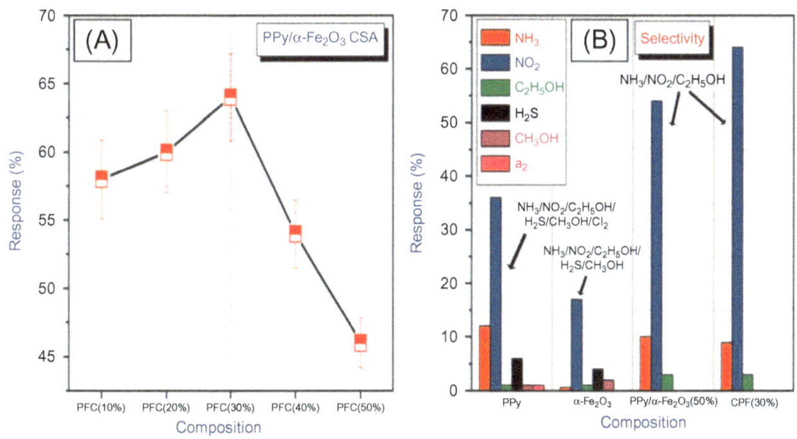

FIG. 6.1 (A) Chemiresistive response of CAS doped PPy/α-Fe_2O_3 hybrid nanocomposites at 100ppm of NO_2, parentheses show the percentage of dopant, CAS. (B) Selectivity of PPy, α-Fe_2O_3, PPy/α-Fe_2O_3, and 30% CAS doped PPy/α-Fe_2O_3 hybrid nanocomposites toward various gases. *(Reproduced with permission from S.T. Navale, G.D. Khuspe, M.A. Chougule, V. B. Patil, Camphor sulfonic acid doped PPy/α-Fe_2O_3 hybrid nanocomposites as NO_2 sensors, RSC Adv. 4 (53) (2014) 27998, The Royal Society of Chemistry.)*

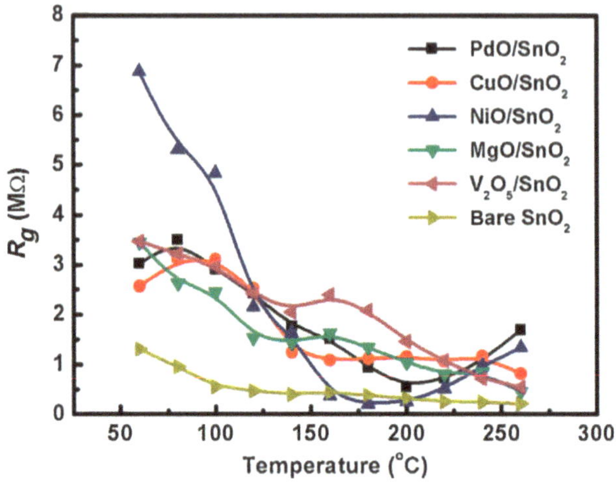

FIG. 6.2 R_g (sensor resistance under exposure to SO_2) of different sensors (PdO/SnO_2, CuO/SnO_2, NiO/SnO_2, MgO/SnO_2, V_2O_5/SnO_2, and bare SnO_2) with respect to temperature. *(Reproduced with permission from P. Tyagi, A. Sharma, M. Tomar, V. Gupta, Metal oxide catalyst assisted SnO_2 thin film based SO_2 gas sensor, Sens. Actuators B 224 (2016) 282, Elsevier.)*

comparison, the bare SnO_2 film exhibits a response (S) of 1.3 at 220°C, while the nanocomposite film has a significantly enhanced response of 56 at a relatively low temperature (180°C) toward 500 ppm of SO_2. Various other metal oxide catalysts are compared in their study, including PdO, CuO, MgO, and V_2O_5 (results as shown in Fig. 6.2). The enhanced resistance change is attributed to the high air resistance related to the formation of the large space charge region and low SO_2 resistance due to the spillover of analyte molecules SO_2 over the uncovered sensor surface.

Hydrogen sulfide (H_2S) is released by biological processes or fuel production. H_2S can influence the human nervous system, and its lethal dose is 800 ppm for 6 min. Pi-Guey et al. fabricated for a room-temperature H_2S gas sensor based on PPy/WO_3 nanocomposite films [20]. Comparative gas sensing results showed a much higher response of the gas sensor based on PPy/WO_3 nanocomposite film than sensors based on PPy or WO_3 films only. The H_2S gas sensor has a linear working range from 100 to 1000 ppb with long-term stability for at least 54 days.

6.2.2 Monitoring of Carcinogenic Gases in Air

Cancer is caused by a cell's DNA change, which can be inherited from parents, or occur upon outside exposure. Carcinogens are the substances that can lead to cancer. Some carcinogens may not be involved in DNA change directly, but through other ways. For example, they may promote cell division at a much faster rate than normal. An increased cell division rate increases the risk of occurrences of genetic errors such as mutations, translocations, and amplifications of oncogenes. The pathogenesis of human cancers may result from the accumulation of such genetic errors [21].

Benzene is a known human carcinogen per the International Agency for Research on Cancer (IARC). Wang et al. reported a gas sensor based on Au nanoparticles functionalized ZnO nanowires (Au-ZnO NWs) for detection of benzene and toluene [22]. ZnO nanowires were first fabricated by a one-pot hydrothermal method, followed by loading Au NPs. The sensor exhibits an enhanced sensitivity with a fast response and short recovery time, which are believed to be due to the strong spillover effect of Au NPs and increased Schottky barriers caused by electronic interaction between ZnO nanowires and Au NPs.

Chlorinated aliphatic hydrocarbons include carbon tetrachloride (CCl_4), chloroform ($CHCl_3$), dichloromethane (CH_2Cl_2), and so forth, which are a group of environmental pollutants. Besides carcinogens, they may also cause other severe health problems such as irritation, headache, lung congestion, or kidney, liver, or brain damage [23]. Kar et al. introduced a chloroform sensor based on a carboxylic acid functionalized multiwalled carbon nanotube doped polyaniline nanocomposite (PANI/c-MWCNT) [24]. The nanocomposite was synthesized by in situ chemical oxidation polymerization of an aniline monomer with ammonium persulfate in the presence of c-MWCNT. Compared with pure PANI and PANI/MWCNT nanocomposites, the PANI/c-MWCNT-based sensor exhibits a better response to chloroform vapor. At 250 ppm chloroform, the response and recovery times of PANI/c-MWCNT (2 wt% of c-MWCNT) are 1.25 and 3.80 min, which is faster than the control PANI/MWCNT (2 wt% of pristine MWCNT) at 5.98 and 9.53 min. The results of the sensing response of the developed sensor toward three different chlorinated methane vapors are shown in Fig. 6.3. The enhanced sensing response could be attributed to a stronger interaction of chloroform molecules with PANI backbones.

6.2.3 Monitoring Organic Gases in Air

Organic gases, known for their volatile organic compounds (VOCs), are a type of greenhouse, flammable gas. Exposure to VOCs can result in acute toxic effects [25]. Most VOCs are released from chemical plants. Take ethylene production as an example; during the cracking and refinement, VOCs including ethylene oxide, vinyl acetate, acrylonitrile, styrene, butadiene, ethanol, and vinyl acetate are released.

Khoang et al. designed a novel hierarchical SnO_2/ZnO nanostructure for scalable production of high performance ethanol sensors [26]. First, single crystalline SnO_2 nanowires (NW) were synthesized via the thermal evaporation method, and they act as the backbones. Subsequently, ZnO nanorods (NR) were grown perpendicularly to the axis of the SnO_2 backbones by the hydrothermal method, and they act as the branches, as shown in Fig. 6.4. Compared with the

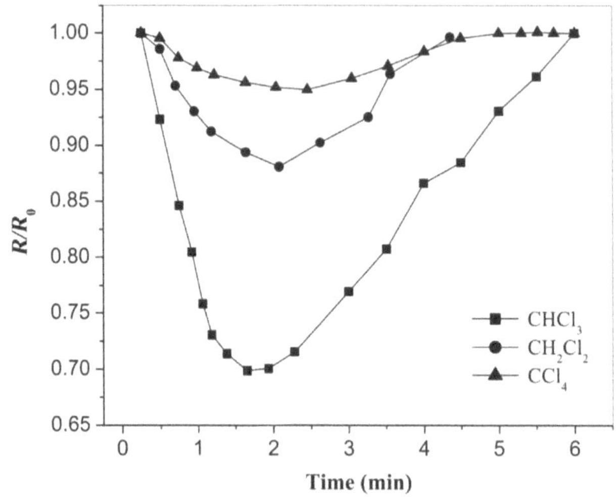

FIG. 6.3 The sensing response of PANI/c-MWCNT sensor toward $CHCl_3$, CH_2Cl_2, and CCl_4. *(Reproduced with permission from P. Kar, A. Choudhury, Carboxylic acid functionalized multi-walled carbon nanotube doped polyaniline for chloroform sensors, Sens. Actuators B. 183 (2013) 25, Elsevier.)*

FIG. 6.4 SEM images of (A) bare SnO_2 nanowires, the SnO_2/ZnO hierarchical nanocomposites grown for (B) 1 h, (C) 2 h, and (D) 4 h. *(Reproduced with permission from N.D. Khoang, D.D. Trung, N. Van Duy, N.D. Hoa, N. Van Hieu, Design of SnO_2/ZnO hierarchical nanostructures for enhanced ethanol gas-sensing performance, Sens. Actuators B 174 (2012) 594, Elsevier.)*

bare SnO_2 NW sensor, the hierarchical nanocomposites showed enhanced ethanol gas response and selectivity for interfering gases, including NH_3, H_2, CO, and CO_2. This could be attributed to the enriched formation of SnO_2/ZnO heterojunctions.

Pramanik et al. recently reported an organic gas sensor based on the polyaniline (PANI) nanofibers modified with bentonite nanohybrid [27]. This nanohybrid was found to form stable dispersions in different organic solvents. Four kinds of organic gases were tested, and the nanohybrid with 0.23 wt% of PANI showed a higher sensitivity to acetone compared with benzene, ethanol, and toluene, than using ruthenium electrodes.

6.3 SOIL MONITORING

Soil resources produce many important goods and services for humans and ecosystems, including the production of food and fiber, recreation, recycling, and assimilation of wastes and by-products [28]. Therefore, the monitoring of soil quality is directly related to human health, food safety, and sustainable economic and social development [29]. Due to global industrialization and urbanization, the amount of soil contaminants such as heavy metals and plastics being produced has significantly increased. Production from sewage irrigation, metal mining and smelting, and processing activities and urban solid waste are substantial. During a 1-year period, 79.6%, 56%, and 63% of the total Cu, Zn, and Cd, respectively, occurring in agricultural soils were caused by agricultural activities. Eighty-five percent, 67.5%, and 43% of Pb, Ni, and Cr, respectively, were originated from atmospheric depositions associated with industrial activities [30]. To protect the soil for a sustainable utilization, an accurate and early detection of soil contaminants is of importance. Many countries have set up official frameworks (e.g., the Soil Monitoring Network, SMN) for the purpose of environmental soil surveys, soil erosion surveys. and monitoring of organic compounds in soil. The analytical methods commonly used for soil monitoring are ICP-AES (Inductively Coupled Plasma-Atomic Emission Spectroscopy), F-AAS (Flame-Atomic Absorption Spectroscopy), and XRF (X-ray Fluorescence Spectroscopy), and so forth. The SMNs and conventional analytical methods have some disadvantages, such as high cost, significant time consumption, off-site determination, and so forth. Nanocomposite-based sensors overcome these problems, and thus have garnered great interest.

Carbendazim is a widely used, broad-spectrum benzimidazole fungicide. It can be uptaken by plants through roots, seeds, and leaves, and subsequently transferred to the whole plants [31]. The benzimidazole ring in carbendazim is difficult to decompose; therefore, it is left in soils for a long time. Luo et al. fabricated a multiwalled carbon nanotube (MWNTs)-graphene oxide (GO) nanocomposite film modified glassy carbon (GC) electrode (MWNTs-GO/GC) to detect the carbendazim using an electrochemical method [32]. The electrochemical behaviors of 3 μM carbendazim at four different electrodes

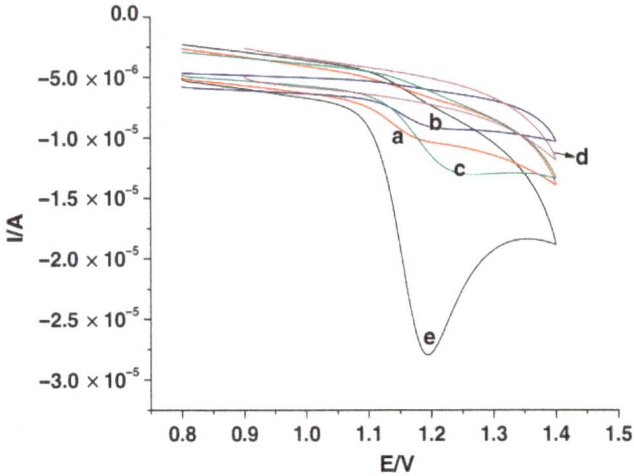

FIG. 6.5 Cyclic Voltammograms of (a) bare GC, (b) MWNTs/GC, (c) GO/GC, (d) GO-MWNTs/GC in 3 μM carbendazim BR buffer solution, (e) GO-MWNTs/GC in pure BR buffer solution. *(Reproduced with permission from S. Luo, Y. Wu, H. Gou, A voltammetric sensor based on GO–MWNTs hybrid nanomaterial-modified electrode for determination of carbendazim in soil and water samples, Ionics 19 (4) (2013) 673, Springer.)*

(bare GC, MWNTs/GC, GO/GC, GO-MWNTs/GC) in a Britton-Robinson buffer solution were investigated using cyclic voltammetry, as shown in Fig. 6.5. GO-MWNTs/GC exhibited a significantly stronger oxidation current peak compared with other electrodes, which should be attributed to the electrochemical activation of carbendazim by GO-MWNTs. The anodic peak current of carbendazim was linear at the range of 10 nM–4 μM, with a detection limit of 5 nM. Soil samples and tap water for detection of carbendazim were successfully verified as well.

Heavy metal pollution is a severe issue in many countries, and influences the quality and yield of crops, the atmosphere, and the hydrological cycle [33]. Platinum group metals (PGMs) are an important group of metals that are being used increasingly in the technologically developing world. PGMs include six elements, Pa, Rh, Pt, Os, Ru, and Ir. They are used for jewelry, catalysts, cancer treatment drugs, and so forth. Horst et al. developed a bismuth-silver bimetallic electrochemical nanosensor for detection of PGMs using a differential pulse adsorptive stripping voltammetry amperometric (DPAdSV) technique [34]. Dimethylglyoxime (DMG) was the chelating agent in a glassy carbon electrode coated with bismuth-silver bimetallic nanoparticles. Various factors were systematically investigated and optimized. The DPAdSV peak current was linear, from the 0.2 to 1 ng/L range (60 s deposition), with the limit of detection (LOC) for Pd (0.19 ng/L), Pt (0.20 ng/L) and Rh (0.22 ng/L), respectively. Reproducibility of the nanosensor was good, and the capability of detecting trace quantities of PGMs in roadside dust and soil samples was demonstrated.

Complex hydrocarbon mixtures include alkanes and aromatic hydrocarbons. Aromatic hydrocarbons such as BTEX are generally more toxic than aliphatic hydrocarbons, and can cause severe short-term health problems such as nervous system disorders, immune system depression, anemia, and long-term hazards such as cancer and damage to the liver, kidneys, and central nervous system. Matin et al. incorporated multiwalled carbon nanotubes (MWCNTs) in polyvinyl chloride (PVC) to prepare PVC/MWCNTs nanocomposites [35]. PVC/MWCNTs was coated on a stainless steel wire as the solid phase microextraction (SPME) fiber. SPME is a popular sample preparation technique in chromatographic analysis, which includes the benefits of being solvent free, easy integration of clean-up and preconcentration steps, and the ability to sample solid samples. The fabricated SPME fiber exhibited a high extraction capability due to MWCNTs, and high mechanical durability due to the stainless steel wire base. PVC is responsible for film formation and adhesion of the coating on the wire. Fig. 6.6 presents a chromatogram of a soil sample using PVC/MWCNTs SPME fiber. The LOC is 0.1–1.10 ng/L (r^2 = 0.9940–0.9994) for water samples, and 0.10–0.77 ng/Kg (r^2 = 0.9946–0.9994) for soil samples showing excellent analytical characteristics in ultratrace detection of complex hydrocarbons as environmental pollutants.

6.4 WATER MONITORING

Water pollution is one of the most severe environmental problems. Though the total water volume on earth is estimated at nearly 1.4 billion km^3, the fresh water that can be readily used only accounts for 0.3% of the total water volume, and is rapidly depleting. Studies have shown that more than 1.2 billion people

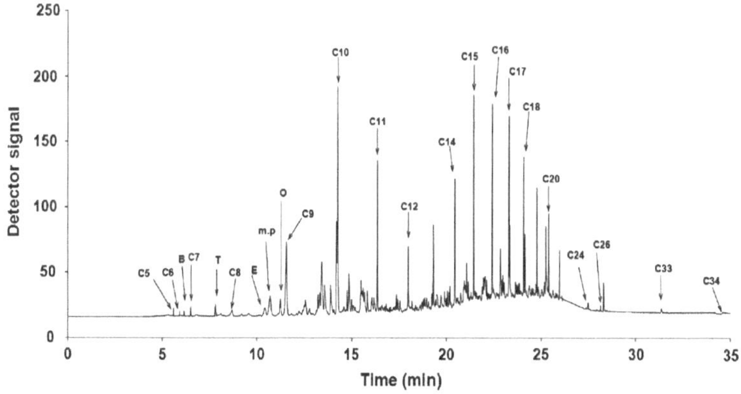

FIG. 6.6 Chromatogram of soil sample using PVC/MWCNTs SPME fiber. *(Reproduced with permission from A.A. Matin, P. Biparva, M. Gheshlaghi, K. Farhadi, A. Gheshlaghi, Environmental monitoring of complex hydrocarbon mixtures in water and soil samples after solid phase microextraction using PVC/MWCNTs nanocomposite fiber, Chemosphere 93 (9) (2013) 1920, Elsevier.)*

lack access to safe drinking water, and 2.6 billion people have been drinking water with little or no sanitation. Millions of people die every year from diseases transmitted from polluted water [36]. The release of water pollutants mainly come from industries. For example, the chloralkali industry releases mercury; the battery industry releases lead, cadmium, and mercury; and persistent toxic organic compounds are released from pesticides, dyes, and the pharmaceutical industry, and so forth.

6.4.1 Monitoring Organic Pollutants in Water

Nanocomposite-based nanosensors have found applications in the detection of numerous organic compounds in aqueous systems such as phenol and its derivatives, thiols, amino acids, pesticides, and so forth [37].

Phenol and its vapors emit environmental pollutants that are corrosive to the eyes, skin, and respiratory tract. Roya et al. proposed a new strategy to determine the concentration of phenol in water [38]. A graphite paste composite modified with *p*-chloranil and carbon nanotubes was used as an electrode for the sensitive and selective voltammetric determination. The detection limit was 2 μmol/L. An interference study and real wastewater samples show that 800-fold of different kinds of interference substances did not affect the selectivity.

Aromatic polyphenols such as gallic acid, pyrogallol, and tannic acid are commonly found in wastewaters. They are the major compounds existing in plants due to their antioxidant properties. Chen et al. have developed a highly selective colorimetric probe for visual detection of aromatic *ortho*-trihydroxy phenols. The sensor was composed of silver nanoparticles capped by chitosan (Ch-Ag NPs) [39]. Due to the hydrogen-bonding recognition and subsequent catalytic oxidation of the aromatic *ortho*-trihydroxy phenols by Ch-Ag NPs, the system produces intensified plasmon absorbance signals and a colorimetric change from light yellow to orange. Fig. 6.7 shows the high selectivity toward gallic acid, pyrogallol, and tannic acid, and other phelnolic compounds.

Paranitrophenol (4-NP) is another severely toxic substituted phenol that can cause headache, drowsiness, and cyanosis through ingestion or inhalation. The United States Environmental Protection Agency (EPA) limit in drinking water is 60 ng/mL. Ying et al. utilized the fluorescent method by fabricating a novel composite of graphene quantum dots (GQDs) and a molecularly imprinted polymer (MIP) for determination of 4-NP [40]. The combination of MIP and GQDs gives a high selectivity. The MIP provides a good recognition specificity of 4-NP, while the resonance energy transfers from GQDs as a donor to 4-NP, and an acceptor efficiently quenches the fluorescence of MIP-coated GQDs. Fig. 6.8 illustrates the preparation of MIP-coated GQDs. The sensor shows a good linearity in the range of 0.02–3 μg/mL, while the detection limit is 9 ng/mL.

The detection of explosives such as trinitrotoluene (TNT) and dinitrotoluene (DNT) in groundwater is also important. Cerutti et al. utilized a phage display

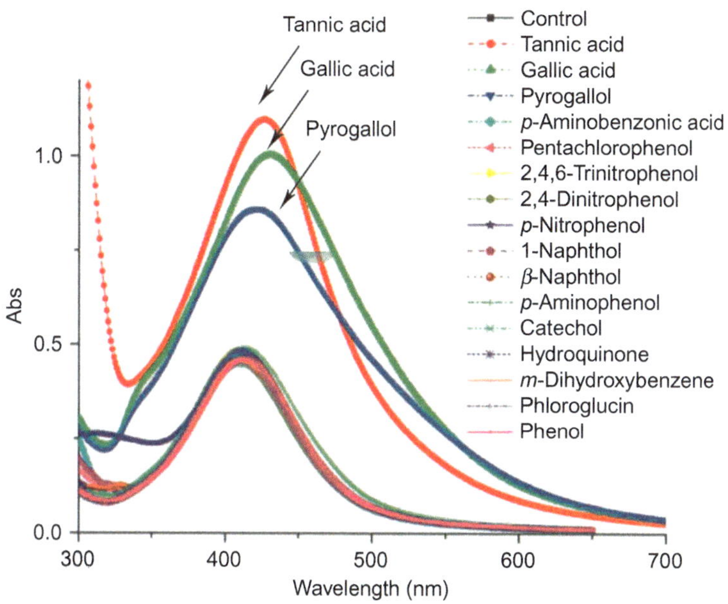

FIG. 6.7 Apparatus used to study sensing performance. *(Reproduced with permission from Z. Chen, X. Zhang, H. Cao, Y. Huang, Chitosan-capped silver nanoparticles as a highly selective colorimetric probe for visual detection of aromatic ortho-trihydroxy phenols, Analyst 138 (8) (2013) 2343, The Royal Society of Chemistry.)*

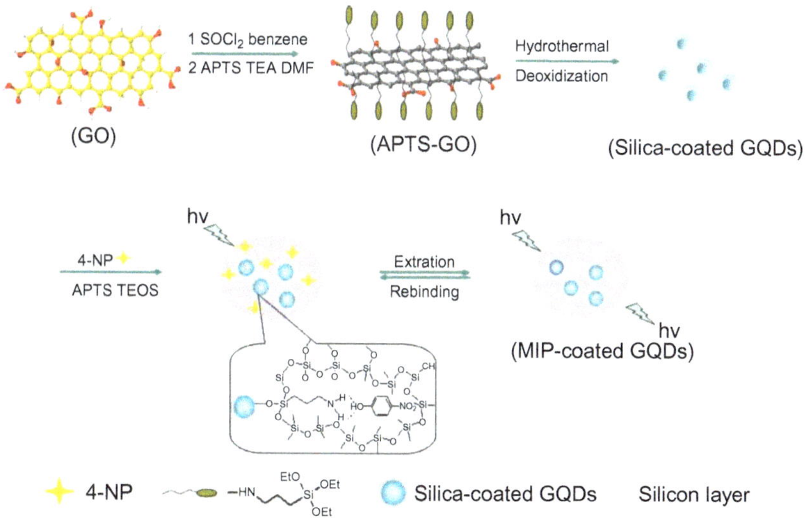

FIG. 6.8 Illustration of the preparation of MIP-coated GQDs. *(Reproduced with permission from Y. Zhou, Z.B. Qu, Y. Zeng, T. Zhou, G. Shi, A novel composite of graphene quantum dots and molecularly imprinted polymer for fluorescent detection of paranitrophenol, Biosens. Bioelectron. 52 (2014) 317, Elsevier.)*

technique to achieve a high selectivity for explosives [41]. An oligopeptide that has selective binding capabilities for TNT was selected as the TNT receptor. Poly(ethylene-*co*-glycidyl methacrylate) (PEGM) was chosen as a polymer matrix to immobilize the TNT receptor, and the composites were coated on a quartz crystal microbalance (QCM). For selectivity tests, a decrease in resonance frequency of the QCM was found in the TNT solution, whereas no change was observed in the DNT solution (see Fig. 6.9).

The detections of amino acid [42], bisphenol A [12], hydroquinone [43], nitrobenzene [44], were also well studied.

6.4.2 Monitoring Heavy Metals in Water

Heavy metals are introduced in aquatic ecosystems via the weathering of rocks, volcanic eruptions, and various human activities, including mining, processing, and so forth [45]. Common heavy metal pollutants in aquatic systems include arsenic (As), cadmium (Cd), chromium (Cr), copper (Cu), nickel (Ni), lead (Pb), and mercury (Hg). Their toxicity depends on many factors such as dose, route of exposure, chemical forms, and the health condition of exposed individuals [46]. Table 6.1 gives the maximum contaminants level (MCL) in National Primary Drinking Water Regulations (NPDWR) set by the United States Environmental Protection Agency (EPA).

In recent years, many novel nanocomposites have been fabricated and are used for sensors, especially optical and electrochemical sensors, because of their high sensitivity and fast response.

A fluorimetric and colorimetric Hg^{2+} chemical sensor based on a porphyrin derivative attached to Au@SiO_2 nanoparticles was developed by Cho et al. [47] The porphyrin derivative was attached onto the core-shell nanocomposites by

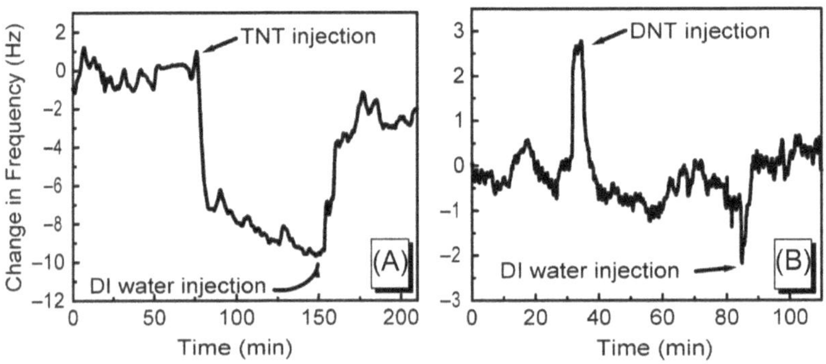

FIG. 6.9 Change in QCM resonance frequency after exposure of (A) TNT solution, (B) DNT solution. *(Reproduced with permission from M. Cerruti, J. Jaworski, D. Raorane, C. Zueger, J. Varadarajan, C. Carraro, S.W. Lee, R. Maboudian, A. Majumdar, Polymer-oligopeptide composite coating for selective detection of explosives in water, Anal. Chem. 81 (11) (2009), 4192, American Chemical Society.)*

TABLE 6.1 Maximum Contaminant Level of Heavy Metals in Drinking Water

Metal	Chemical Symbol	Maximum Contaminant Level (mg/L)
Arsenic	As	0.010
Cadmium	Cd	0.005
Chromium	Cr	0.1
Lead	Pb	0.015
Mercury	Hg	0.002
Selenium	Se	0.05
Copper	Cu	1.3

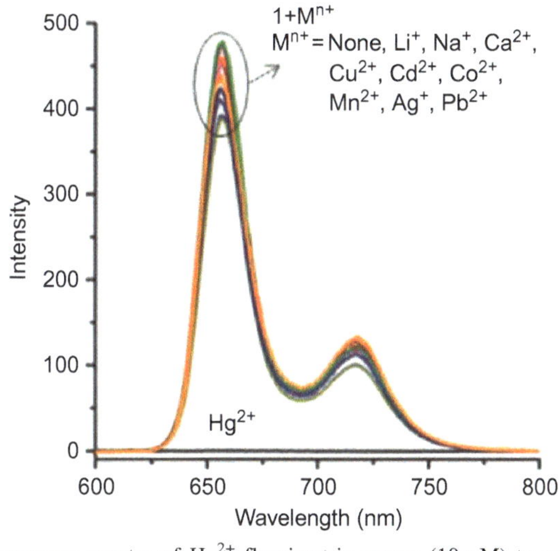

FIG. 6.10 Fluorescence spectra of Hg^{2+} fluorimetric sensor (10mM) to various metal ions. *(Reproduced with permission from Y. Cho, S.S. Lee, J.H. Jung, Recyclable fluorimetric and colorimetric mercury-specific sensor using porphyrin-functionalized Au@SiO$_2$ core/shell nanoparticles. Analyst 135 (7) (2010) 1551, The Royal Society of Chemistry.)*

covalent bonds, and showed a red color with strong fluorescence. Meanwhile, it also acted as an Hg^{2+} receptor. In the presence of Hg^{2+}, the color changed from red to green within 10s, and the fluorescence signal became weak. Excellent selectivity was shown by testing the fluorescence change of the nanocomposite toward other metal ions, including Li^+, Na^+, Ca^{2+}, Cd^{2+}, Co^{2+}, Ag^+, Pb^{2+}, and so forth, by fluorophotometry, as shown in Fig. 6.10. The detection limit is 1.2ppb, which is way below the MCL set by the EPA. Furthermore, the fluorescence response was constant over a wide range of pH from 4 to 10, and the sensor is reversible simply by treating it with an EDTA solution to strip Hg^{2+} ions off from the porphyrin derivative. Techniques such as DNA-DNA hybridization, protein-DNA recognition, and so forth were also used for heavy metal analysis. Wang et al. reported an ultrasensitive electrochemical biosensor using gold nanoparticles decorated with a bovine serum albumin reduction of graphene oxide (AuNP-BSA-rGO) for detection of Hg^{2+} in water [48]. The triple helix DNA containing a thiol labeled single-strand DNA, and a thymine-rich DNA will form a stable thymine-Hg^{2+}-thymine complex with Hg^{2+}. This biosensor was found to have a wide linear range from 0.1 to 130 nM, with a detection limit of 0.03 nM.

A small amount of Pb^{2+} and Cd^{2+} ingestion may cause nervous system dysfunction, anemia, kidney problems, or high blood pressure; and a high dose can result in death [49]. Niu et al. fabricated screen-printed electrodes (SPEs) that were made of bismuth nanoparticles' porous carbon nanocomposite [50]. The resulting electrochemical sensors were sensitive to Pb^{2+} and Cd^{2+} in water samples. The porous structure of the porous carbon matrix and high surface area of the bismuth nanoparticles enabled the nanosensor to have a detection limit of 4 ppb in tap drinking water and waste water samples. Dai et al. developed an electrochemical sensor based on phytic acid (PA) functionalized polypyrrole/graphene oxide (PPy/GO) nanocomposites for simultaneously detecting Pb^{2+} and Cd^{2+} ions in water [51]. The nanocomposites were synthesized by in situ chemical polymerization following functionalization of PA by electrostatic attraction. The PA/PPy/GO electrode showed a higher conductivity compared with PPy/GO and PA/GO electrodes, and was used for sensing metal ions by differential pulse voltammetry (DPV). This nanosensor exhibited a wide linear working range from 5 to 150 μg/L, as shown by Fig. 6.11.

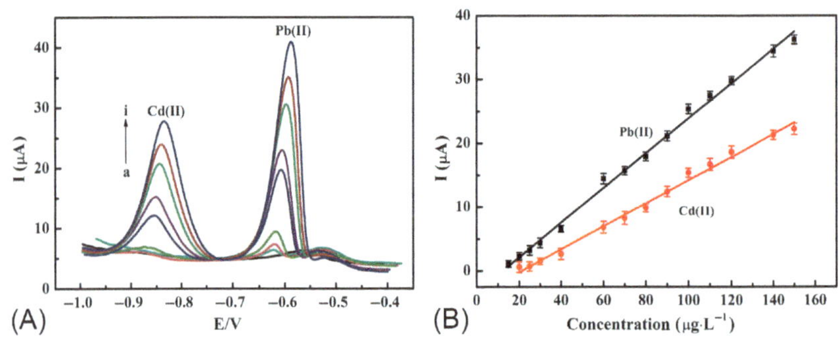

FIG. 6.11 Differential pulse voltammetric detection of Cd^{2+} and Pb^{2+} (A) and corresponding calibration plot (B) toward metal ions at different concentrations. *(Reproduced with permission from H. Dai, N. Wang, D. Wang, H. Ma, M. Lin, An electrochemical sensor based on phytic acid functionalized polypyrrole/graphene oxide nanocomposites for simultaneous determination of Cd (II) and Pb (II), Chem. Eng. J. 299 (2016) 150, Elsevier.)*

Simultaneous detection of heavy metals with high sensitivity and selectivity is a challenge for nanocomposite-based nanosensors. Veerakumar et al. successfully developed an electrochemical sensor platform using palladium nanoparticles (Pd NPs)/porous activated carbons (PACs) modified by glassy carbon electrodes (Pd/PAC-GCE) to detect Cd^{2+}, Pb^{2+}, Cu^{2+}, and Hg^{2+} simultaneously [52]. The electrochemical sensor exhibited a superior linear response of 0.5–5.5, 0.5–8.9, 0.5–5.0, and 0.24–7.5 μM with the detection limits of 41, 50, 66, and 54 nM for Cd^{2+}, Pb^{2+}, Cu^{2+}, and Hg^{2+} ions, respectively. The enhanced sensitivity and selectivity were believed to be due to the synergistic effect between the Pd NPs and PACs.

6.5 CONCLUSION

Human activities have greatly contaminated ecosystems around the world, especially in the past two centuries. Toxic chemicals can be found in the most pristine rainforest, the middle of the Pacific Ocean, and in the blood of arctic animals. To protect the environment, early monitoring of pollutants is essential. With the development of nanotechnology, nanocomposites offer tremendous opportunity and possibilities for applications in environmental monitoring. Most of the traditional analytical methods, including UV-Vis spectroscopy, LC (Liquid Chromatography), GC (Gas Chromatography), ICP-AES (Inductively Coupled Plasma-Atomic Emission Spectroscopy), F-AAS (Flame-Atomic Absorption Spectroscopy), and XRF (X-ray Fluorescence Spectroscopy), and so forth have disadvantages such as high cost, and requiring a lot of time, off-site determination, and technical trainingTherefore, the nanocomposite-based sensor is of great interest due to its small size, simplicity, low cost, on-site determination, high sensitivity, high selectivity, fast response time, and so forth. Most of the nanosensors reported are chemical sensors, optical sensors, electrochemical sensors, and resistive sensors. Despite the fast progress in developing nanocomposite sensors, there are still some disadvantages in the most recently reported sensors: (i) Distinct and precise control of morphology is hardly guaranteed. Multidimensional heterostructures composed of multiple compositions greatly affect the overall performance. (ii) Though the sensitivity is impressive, most sensors failed to have a high selectivity, especially gas sensors. In the real world, unknown interference can alter the result. (iii) Scalable fabrication of nanosensors still remains challenging. (iv) Nanosensors are not as durable as bulk sensors. For example, most colorimetric sensors are humidity- and temperature-sensitive. Semiconducting metal oxide-based sensors are temperature sensitive and need a power supply. Fluorescence sensors offer low photo-stability for chromophores in air.

Following are some keys in designing nanocomposite-based sensors: (i) Size and shape of the nanocomposites should be reproducible and of high performance. (ii) The specific detection molecule or nanomaterials should be active sites with a high affinity to the analyte. (iii) High conductive, electrochemical active, mechanical stable, and high-surface-area substrates enhance the sensitivity and robustness of the sensor. (iv) Design of the nanocomposite interphase should be reinforced to the nanocomposite, and immobilize the active sites. (v) There should be an easy way to revert to the state before the test to make it reusable. Meeting some or all of these criteria while overcoming the common disadvantages discussed earlier can help in developing an ideal environmental nanocomposite sensor.

This chapter describes recent developments of nanocomposite sensors in environmental monitoring of air, soil, and water. It is expected that the sensors based on nanocomposites will be more sensitive, selective, accurate, comprehensive, stable, and robust with continuous effort in their advancement; and the full commercialization of nanosensors for environmental monitoring will be realized in the near future.

REFERENCES

[1] D. Hoornweg, P. Bhada-Tata, What a Waste: A Global Review of Solid Waste Management, Urban Development Series Knowledge Papers, vol. 15, (2012) p.1.

[2] D. Hoornweg, P. Bhada-Tata, C. Kennedy, Environment: waste production must peak this century, Nature 502 (7473) (2013) 615.

[3] G. Maduraiveeran, W. Jin, Nanomaterials based electrochemical sensor and biosensor platforms for environmental applications, Trends Environ. Anal. Chem. 13 (2017) 10.

[4] J.R. Potts, D.R. Dreyer, C.W. Bielawski, R.S. Ruoff, Graphene-based polymer nanocomposites, Polymer 52 (1) (2011) 5.

[5] J. Zhu, H. Deng, B. Zhu, W. Dong, W. Zhang, J. Li, X. Bao, Polymer-assistant ceramic nanocomposite materials for advanced fuel cell technologies, Ceram. Int. 43 (7) (2017) 5484.

[6] C. Zhan, G. Yu, Y. Lu, L. Wang, E. Wujcik, S. Wei, Conductive polymer nanocomposites: a critical review of modern advanced devices, J. Mater. Chem. C 5 (7) (2017) 1569.

[7] N. Fukata, M. Mitome, Y. Bando, W. Wu, Z.L. Wang, Lithium ion battery anodes using Si-Fe based nanocomposite structures, Nano Energy 26 (2016) 37.

[8] N. Yousefi, M.M. Gudarzi, Q. Zheng, X. Lin, X. Shen, J. Jia, F. Sharif, J.-K. Kim, Highly aligned, ultralarge-size reduced graphene oxide/polyurethane nanocomposites: mechanical properties and moisture permeability, Compos. A: Appl. Sci. Manuf. 49 (2013) 42.

[9] Y. Wang, H. Wei, Y. Lu, S. Wei, E. Wujcik, Z. Guo, Multifunctional carbon nanostructures for advanced energy storage applications, Nanomaterials 5 (2) (2015) 755.

[10] H. Hu, X. Wang, F. Liu, J. Wang, C. Xu, Rapid microwave-assisted synthesis of graphene nanosheets–zinc sulfide nanocomposites: optical and photocatalytic properties, Synth. Met. 161 (5) (2011) 404.

[11] V.M. Kariuki, S.A. Fasih-Ahmad, F.J. Osonga, O.A. Sadik, An electrochemical sensor for nitrobenzene using pi-conjugated polymer-embedded nanosilver, Analyst 141 (7) (2016) 2259.

[12] L. Hu, C.C. Fong, X. Zhang, L.L. Chan, P.K. Lam, P.K. Chu, K.Y. Wong, M. Yang, Au nanoparticles decorated TiO_2 nanotube arrays as a recyclable sensor for Photoenhanced electrochemical detection of Bisphenol A, Environ. Sci. Technol. 50 (8) (2016) 4430.

[13] L. Rassaei, F. Marken, M. Sillanpää, M. Amiri, C.M. Cirtiu, M. Sillanpää, Nanoparticles in electrochemical sensors for environmental monitoring, TrAC Trends Anal. Chem. 30 (11) (2011) 1704.

[14] M. Kampa, E. Castanas, Human health effects of air pollution, Environ. Pollut. 151 (2) (2008) 362.

[15] K. Wetchakun, T. Samerjai, N. Tamaekong, C. Liewhiran, C. Siriwong, V. Kruefu, A. Wisitsoraat, A. Tuantranont, S. Phanichphant, Semiconducting metal oxides as sensors for environmentally hazardous gases, Sens. Actuators B 160 (1) (2011) 580.
[16] H. Bai, G. Shi, Gas sensors based on conducting polymers, Sensors 7 (3) (2007) 267.
[17] D.W. Hatchett, M. Josowicz, Composites of intrinsically conducting polymers as sensing nanomaterials, Chem. Rev. 108 (2) (2008) 746.
[18] S.T. Navale, G.D. Khuspe, M.A. Chougule, V.B. Patil, Camphor sulfonic acid doped PPy/α-Fe2O3 hybrid nanocomposites as NO2 sensors, RSC Adv. 4 (53) (2014).
[19] P. Tyagi, A. Sharma, M. Tomar, V. Gupta, Metal oxide catalyst assisted SnO_2 thin film based SO_2 gas sensor, Sens. Actuators B 224 (2016) 282.
[20] P.-G. Su, Y.-T. Peng, Fabrication of a room-temperature H2S gas sensor based on PPy/WO3 nanocomposite films by in-situ photopolymerization, Sens. Actuators B 193 (2014) 637.
[21] S. Preston-Martin, M.C. Pike, R.K. Ross, P.A. Jones, B.E. Henderson, Increased cell division as a cause of human cancer, Cancer Res. 50 (23) (1990) 7415.
[22] L. Wang, S. Wang, M. Xu, X. Hu, H. Zhang, Y. Wang, W. Huang, A Au-functionalized ZnO nanowire gas sensor for detection of benzene and toluene, Phys. Chem. Chem. Phys. PCCP 15 (40) (2013) 17179.
[23] M. La Farre, S. Pérez, L. Kantiani, D. Barceló, Fate and toxicity of emerging pollutants, their metabolites and transformation products in the aquatic environment, TrAC Trends Anal. Chem. 27 (11) (2008) 991.
[24] P. Kar, A. Choudhury, Carboxylic acid functionalized multi-walled carbon nanotube doped polyaniline for chloroform sensors, Sens. Actuators B 183 (2013) 25.
[25] R. Gardner, Use of the reciprocal calculation procedure for setting workplace emergency action levels for hydrocarbon mixtures and their relationship to lower explosive limits, Ann. Occup. Hyg. (2011) mer098.
[26] N.D. Khoang, D.D. Trung, N. Van Duy, N.D. Hoa, N. Van Hieu, Design of SnO_2/ZnO hierarchical nanostructures for enhanced ethanol gas-sensing performance, Sens. Actuators B 174 (2012) 594.
[27] S. Pramanik, G. Das, N. Karak, Facile preparation of polyaniline nanofibers modified bentonite nanohybrid for gas sensor application, RSC Adv. 3 (14) (2013) 4574.
[28] D. Arrouays, B. Marchant, N. Saby, J. Meersmans, T. Orton, M. Martin, P. Bellamy, R. Lark, M. Kibblewhite, Generic issues on broad-scale soil monitoring schemes: a review, Pedosphere 22 (4) (2012) 456.
[29] Y. Liu, C. Wen, X. Liu, China's food security soiled by contamination, Science 339 (6126) (2013) 1382.
[30] L. Luo, Y. Ma, S. Zhang, D. Wei, Y.-G. Zhu, An inventory of trace element inputs to agricultural soils in China, J. Environ. Manage. 90 (8) (2009) 2524.
[31] M.G. de la Huebra, P. Hernández, O. Nieto, Y. Ballesteros, L. Hernández, Determination of carbendazim in soil samples by anodic stripping voltammetry using a carbon fiber ultramicroelectrode, Fresenius J. Anal. Chem. 367 (5) (2000) 474.
[32] S. Luo, Y. Wu, H. Gou, A voltammetric sensor based on GO–MWNTs hybrid nanomaterial-modified electrode for determination of carbendazim in soil and water samples, Ionics 19 (4) (2013) 673.
[33] E. Solgi, A. Esmaili-Sari, A. Riyahi-Bakhtiari, M. Hadipour, Soil contamination of metals in the three industrial estates, Arak, Iran, Bull. Environ. Contam. Toxicol. 88 (4) (2012) 634.
[34] C. van der Horst, B. Silwana, E. Iwuoha, V. Somerset, Bismuth–silver bimetallic nanosensor application for the voltammetric analysis of dust and soil samples, J. Electroanal. Chem. 752 (2015) 1.

[35] A.A. Matin, P. Biparva, M. Gheshlaghi, K. Farhadi, A. Gheshlaghi, Environmental monitoring of complex hydrocarbon mixtures in water and soil samples after solid phase microextraction using PVC/MWCNTs nanocomposite fiber, Chemosphere 93 (9) (2013) 1920.

[36] M.A. Shannon, P.W. Bohn, M. Elimelech, J.G. Georgiadis, B.J. Marinas, A.M. Mayes, Science and technology for water purification in the coming decades, Nature 452 (7185) (2008) 301.

[37] S. Thatai, P. Khurana, J. Boken, S. Prasad, D. Kumar, Nanoparticles and core–shell nanocomposite based new generation water remediation materials and analytical techniques: a review, Microchem. J. 116 (2014) 62.

[38] R. Sadeghi, H. Karimi-Maleh, M.A. Khalilzadeh, H. Beitollahi, Z. Ranjbarha, M.B. Zanousi, A new strategy for determination of hydroxylamine and phenol in water and waste water samples using modified nanosensor, Environ. Sci. Pollut. Res. Int. 20 (9) (2013) 6584.

[39] Z. Chen, X. Zhang, H. Cao, Y. Huang, Chitosan-capped silver nanoparticles as a highly selective colorimetric probe for visual detection of aromatic ortho-trihydroxy phenols, Analyst 138 (8) (2013) 2343.

[40] Y. Zhou, Z.B. Qu, Y. Zeng, T. Zhou, G. Shi, A novel composite of graphene quantum dots and molecularly imprinted polymer for fluorescent detection of paranitrophenol, Biosens. Bioelectron. 52 (2014) 317.

[41] M. Cerruti, J. Jaworski, D. Raorane, C. Zueger, J. Varadarajan, C. Carraro, S.W. Lee, R. Maboudian, A. Majumdar, Polymer-oligopeptide composite coating for selective detection of explosives in water, Anal. Chem. 81 (11) (2009) 4192.

[42] L. Wang, Q. Zhang, S. Chen, F. Xu, S. Chen, J. Jia, H. Tan, H. Hou, Y. Song, Electrochemical sensing and biosensing platform based on biomass-derived macroporous carbon materials, Anal. Chem. 86 (3) (2014) 1414.

[43] D.-M. Zhao, X.-H. Zhang, L.-J. Feng, L. Jia, S.-F. Wang, Simultaneous determination of hydroquinone and catechol at PASA/MWNTs composite film modified glassy carbon electrode, Colloids Surf. B Biointerfaces 74 (1) (2009) 317.

[44] G. Maduraiveeran, R. Ramaraj, Potential sensing platform of silver nanoparticles embedded in functionalized silicate shell for nitroaromatic compounds, Anal. Chem. 81 (18) (2009) 7552.

[45] H.A. Naser, Assessment and management of heavy metal pollution in the marine environment of the Arabian gulf: a review, Mar. Pollut. Bull. 72 (1) (2013) 6.

[46] P.B. Tchounwou, C.G. Yedjou, A.K. Patlolla, D.J. Sutton, Molecular, Clinical and Environmental Toxicology, Basel: Springer, 2012.

[47] Y. Cho, S.S. Lee, J.H. Jung, Recyclable fluorimetric and colorimetric mercury-specific sensor using porphyrin-functionalized Au@SiO2 core/shell nanoparticles, Analyst 135 (7) (2010) 1551.

[48] H. Wang, Y. Zhang, H. Ma, X. Ren, Y. Wang, Y. Zhang, Q. Wei, Electrochemical DNA probe for Hg 2+ detection based on a triple-helix DNA and multistage signal amplification strategy, Biosens. Bioelectron. 86 (2016) 907.

[49] R. Singh, N. Gautam, A. Mishra, R. Gupta, Heavy metals and living systems: an overview, Indian J. Pharm. 43 (3) (2011) 246.

[50] P. Niu, C. Fernández-Sánchez, M. Gich, C. Navarro-Hernández, P. Fanjul-Bolado, A. Roig, Screen-printed electrodes made of a bismuth nanoparticle porous carbon nanocomposite applied to the determination of heavy metal ions, Microchim. Acta 183 (2) (2016) 617.

[51] H. Dai, N. Wang, D. Wang, H. Ma, M. Lin, An electrochemical sensor based on phytic acid functionalized polypyrrole/graphene oxide nanocomposites for simultaneous determination of Cd (II) and Pb (II), Chem. Eng. J. 299 (2016) 150.

[52] P. Veerakumar, V. Veeramani, S.M. Chen, R. Madhu, S.B. Liu, Palladium nanoparticle incorporated porous activated carbon: electrochemical detection of toxic metal ions, ACS Appl. Mater. Interfaces 8 (2) (2016) 1319.

Chapter 7

Nanocomposites for Biomedical Applications

Natali Gendelberg, David Bird and Nuggehalli M. Ravindra
Interdisciplinary Program in Materials Science & Engineering, New Jersey Institute of Technology, Newark, NJ, United States

Chapter Outline

7.1 INTRODUCTION

Composites consist of two or more materials, usually consisting of a polymer, an inorganic material, a filler or organic/inorganic fiber/particulate form filler, and a resin. Usually, the composite simply consists of filler dispersed into a polymer. The requirements for composite design are [1]:

1. The material phases must differ physically and chemically.
2. The material phases must be distinct from each other on some scale, whether it is nanometers or millimeters.
3. Unlike the constituents of alloys, the materials retain their individual identities and characteristics.
4. Despite the third requirement, the composite, as a whole, exhibits bulk properties that are different from those of the individual components. Often times, these bulk properties result in synergetic enhancements to the individual properties of the components.
5. Composites are obtained by uniformly dispersing one material, referred to as the dispersed phase, into another material that is referred to as the continuous phase, or the matrix phase.

The filler properties impact those of the composite based on their inherent properties, physical dimensions, and shape. For example, fiberglass is widely used to enhance the mechanical properties of the composite, as the fibers have a high

Polymer-Based Multifunctional Nanocomposites and Their Applications
https://doi.org/10.1016/B978-0-12-815067-2.00007-X

strength-to-weight ratio and good dimensional stability. Longer fibers that are used to reinforce polymers more significantly impact the viscosity of the polymer at low shearing rates [1]. This phenomena results in the formation of an aligned and continuous order of the fibers, which gives it significant strength, but results in an anisotropic character. Whereas short fibers are discontinuous, irrespective of whether they are aligned or randomly dispersed, it results in a weaker, but isotropic, composite. In other words, longer fibers handle stress more efficiently because tensile stress is transferred to the fiber from the matrix by a shear stress at the interface. If the fiber is too short or near the ends of the fiber, there is not enough surface area to transfer the full stress. Longer fibers reduce/hinder the "end effect."

Not all fillers are compatible with the polymer resin in which they are to be dispersed. This property is especially sensitive when the filler is inorganic. Negative effects of incompatible dispersion of filler in the matrix include the following:

1). Reduced wetting of the filler by the polymer.
2). Agglomeration of the filler.
3). Inadequate dispersion of the filler in the polymer.
4). Significant increase in the viscosity of the polymer with the filler content.

Coupling agents are chemicals that can react with the reinforcement and the resin matrix of a composite material and act as an interface between the resin and glass fiber/mineral filler to form a chemical bridge. This allows for the enhancement of performance properties of the composite, and may improve production or melt-processing by optimizing the melt flow.

The performance of composites is influenced by the adhesive strength of the filler-matrix interface. The presence of water, absorbed by the filler surface, and the thermal stresses, generated by the differential thermal coefficients of linear expansion between the filler and matrix materials, reduce the interfacial adhesive strength. Polymers have relatively higher linear expansion coefficients than filler materials. In practice, interfacial adhesive strength is enhanced by the use of coupling agents, which are usually low-molecular-weight organofunctional silanes, stereates, or titanates that act as a bridge between the matrix resin and fiber material. The coupling agents can modify the surface of the filler and polymer, as well [1].

The classification of man-made, multiphase composites began around the middle of the 20th century. Fiberglass-reinforced polymers were the first to fall into these classifications [2]. Some of the applications of composite material development include the military sector, aerospace technologies, aviation technologies, automotive industry, and more recently, the development of scalable and affordable production materials for biomedical applications. Even though composite classification is quite recent, composites have been implemented in nature, and known by humans, for centuries. Some examples of composites include wood trunks, seashells, and clay construction bricks reinforced with straw and bones.

For example, aircraft wings were initially made of wood that proved to be a liability due to its directional mechanical properties and wing warping. Following this experience, prototypes were made of steel tubing and aluminum, which were not fuel efficient when designed in the proper dimensions to hinder the drag effects on the wind. Current airplane wings are made of composite materials such as carbon fibers in order to provide the wings with the necessary stability and strength, as well as low density, which results in lighter wings with a large surface area to facilitate higher speed. The Wright brothers began designing aircrafts in the early 1900s, but the core investment into the optimization of wing design was done in the 1920s and 1930s in response to the United States' participation in WWI and WWII. From this example, it is evident that most of the composite material developments have been in response to industry and government needs. As will be discussed in this chapter, when the classified composites began to be researched for medical use, many side effects resulting from immune response and toxicity issues arose [3].

It is ironic that investment into the development of composites for medical purposes began considerably late, as many parts of the human body are composites, and it is only logical that treatment using composites could result in better forms of treatment and improved healing potential. An example of a composite in the body that has served as the subject of much of the biomedical composite research to date is the bones. The bones in the human body fall under two categories: cortical bone and trabecular bone. Cortical bone acts as the load-bearing bone and constitutes the majority of the bone of the body, and its presence increases with age. Trabecular bone has shock absorbing properties and is less dense than cortical bone [4].

Both cortical bone and trabecular bone form a composite structure, with their difference in functionality resulting from the arrangement and infrastructure formed by the osteons. The osteons are horizontally connected cylindrical structures that form the larger cylindrical structure surrounding the trabecular bone, that is, the cortical bone. Due to this arrangement, cortical bone is high in density (Fig. 7.1), whereas the trabecular bone forms a system of trabecular plates and rods that are dispersed in the bone marrow. This porous composite geometry that occurs in the trabecular bone alone allows for the passage of nutrients both vertically and laterally in the bone, while facilitating multidirectional shock absorption and bone development (Fig. 7.1).

The cortical bone adds a phase to the composite formed within the trabecular bone, which amplifies the range of endurance of the bone. Had there been cortical bone alone, bone fractures would be more prevalent due to the bone's high rigidity. On the other hand, trabecular bone alone would not be able to support the body's weight and vertical posture. Bone implants such as the hip implant are one of the most investigated applications in which several materials have been combined to provide treatments that result in outcomes compatible with the bone's true properties. The basic definition of composites will be discussed and compared with one of the most widely used forms of implants—the total hip arthoplasty (THA)—and its evolution over time from a single component

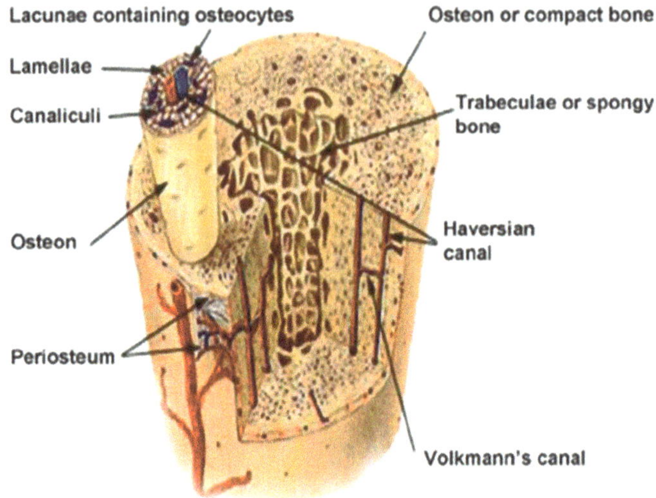

FIG. 7.1 Cortical bone versus trabecular bone [5].

structure, to a multi-component structure that showcases the advantages of combining several materials into a composite. These advantages will be further showcased by discussing the osteointegration potential posed by the use of composites, as will be defined in the next section.

The function of the hip is to support the body mass and balance the body in movement and at rest. The hip is really a joint coming out of the femur and forming a ball-in-socket complex with the pelvis. This structure is made up of the femoral neck that connects the femur to the femoral head (ball component). The femoral head is inserted into an indent in the pelvis, the acetabulum (the socket). The femoral neck angular extension from the femur provides a movement arm that eases the movement of the femur with respect to the pelvis when one is walking. In order to reduce friction from the constant femur-to-acetabulum contact, the femoral head is covered by a layer of smooth cartilage; there is synovial fluid (lubricant) between the femoral head and acetabulum, and the acetabulum is lined with an additional layer of smooth cartilage. Behind the aforementioned layer of acetabular cartilage, there is an extra layer of cartilage that acts as a weight/shock absorber to alleviate the force impact on the joint.

Hip implants date as far back as 1891 and have materialistically and mechanically advanced all through the 1960s, when the basis for the THA models used today was made. As of then, most improvements have been, and are still focused on extending the lifespan of the implant, improving ease of removal and re-implantation, homogenization of surgical techniques, and adaptation to the dynamic human lifestyle. This shift in focus is due to the fact that the human lifespan has significantly increased, making the average 10–20 year lifespan of the implant insufficient [6,7].

The human body is a very hostile environment to any foreign object that is introduced to it; materials must be very specific and must undergo certain processes in order to adapt well to the body. Materials that are inserted into the body must be modified, in many cases, to be suitable for implant, where there is a balance of cost with respect to the demand of the implant and its viable substitutions—it is a balance between cost and quality. This is expressed in the alloying of cobalt with chromium because of cobalt's high monetary value; but this alloy is chosen because it has a high resistance to wear corrosion with a reduced cost. Although titanium is very corrosion and heat resistant, for many years it was not used because of its high cost, until its more economical version was developed by alloying it with aluminum and vanadium. Alumina and zirconia (ceramics) are especially engineered for orthopedics to reduce brittleness and increase homogeneity with reduced wear [6]. Still, some debris from materials used can result, as in the case of friction being applied to polyethylene. The debris from polyethylene is inert, so it does not trigger an immune response, and in small amounts is not an issue. In large amounts, it can result in inflammation. In order to overcome this obstacle, the polyethylene is gamma-irradiated to create cross links and increase its molecular weight and density, thus resulting in ultra-high molecular weight polyethylene. Even though this is a good solution, the increased brittleness resulting from the increased strength does pose some risk for fracture.

Fig. 7.2 demonstrates the three most predominant hip implant models, which are: (1) metal-on-polyethylene (M-on-PE), (2) metal-on-metal (M-on-M) and (3) ceramic-on-ceramic (C-on-C). The general structure of the implant is made up of a titanium, or Co-Cr, femoral stem that inserts into the femur from one side, and from its other side into a Co-Cr ball portion (ceramic in the case of C-on-C) that is positioned in either a polyethylene, metal, or ceramic liner that lines a metal cup (all make up the acetabular component) that is placed in the

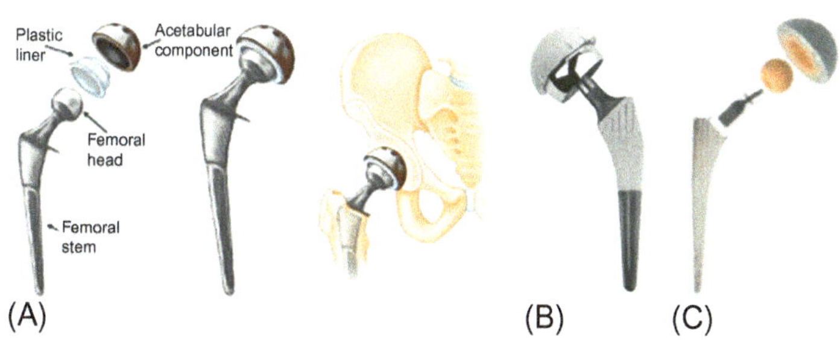

FIG. 7.2 The three predominant hip implant designs. (A) M-on-PE broken down, assembled and inserted into the body [7]; (B) M-on-M [6]; (C) C-on-C [6].

acetabulum. As part of the surgery, the femoral stem is either press fitted or cemented (with poly-methyl-methacralate—PMMA) into the femur after some bone is removed from inside the femur to make space for the femoral stem. When used, the PMMA is not used as an adhesive, but rather to improve the form fit or snug fit of the femoral stem. In general, the press-fit method is more often used because PMMA introduces an extra interface into the implant, and a greater number of interfaces increase the possibilities for implant failure. The popularity of the use of titanium as the metal component inserted into the bone can be attributed to its ability to osseointegrate into the bone, and the fact that its elastic modulus is comparable to that of the bone [6,7].

The evolution of materials used for tooth implants, crown development, and fillings has followed a similar trend to that used in hip replacements. In recent years, these sorts of implants have also experienced a shift toward polymer, glass, and ceramic composites due to their preferred appearance, and milder byproducts and handling techniques [8].

Carbon fiber reinforced polymers have the potential to resolve many of the issues that arise from the utilization of metal implants, including: harmful wear debris, cold-welding that results from the titanium locking screw construct, corrosion and its radiodensity that results in inaccurate visualization of fracture healing, infection or tumors in CTs and MRIs. An example of a carbon reinforced polymer used for anchoring bone plates is carbon-fiber-reinforced polyestheretherketone (CFR-PEEK) (Fig. 7.3). The elastic modulus of the CFR-PEEK implant can be fine-tuned with respect to the bone it is to be attached to by varying the orientation of the carbon fibers and the number of layers of the carbon fibers [10].

In Table 7.1, a comparison of the mechanical properties of various metal implants and CFR-PEEK to that of cortical and trabecular bone is presented. It is evident from Table 7.1 that CFR-PEEK falls within the composite modulus

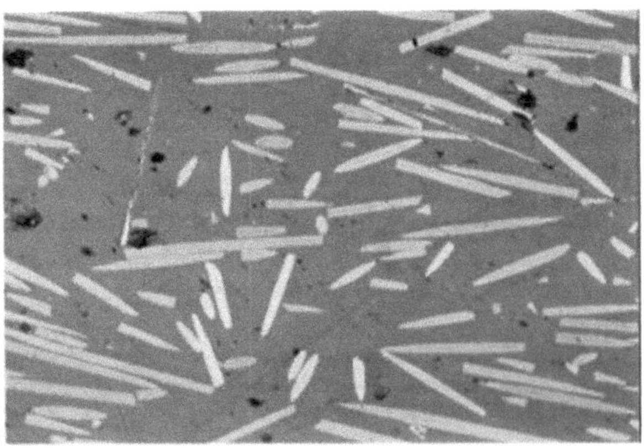

FIG. 7.3 Mircroscopic structure of CFR-PEEK captured with reflected polarized light microscopy (×200 magnification) [9].

TABLE 7.1 Comparison of the Mechanical Properties of Implants with Cortical/Trabecular Bone [10]

Material	Modulus of Elasticity (GPa)
Stainless steel	230
Cobalt-chrome	210
Titanium alloy	106–155
CFR-PEEK	3.5
Cortical bone	12–20
Trabecular bone	1

of elasticity of the cortical and trabecular bone. This serves the advantage of preventing stress shielding, in which the component with the higher elastic modulus carries the load. This property is important because stress shielding can cause the formation of weak bone during the bone repair process, whereas an implant with too low of a modulus of elasticity would not serve its purpose in preventing further injury to the affected area [10].

Besides its mechanical compatibility with the bone, CFR-PEEK has exquisite radiolucent properties and is highly compatible with MRIs. These properties allow for more accurate imaging of implant areas, which are normally occluded due to the optical properties of metals. The advantages of accurate implant site visual capturing provide insight into bone healing, osseointegration, bone tumor surveillance, and bone infection site surveillance. The last two items are attributed to some bone disease treatments, including the insertion of a therapeutic coated screw into the bone; thus the effectiveness of such treatment can be accurately tracked [9].

One disadvantage of CFR-PEEK is its inability to be contoured, which can be compensated for by utilizing different lengths of screws where needed. A second disadvantage is that increased flexibility of the material mat leads to pseudoarthrosis. Finally, despite its theoretical reduced risk of fatigue failure, it is difficult to visualize potential fatigue failure in the screw holes due to the high radiolucency of CFR-PEEK. This has been resolved by adding radiopaque markers that identify the intramedullary nail interlocking screw holes [9].

As has been discussed so far, the predecessors of composites in biomedical applications were metal implants, followed by structural composites. The discussed progression leading toward composite development shows that as the composite design delves to the smaller scales, the mechanical and physical properties of materials can be better maneuvered, especially in biological applications. This has to do, in part, with the scale in which natural intricacies in bodily composites occur. The average cell size is approximately 100 μm and the molecular interactions occur on the nanoscale. This leads to the next topic of biomedical composite research—bionanocomposites.

7.2 BIONANOCOMPOSITES

Bionanocomposites introduce a new and captivating area of research that encompasses all the innovative areas of biology, materials science, and nanotechnology with the potential to have a huge impact in biomedical sciences [11–20]. Generally, these nanocomposites are formed by the incorporation of inorganic/organic fillers into a biopolymer matrix, and the molecular bridging interactions between filler and biopolymer determine the mechanical properties of the polymer matrix [11]. The biopolymers of interest are natural or synthetic biodegradable polymers and have traditionally included polysaccharides, aliphatic polyesters, polyethylene glycols, polypeptides and proteins, and polynucleic acids, whereas the fillers have included clays, hydroxyapatite, and metal nanoparticles. The key to being able to introduce these materials into the body is based on biomimetics, and the ability of the human body and its constituents to accept these non-corrosive, non-toxic materials with no harm or cytotoxic effects. Therefore, materials that biodegrade hydrolytically, or can be eliminated via metabolism, are of key interest to biotechnologies such as tissue engineering, bone replacement/repair, dental applications, and controlled drug delivery [11]. Table 7.2 lists some of the polymers and their respective structures that are currently utilized in biomedical applications.

Hydroxyapatite (HAP)-polymer nanocomposites have been extensively researched and used in orthopedic surgery [11]. However, HAP [$Ca_{10}(PO_4)_6(OH)_2$] is problematic due to brittleness, lack of flexibility, migration, aggregation, and trouble with dispersion. Polysaccharides and polypeptidic matrices have been utilized to overcome these issues for composite formation. Chitosan-HAP nanocomposites are one such example in which researchers utilized a flexible biopolymer with resistance to heat and were able to synthesize a mechanically flexible nanocomposite via the coprecipitation method. Other researchers have looked at collagen-based polypeptidic gelatin, which is attractive due to the high number of functional groups in the polymer, the flexibility, and cost-effectiveness when combined with the osteoconductivity of HAP. The porous scaffolds that are formed in polymer solution suffer no aggregation and promote blood circulation and cell ingrowth due to the well-developed pore configuration. Porosity is considered to be critical for tissue engineering as it enables the diffusion of cellular byproducts and induces cell movement. Fig. 7.4 exemplifies a similar uniform scaffold prepared by

TABLE 7.2 Biopolymers Commonly Used in Biomedical Applications [11]

Polymer or Biopolymer	Structure
Poly(ethylene glycol)—PEG	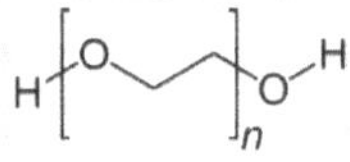
Chitosan	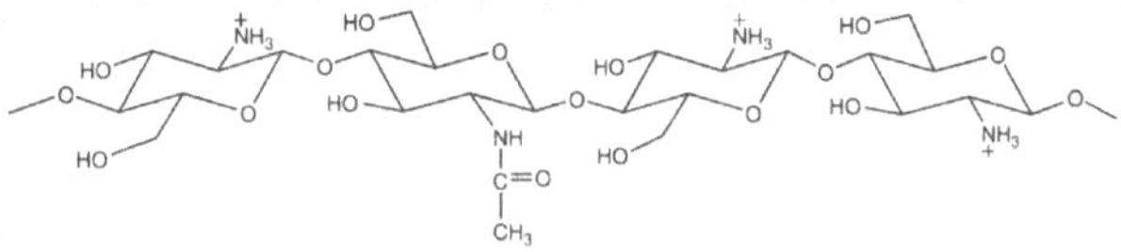
Collagen	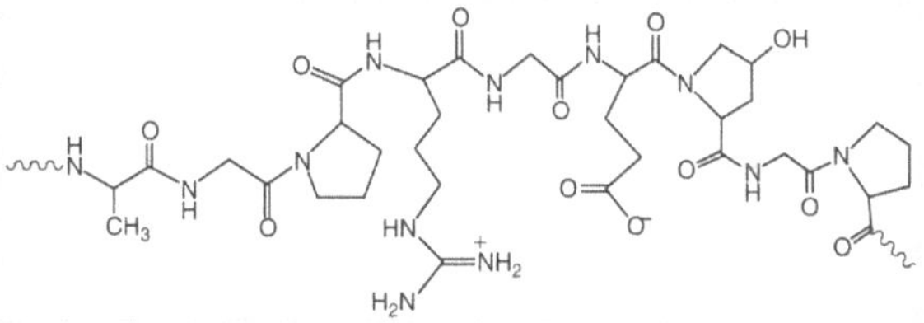
Poly(lactic acid)—PLA	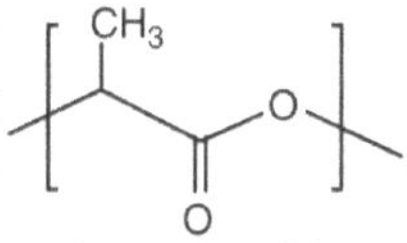
Poly (ε-caprolactone)—PCL	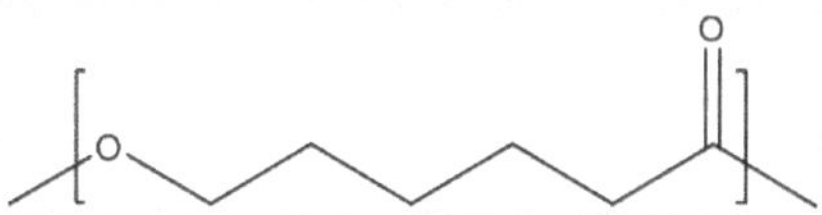
Poly(L-lysine)—PLL	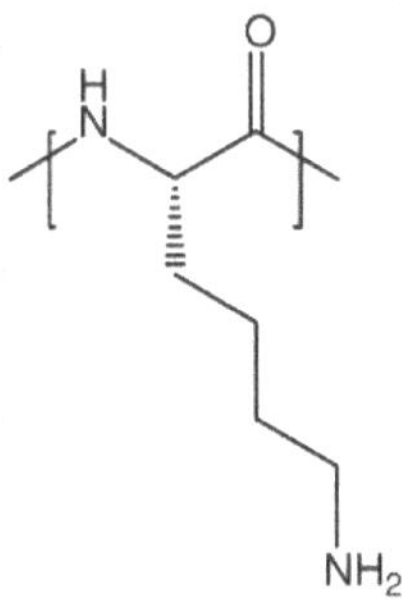

HAP/chitosan-gelatin/acetic acid. HAP has also been utilized to strengthen the sponge structure of polysaccharides such as alginates.

Polypeptide nanocomposite matrices have features that are unique to biomedical applications because the polymers are not synthetic, and they have

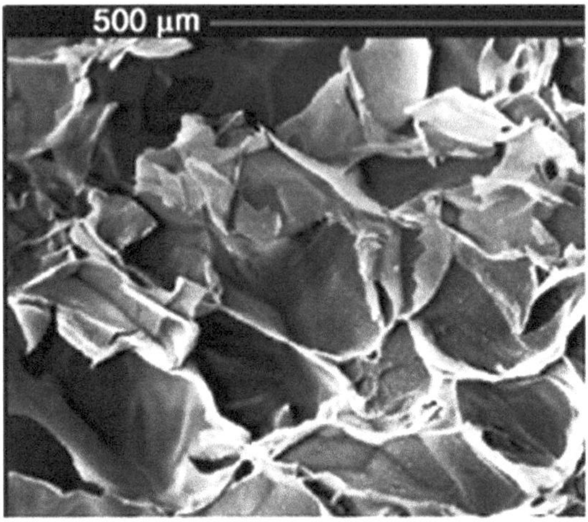

FIG. 7.4 SEM image showing the uniform porosity formed from HAP/chitosan-gelatin composite [11].

the ability to conform to a variety of structures: random coil, α-helix, or β-sheets in aqueous solution [11]. Nanocomposites based on polypeptide examples predominantly include poly(L-lysine) (PLL) as the matrix and montmorillonite clay (MMT) and fibrils. Such nanocomposites exhibit enhanced mechanical and thermal properties. Further investigations have focused on the effect of molecular weight on final conformations and transitions that are influenced by environmental conditions such as pH, temperature, salt concentration, or alcohol content. Fourier Transform Infrared (FTIR) Spectroscopy, Raman Spectroscopy and Circular Dichroism (CD) have been used to analytically map structural transitions in liquid solutions displaying a kinetic dependence of conformation on temperature.

Aliphatic polyesters represent a class of polymers that have been widely studied and show promise in biomedical applications [11]. These polymers, notably poly(L-lactic acid) (PLLA), have proven to be useful in degradable drug delivery systems, prosthetics, scaffolds, vascular grafts, bone screws, pins, and plates for temporary internal fixation due to their good mechanical properties and degradation into non-toxic properties. PLLA is hard, water insoluble, and crystalline with a melting point between 170°C and 180°C, and a glass transition temperature (T_g) of ~60–70°C. PLLA lends itself as a suitable polymer for mixing to form nanocomposites and for homopolymerization or copolymerization, depending on the desired properties of the polymer. PLA, the racemic polymer, poly(D, L-lactide) is also popular in biomedical studies and is amorphous with a T_g of ~50–60°C. PLLA reinforced with fibers, HAP, and clays are ways to compensate for the subpar mechanical properties, and such nanocomposites have been reported to have increased flexural modulus, strength, and moduli values comparable to bone replacement implants. Electrospinning has been used for scaffold fabrication and is a proven method for tailoring the mechanical, biological, and kinetic properties by optimizing the polymer solution and processing conditions. PLLA with montmorillonite (MMT) solutions that were spun into woven mats had improved mechanical and biodegradation characteristics with the desired fiber diameters and pore sizes to promote cell growth.

Poly(ε-caprolactone) is a candidate aliphatic polyester for controlled release and soft tissue engineering due to its rubbery state at body temperature due to its T_g of −60°C [11]. Diffusion of small molecules is possible through encapsulation and thin films of this polymer, of which the rate of diffusion, mechanical properties, and other physicochemical characteristics can be controlled via copolymerization and the introduction of nanofillers. Poly(urethane urea) PUU segmented block copolymers have applications in ventricular assist devices and total artificial hearts as blood sacs, but have disadvantages due to high permeability of air and water vapor through the soft segments of poly(tetramethylene oxide) in the block copolymer. Polymer hydrogels based on polypeptides, (meth)acrylated gelatins, carboxymethyl guar gum, poly(acrylamide) (PAAm), poly(vinyl alcohol) (PVA), and aliphatic polyesters with varying nanoparticles are extensively used in biomedical applications [11,12].

Carbon nanotubes (CNT) and graphene represent most of the carbon-based fillers being studied for nanocomposite hydrogels (NCHs). The applications of these carbon-based nanomaterials are actuators, biosensors, tissue engineered scaffolds, drug delivery, and biomedical devices [12–16]. CNTs can exist in multi-wall or single-wall forms, and are highly hydrophobic and self-aggregate in solution due to their sp^2-hybridization. These physicochemical characteristics limit their biomedical applications in hydrogels. However, surface modifications with polar moieties such as amine, hydroxyl, and carboxyl groups have circumvented these issues establishing NCHs reinforced with CNTs as potential conductive tissues such as nerves, muscles, and cardiac tissues. Similarly, graphene has been functionalized to graphene oxide (GO) to amend solubility issues and surface energy interactions with hydrophilic polymers leading to GO as an efficient nanocarrier for applications ranging from drug/gene delivery, biological sensing and imaging, antibacterial materials, to biocompatible scaffold for cell culture.

Exploring graphene for cancer therapy has investigated in vivo tumor uptake and photothermal therapy with PEGylated GO using xenograft tumor mouse models. The efficient uptake into tumor cells of PEGylated GO is caused by the EPR (enhanced permeability and retention effect), and could be explained by PEGylating the highly hydrophobic GO and functionalizing it to provide it with the proper physiological stability and biocompatibility [16]. Subsequent irradiation with low-power, near-infrared (NIR) lasers has been highly destructive to the tumor. Further approaches have investigated the photothermal anti-cancer activity and photodynamic therapy applications of coated graphene/CNT nanoparticles with polyvinylpyrrolidone (PVP), and folic or sulfonic acid conjugated GO with photosensitizers (PS), respectively. These structural and processing modifications of GO show promising opportunities to load and deliver drugs, genes, PS, and so forth for cancer and disease treatment.

Biosensing utilizes pristine graphene, GO, doped graphene, and chemically reduced GO (rGO) [16]. GO-based biosensors presently used are fluorescence resonance energy transfer (FRET), field effect transistors (FET), self-assembling graphene-biomolecules for detection of DNA and other molecules, graphene based nanoplatforms for matrix-assisted laser desorption/ionization time-of-flight mass spectrometry (MALDI-TOF MS), and novel biosensors that utilize the large surface area, good electrical conductivity, and capability for loading biomolecules into the nanomaterial. Bioimaging with graphene has mostly involved bottom-up approaches in the preparation of graphene quantum dots (GQD) that possess a size of 10 nm or less. GQDs exhibit intrinsic fluorescence and have the potential for cellular imaging due to their excellent biocompatibility, physiological solubility, and lack of surface processes/modifications required for use. Fig. 7.5 highlights the possibilities of GQDs for cellular imaging.

Hydrogels lend themselves as a new and attractive polymeric matrix for biomimetics due to their ability to mimic physical, chemical, electrical, and biological tissues. The introduction of nanoparticles or nanostructures to obtain nanocomposite hydrogels (NCHs) with improved mechanical properties, multiple functionalities, responsiveness to external stimuli, and capabilities for dynamic interactions between matrices and cells embodies most of the current research trends in biomedical applications [11–18]. Hydrogels can be either synthetic or natural, with physical or chemical crosslinking and the presence of hydrophilic groups such as amido, amino, carboxyl, and hydroxyl groups along the polymer backbone, enabling high water retention. Fig. 7.6 iterates the current research focus, the processing parameters, and the influence on structure and the properties of NCH and their resulting performance in the medical field.

Biomimetic NCHs have been utilized in the development of scaffolds for tissue engineering through preparation of analogue extracellular matrices (ECMs) synthesized from methacrylated gelatin (GelMA) and CNTs [12]. Cell differentiation can be steered via stiffness; as an example, multi-wall COOH-functionalized CNTs have been coated with GelMA to mimic the biological environment of the ECM from the perspective of mechanical properties. Favorable surface energy interactions between the hydrophobic polypeptide chains lead to a coating of GelMA on the sidewalls of CNTs, enabling the synthesis of CNT-GelMA hybrid hydrogels. CNT concentrations, as low as 0.5%, lead to a three-fold increase in the tensile modulus of the hydrogel, and the presence

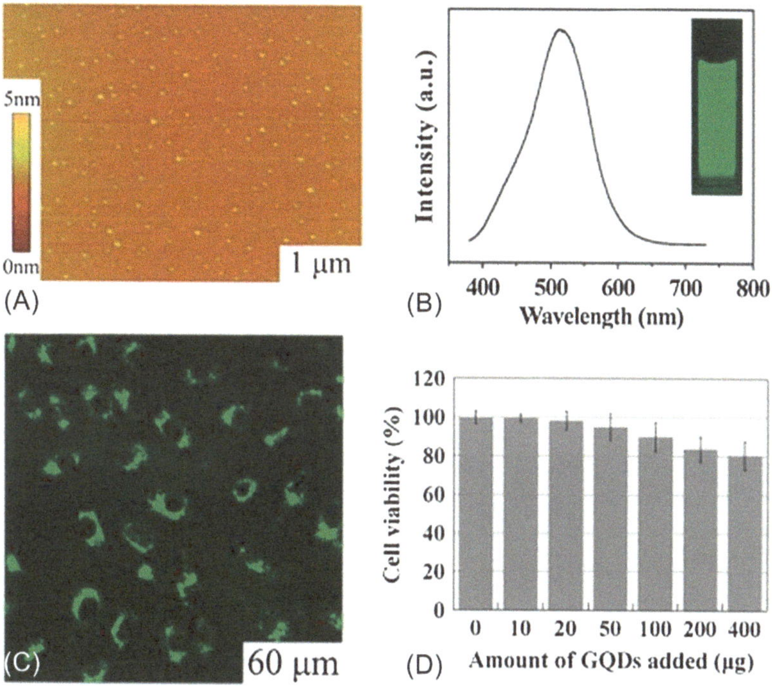

FIG. 7.5 (A) Atomic Force Microscopy (AFM) image of GQDs; (B) Fluorescence spectra of aqueous solution of GQDs excited at 375 nm. Inset: aqueous solution of GQDs under UV light; (C) Cellular imaging of GQDs imaged under 405 nm; (D) Cytotoxicity of GQDds [16].

of CNTs reduce the external voltage necessary to induce cell beating due to the increase in electrical conductivity attributed to CNTs.

Polymeric nanoparticles, including NCHs, are ideal polymer systems that can be tuned for obtaining enhanced mechanical properties and/or for controlled drug delivery [11–18]. The polymer delivery system can include micelles, core-shell particles, dendrimers, and hyper-branched polymers, as illustrated in Fig. 7.7. An example of micelles that will enhance the solubility of hydrophobic drugs involves encapsulating erythromycin (a hydrophobic antibiotic) with a diacrylate macromer based on triblock copolymers of poly(ethylene oxide)-poly(propylene oxide)-poly(ethylene oxide) (PEO-PPO-PEO). The hydrophobicity of PPO allows for efficient loading of the hydrophobic drug, while enhancing water solubility with the hydrophilic PEO domains on the polymer backbone. Photopolymerization under UV light has been used to form the final NCH with the acrylate functionalities at the site where polymerization occurs via chain extension.

Dendritic polymers such as poly(amidoamine) (PAMAM), with their amide and amine repetitively branched subunits, resemble the chemical structure of

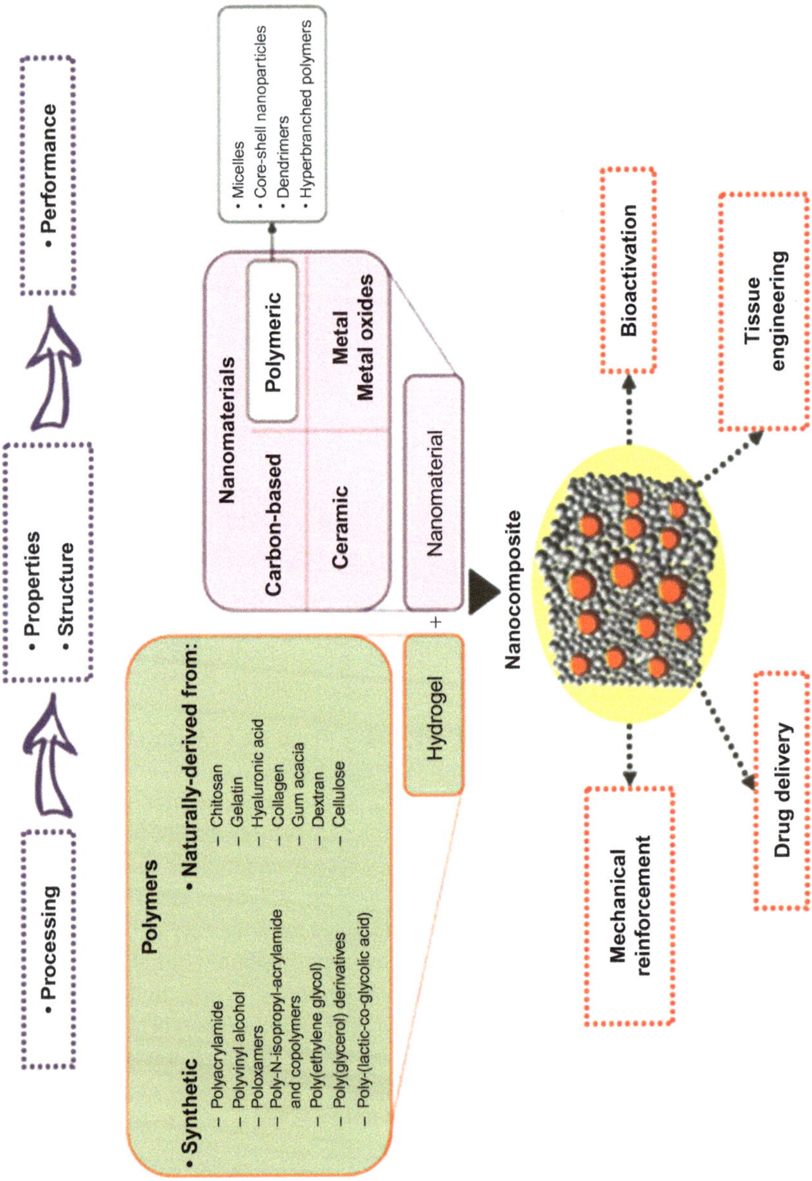

FIG. 7.6 NCHs and their performance in the medical field [12].

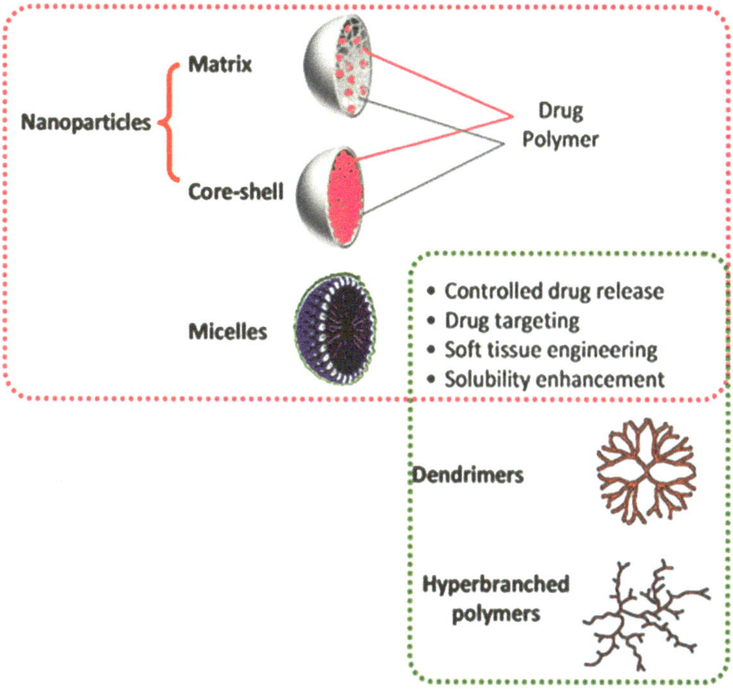

FIG. 7.7 Hydrogels represent a delivery system for controlled release of drugs and an opportunity to increase the mechanical properties. Amphiphilicity can enhance the solubility of sparingly soluble drugs [12].

globular proteins [12]. The introduction of PAMAM nanoparticles into collagen scaffolds has been reported to increase the mechanical properties while subsequently enhancing the proliferation of human conjuctivital fibroblasts in the hydrogels. Hyperbranched polyesters have demonstrated their ability to produce NCHs by UV irradiation, forming a porous and controlled polymer encapsulation of hydrophilic dexamethasone acetate. The amphiphilic encapsulation shows high efficacy with greater control than conventional controlled release hydrogels. The porous structure, coupled with the greater controlled release properties, makes these ideal candidates for scaffolds in tissue engineering. Other examples of hydrogel scaffolds that have proven useful for cartilage repair include tri-block copolymers consisting of a PEG core and methacrylated poly (glycerol succinic acid) dendrimer terminal block. The dendrimer concentration lends itself to tailoring the hydrogel stiffness and hydration/degradation features of the polymer. Further testing of mechanical properties portray a higher dendrimer concentration that corresponds to a transition from elastic to viscoelastic behavior, as evidenced by the loss angle (δ) increasing from 6 degree at the lowest concentration to 71 degree at the highest concentration.

Lefnaoui et al. [17] have developed antihistaminic transdermal films from alginate-chitosan polyelectrolyte complexes (PEC) and studied the transdermal drug delivery systems based on varying the ratios of chitosan and sodium alginate in the hydrogel. The goal is to optimize critical parameters for transdermal therapeutic systems, which include formulation (polymer ratio, plasticizers, permeability enhancers), skin compatibility and bioadhesion, drug release characteristics, and its permeation through skin. The effects of the polymers on the physico-mechanical properties, as well as the drug performance and permeation characteristics, have been subsequently studied. The authors find that the hydrophilic polymer concentration is directly proportional to the moisture content of the resulting films. It is important to keep the moisture content low for stability and protection from possible microbial contamination. Importantly, the researchers utilized FTIR to demonstrate the compatibility between Ketotifen fumarate (KF) (antihistamine) and the sodium alginate, or chitosan, when mixed together, as illustrated in Fig. 7.8. The IR spectra indicate no change in the major peak wave numbers of KF (a) when mixing with sodium alginate (b) or chitosan (c). This is important to monitor to ensure that the drug efficacy remains unchanged in the delivery system.

Tensile testing indicates an increase in chitosan concentration that correlates to an increase in tensile strength due to its rigidity and lack of elasticity. Mechanical strength and flexibility can be improved by PEC as it promotes additional bonding between polysaccharide chains [17]. Bioadhesion is decreased when the formulated films incorporate more sodium alginate, as it is postulated that the surface amino groups of chitosan would interact with the carboxylic ion groups of sodium alginate; hence the lack of surface amino groups (with higher concentrations of sodium alginate) correspond to less bioadhesive strength as the protonated (cationic) chitosan (see Table 7.2) electrostatically interacts with negatively charged skin, leading to greater bioadhesion strength.

During in vitro studies of controlled release in films, it has been observed that the biopolymer concentration in the formulation has a direct impact on the release rate, as the greater chitosan concentrations are observed to swell in the surrounding medium, accelerating release of KF by diffusion through the porous network of the viscous gel layers [17]. Additionally, equal concentrations of the two biopolymers, formed by PEC, are found to sustain drug release, and it is believed that a denser and diffusion resistant polymer network would control the release rate and deliver the drug at ideal therapeutically effective concentrations for prolonged periods. Fig. 7.9 exemplifies the influence of the composition of biopolymer films on release rate. Permeation enhancers (surfactants that form micelles) and plasticizer propylene glycol added to the formulations are found to increase drug release, which is best explained by a diffusion-partition-solubility theory [18].

The incorporation of nanomaterials into biopolymers is an attractive, yet challenging, approach to tailor the mechanical properties and design new

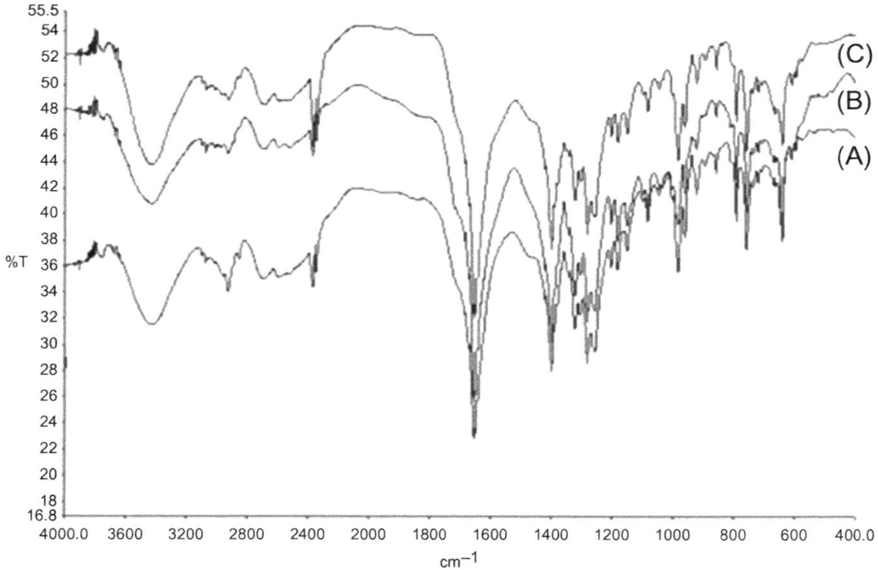

FIG. 7.8 FTIR spectra showing (a) KF drug, (b) Physical mixture of KF and sodium alginate, (c) Physical mixture of KF and chitosan [17]. The IR spectra indicate that the drug and polymers are compatible as no changes in peak positions are observed when KF is mixed with the polymers.

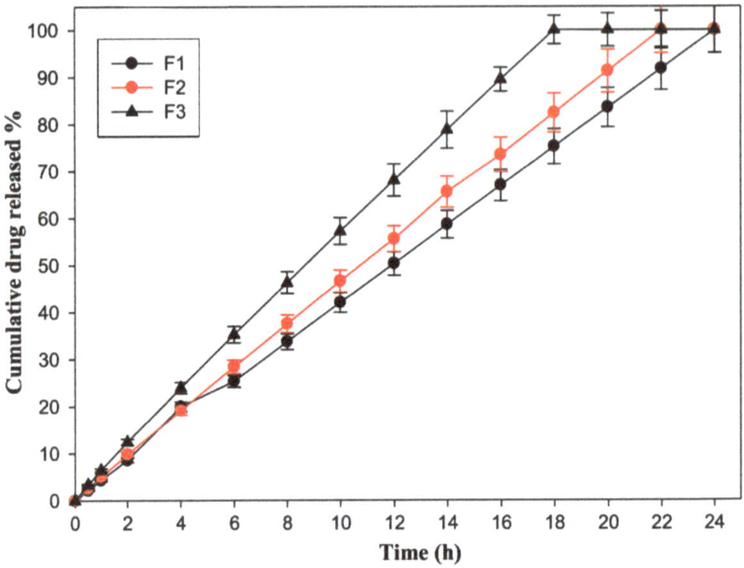

FIG. 7.9 Drug release from transdermal film formulations. F1 corresponds to a biopolymer ratio (chitosan:alginate) of 1:1, F2 corresponds to a ratio of 1:2, and F3 corresponds to a ratio of 2:1 [17].

innovative biomimetic materials with responsiveness to mechanical, thermal, magnetic, and electrical stimuli [11–18]. Researchers in the area of bionanocomposites have synthesized and functionalized various materials, as a proof of concept, with in vitro and in vivo testing. Much work will now be focused on carbon based nanomaterials, such as graphene/GO-based scaffolds for cell culture, drug delivery systems, understanding the exceptional physicochemical properties of the materials and their respective functionalization, controllable modifications and synthesis routes and techniques. Biosensing and bioimaging have massive potential, given the biocompatibility, the possibilities of tuning the electrical properties, and the integration of GQDs [16].

7.3 SMART BIOPOLYMERS, SHAPE MEMORY POLYMERS, AND SELF-HEALING MATERIALS

In addition to biopolymers being biodegradable and biocompatible, smart biopolymers are gaining attention due to their capability to undergo fast and reversible changes that are triggered by changes in temperature, pH, humidity, intensity of light, and electrical or magnetic fields [11–19]. Furthermore, some of these smart biopolymers exhibit self-healing properties, or are shape memory polymers (SMPs) [20–27]. Smart biopolymers are classified based on the type of stimuli and functionality of the polymers, and are grouped as pH-sensitive, thermosensitive, and stimuli responsive smart polymers. Specifically, the SMP phenomena arises from a dual-segment system, where one segment is highly elastic and the other segment is capable of reducing its stiffness in the presence of external stimuli, as listed herein. SMPs have the capability in the biomedical field for novel medical devices such as smart sutures, whose temporary shape can be obtained externally with minimal invasion to the body, then achieve the desired knot in response to stress and temperature, as the suture will shrink and tighten, as exemplified in Fig. 7.10. SMPs have various other applications in tissue engineering, controlled drug delivery systems, implants, and wound covers [20–27].

Self-healing polymers for biomedical areas have focused on all classes of polymers, from thermosets to thermoplastics to elastomers [20]. Self-healing is a synthetic route that attempts to repair the inevitable degradation, damage, and failure that biological tissues are subjected to with faster response times than the natural response of biological systems (Fig. 7.11). The three conceptual approaches include: capsule-based healing systems, vascular healing systems, and intrinsic healing polymers, with each approach having benefits and challenges. The capsule-based approach is easily integrated into a polymer, but suffers limitations from capsules being depleted after a single damage effect. Vascular approaches hold immense potential for large damage volumes over

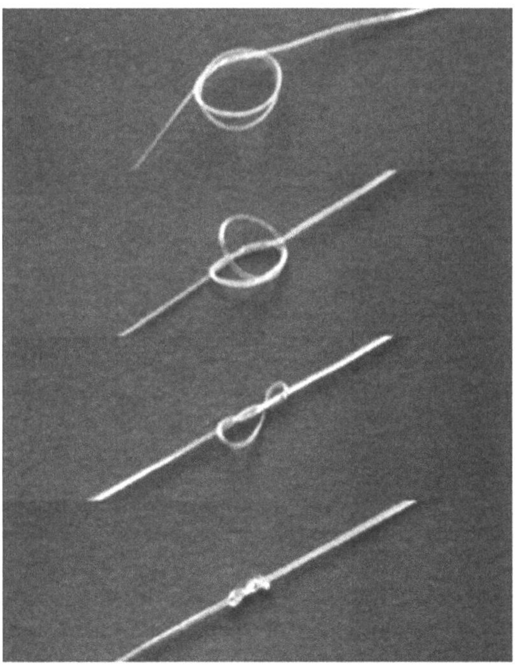

FIG. 7.10 Fiber of thermoplastic shape-memory polymer with the potential as a surgical suture. The suture can be processed at room temperature, but will shrink and tighten the knot above its transition temperature (T_{trans}) [19].

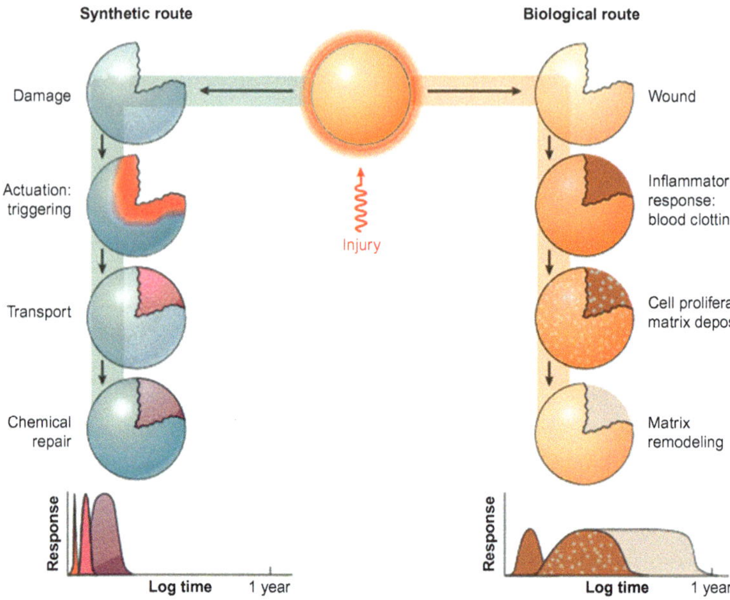

FIG. 7.11 Details of the synthetic route (*left*) and the biological route (*right*) to damage healing. The long-term response of matrix remodeling observed in biological systems is postulated to be shortened through the introduction of synthetic materials (e.g., the rupture of embedded microscapusles), which involves transporting chemical species to the damage site at a rapid rate for repair [20].

multiple events, but suffer challenges of integration into existing materials. Intrinsic healing of the polymer matrix is less complex, as the matrix is inherently self-healing, and does not require integration of the healing-agent compatibility; but meeting the desired mechanical, chemical, and optical properties performance is often challenging, and requires intimate material contact for healing. Intrinsic self-healing approaches include those based on:

- Reversible reactions—such as transformation from monomeric to cross-linked state via external energy.
- Meltable thermoplastic additives—additives dispersed within a matrix material, which subsequently melts and fills the damage, cracks, voids, and so forth.
- Ionomeric self-healing—ionic segments of copolymers that reversibly cross-link by stimulation from temperature or UV irradiation.
- Supramolecular self-healing materials—elastomeric polymers that form reversible hydrogen bonds due to their rubbery nature, allowing for close contact of molecules after damage that can reform hydrogen bonds.

- Self-healing via molecular diffusion—time, temperature, and pH dependence of compounds on mobility has been used to prevent corrosion and induce void closure, surface interaction, and molecular entanglement between damaged faces.

Recently, significant research has focused on hydrogels that are constructed by dynamic chemical linkages due to higher mechanical strength and properties that are similar to human soft tissues [21–27]. Self-healing hydrogels restore their integrity and property/function after damage and lend themselves as ideal candidates to be injected in vivo to deliver high concentrated drugs or cells avoiding delivery failure or unwanted release. Li et al. [21] have developed a series of chitosan-based self-healing hydrogels using a benzaldehyde terminated PEG to crosslink chitosan and its derivatives. The aldehyde groups react reversibly with the amine groups on chitosan to form a Schiff base (imine), resulting in a multi-responsive (pH, amino acids, enzymes, and vitamin B6 derivatives) self-healing hydrogel.

Yang et al. [24] have used glycol-chitosan (enhanced hydrophilicity) and reacted it with the benzaldehyde terminated PEG to investigate the hydrogel as a carrier system for tumor chemotherapy. The in vivo evaluation confirms that a highly concentrated anti-tumor drug (Taxol) was delivered with a steady release rate at a safe level. The self-healing hydrogel can adapt to the intratumor structure and is capable of regeneration, circumventing the accelerated release that is associated with the destruction of the hydrogel upon injection. The regenerated hydrogel presents a similar storage modulus to the native hydrogel and outperforms controls (direct injection or non-self-healable thermal-sensitive hydrogel carrier), indicating that this smart carrier could be a promising solution for tumor chemotherapy due to its high efficiency, low toxicity, and requirement of fewer injections (Fig. 7.12).

Buono et al. [25] have reported on the preparation and characterization of self-healing materials based on lignin. Lignin, obtained by a soda process (soda lignin—SL), was esterified with 11-maleimidoundecylenic acid and polymerized through Diels-Alder cycloaddition reactions with different polyfunctional furan linkers (Fig. 7.13). This allowed for on-demand depolymerization-repolymerization, setting the stage for thermo-reversible and self-healing materials with linker functionality and mobility that is responsible for mechanical properties and reprocessing of the polymer. The polymers were processed in films and analyzed by thermogravimetric analysis (TGA), differential scanning calorimetry (DSC), dynamic mechanical analysis (DMA), tensile testing, and optical microscopy.

Khamrai et al. [26] prepared self-healable polyelectrolyte films by crosslinking chitosan with bacterial cellulose and grafting poly(acrylic acid) onto the surface. Self-healing of the non-toxic films at physiological pH was established via ionic interactions between the matrix and cellulose. Further, the composite films demonstrated the capability to encapsulate the antimicrobial and

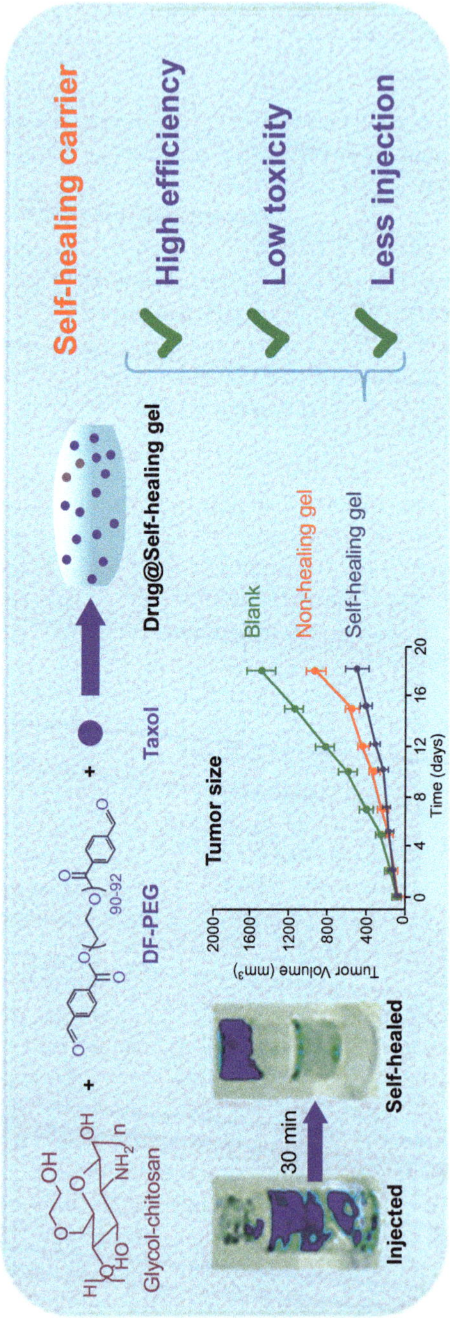

FIG. 7.12 Self-healing hydrogel demonstrates as a good carrier system for anti-tumor drug [24].

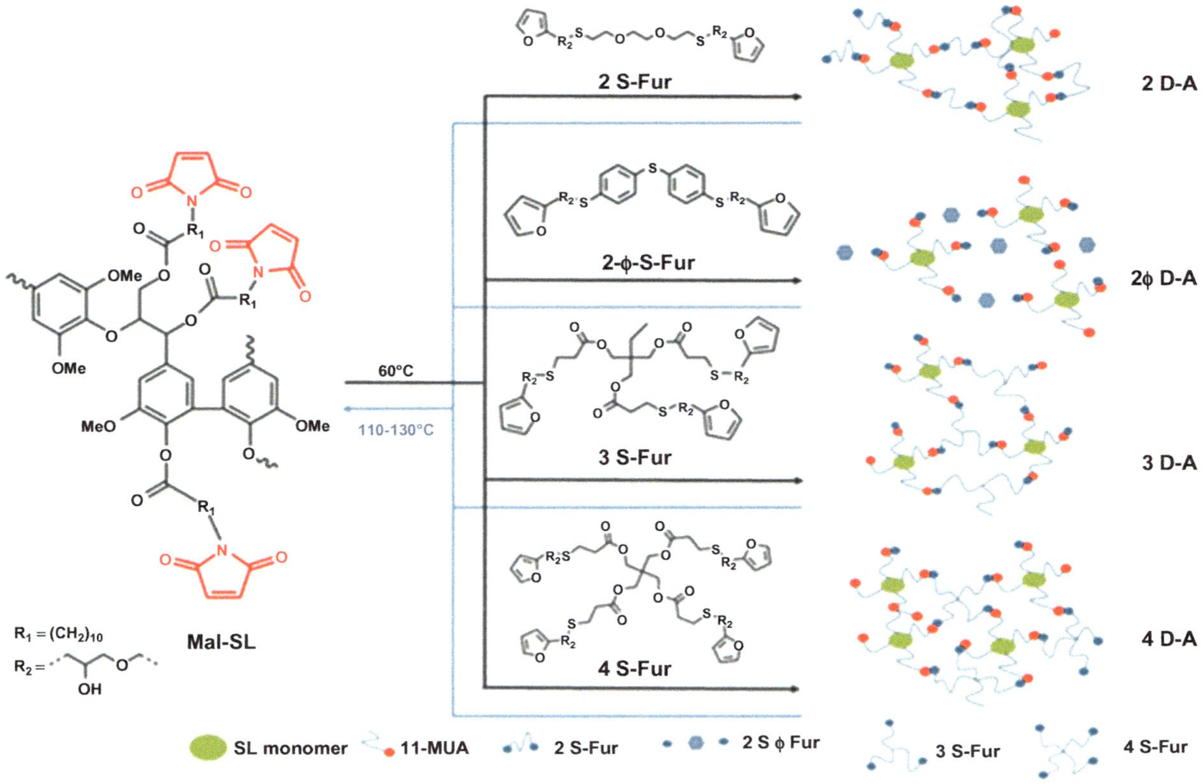

FIG. 7.13 Thermo-reversible D-A polymerization of Mal-SL with the polyfunctional furan linkers [25].

wound healing drug, curcumin, demonstrating efficiency in wound dressing applications. Bankoti et al. [27] investigated the effect of phase separation utilizing polyurethane diol dispersion and anti-bacterial chitosan microporous hydrogel scaffolds (MHS). The mechanical properties and in vitro degradation rendered the MHS suitable for wound healing, with accelerated healing that has been demonstrated in vivo with Wistar rates compared to a commercial dressing, Tegaderm.

7.4 CONCLUSIONS

Biomaterials will continue to play a key role in addressing the growing demands of enhanced human life expectancy and quality of life. Polymer composites, with their attractive properties such as excellent biocompatibility, reliable and reproducible mechanical properties, and self-healing, offer significant opportunities in a variety of human health-related applications.

REFERENCES

[1] R.O. Ebewele, Polymer Science and Technology, CRC Press, Boca Ratón, FL, 2000. Ch. 9.
[2] W.D. Callister, D.G. Rethwisch, Materials Science and Engineering: An Introduction, eighth ed., John Wiley & Sons, Hoboken, NJ, 2014. Ch. 16.
[3] Anderson J. 2011 "WINGS: From the Wright Brothers to the Present" Smithsonian National Air and Space Museum. https://airandspace.si.edu/stories/editorial/wings-wright-brothers-present (Accessed 8 January 2018).
[4] B. Clarke, Normal bone anatomy and physiology, Clin. J. Am. Soc. Nephrol. 3 (Suppl 3) (2008) S131–9.
[5] S.C. Cowin, L. Cardoso, Blood and interstitial flow in the hierarchical pore space architecture of bone tissue, J. Biomech. 48 (5) (2015) 842–854.
[6] S.R. Knight, R. Aujla, S.P. Biswas, Total hip arthroplasty—over 100 years of operative history, Orthop. Rev. 3 (2) (2011) e16. 72–74.
[7] AAOS. 2014. Revision Total Hip Replacement-OrthoInfo—AAOS. http://orthoinfo.aaos.org/topic.cfm?topic=A00774 (Accessed 8 January 2018).
[8] C.M. Abraham, A brief historical perspective on dental implants, their surface coatings and treatments, Open Dent. J. 8 (1) (2014) 50–55.
[9] G.I. Howling, H. Sakoda, A. Antonarulrajah, H. Marrs, T.D. Stewart, S. Appleyard, B. Rand, J. Fisher, E. Ingham, Biological response to wear debris generated in carbon based composites as potential bearing surfaces for artificial hip joints, J. Biomed. Mater. Res. 67B (2) (2003) 758–764.
[10] D.J. Hak, C. Mauffrey, D. Seligson, B. Lindeque, Use of carbon-fiber-reinforced composite implants in orthopedic surgery, Orthopedics 37 (12) (2014) 825–830.
[11] R.A. Hule, D.J. Pochan, Polymer nanocomposites for biomedical applications, MRS Bull. 32 (2007) 354–358.
[12] M. Biondi, A. Borzacciello, M. Laura, L. Ambrosio, Nanoparticle-integrated hydrogels as multifunctional composite materials for biomedical applications, Gels 1 (2015) 162–178.

[13] N.S. Sambudi, M.G. Kim, S.B. Park, The formation of web-like connection among electrospun chitosan/PVA fiber network by the reinforcement of ellipsoidal calcium carbonate, Mater. Sci. Eng. C 60 (2016) 518–525.

[14] A.K. Gaharwar, N.A. Peppas, A. Khademhosseini, Nanocomposite hydrogels for biomedical applications, Biotechnol. Bioeng. 111 (3) (2014) 441–453.

[15] V.I. Kalita, D.I. Komlev, V.S. Komlev, A.A. Radyuk, The shear strength of three-dimensional capillary-poroustitanium coatings for intraosseous implants, Mater. Sci. Eng. C 60 (2016) 255–259.

[16] H. Shen, M. Liu, H. He, L. Zhang, J. Huang, Y. Chong, J. Dai, Z. Zhang, PEGylated graphene oxide-mediated protein delivery for cell function regulation, ACS Appl. Mater. Interfaces 4 (11) (2012) 6317–6323.

[17] S. Lefnaoui, N. Moulai-Mostefa, M.M. Yahoum, S.N. Gasmi, Design of antihistaminic transdermal films based on alginate-chitosan polyelectrolyte complexes: characterization and permeation studies, Drug Dev. Ind. Pharm. (2017) 1–12.

[18] M.E. Lane, Skin penetration enhancers, Int. J. Pharm. 447 (1-2) (2013) 12–21.

[19] N.A. Pattanashetti, G.B. Heggannavar, M.Y. Kariduraganavar, Smart biopolymers and their biomedical applications, Procedia Manuf. 12) (2017) 263–279.

[20] B.J. Blaiszik, S.L.B. Kramer, S.C. Olugebefola, J.S. Moore, N.R. Sottos, S.R. White, Self-healing polymers and composites, Annu. Rev. Mater. Res. 40 (2010) 179–211.

[21] Y. Li, X. Wang, Y. Wei, L. Tao, Chitosan-based self-healing hydrogel for bioapplications, Chin. Chem. Lett. 28 (11) (2017) 2053–2057.

[22] L. Yang, Y. Li, Y. Gou, X. Wang, X. Zhao, L. Tao, Improving tumor chemotherapy effect using an injectable self-healing hydrogel as drug carrier, Polym. Chem. 8 (2017) 5071–5076.

[23] D.L. Taylor, M. In Het Panhuis, Self-healing hydrogels, Adv. Mater. 28 (41) (2016) 9060–9093.

[24] Z. Wei, J.H. Yang, J. Zhou, F. Xu, M. Zrínyi, P.H. Dussault, Y. Osada, Y.M. Chen, Self-healing gels based on constitutional dynamic chemistry and their potential applications, Chem. Soc. Rev. 43 (23) (2014) 8114–8131.

[25] P. Buono, A. Duval, L. Averous, Y. Habibi, Thermally healable and remendable lignin-based materials through diels-alder click polymerization, Polymer 133 (2017) 78–88.

[26] M. Khamrai, S.L. Banerjee, P.P. Kundu, Modified bacterial cellulose based self-healable polyeloctrolyte film for wound dressing application, Carbohydr. Polym. 174 (2017) 580–590.

[27] K. Bankoti, A.P. Rameshbabu, S. Datta, P.P. Maity, P. Goswami, P. Datta, S.K. Ghosh, A. Mitra, S. Dhara, Accelerated healing of full thickness dermal wounds by macroporous waterborne polyurethane-chitosan hydrogel scaffolds, Mater. Sci. Eng. C 81 (2017) 133–143.

Chapter 8

Polymer-Based Nanocomposites with High Dielectric Permittivity

Shaohua Jiang*,[1], Li Jin[†], Haoqing Hou[‡] and Lin Zhang[†,1]
[*]*College of Materials Science and Engineering, Nanjing Forestry University, Nanjing, China,* [†]*Electronic Materials Research Laboratory, Key Laboratory of the Ministry of Education & International Center for Dielectric Research, School of Electronic and Information Engineering, Xi'an Jiaotong University, Xi'an, China,* [‡]*Department of Chemistry and Chemical Engineering, Jiangxi Normal University, Nanchang, China*
[1]*Corresponding authors: shaohua.jiang@njfu.edu.cn; zhanglin.materials@gmail.com*

Chapter Outline

8.1 INTRODUCTION OF DIELECTRIC MATERIALS

Dielectrics have been widely used in current electronic and electrical industries, in capacitors, insulation, sensors, actuators, sonars, gate dielectrics,

Polymer-Based Multifunctional Nanocomposites and Their Applications
https://doi.org/10.1016/B978-0-12-815067-2.00008-1

high-frequency transducers, electro-optical devices, and microelectromechanical systems. For a certain dielectric, the performance is characterized by its dielectric permittivity, dielectric loss, breakdown field (i.e., dielectric strength, E_b), and polarization-electric field (P-E) relationship. In this section, the fundamentals of dielectrics and well known dielectric materials are introduced. In order to make the best use of different dielectric materials, polymer-based 0-3 dielectric nanocomposites have been extensively studied, because nanocomposites offer the advantages of being compact, low-cost, lightweight, flexible devices. According to the different types of filler selected, polymer-based nanocomposites can be classified into two types: dielectric-dielectric nanocomposites, and conductor-dielectric nanocomposites. They are discussed as follows.

8.1.1 Fundamentals of Dielectrics

8.1.1.1 Capacitance and Dielectrics

When two metal plates with the area of A separated by a distance d in a vacuum are connected with potential V, the surface charge density σ_{vac} and capacitance C at equilibrium status can be expressed as [1,2]:

$$\sigma_{vac} = \frac{Q_{vac}}{A} = \frac{\varepsilon_0 V}{d} = \varepsilon_0 E \quad (8.1)$$

$$C_{vac} = \frac{\varepsilon_0 A}{d} \quad (8.2)$$

where Q is the net amount of charge in one of the two metal plates (the unit is C), ε_0 is the dielectric permittivity in vacuum ($8.8542 \times 10^{-12}\,C^2/Jm$), E is the external electric field applied on the space between two plates, which can be obtained by V/d (the unit is V/m). C is the capacitance expressed in farad (F). A dielectric is an electrical insulator that can respond to an external electric stimulation. When a dielectric is placed between two metal plates, the surface charge density will be changed as:

$$\sigma_{die} = \frac{Q_{die}}{A} = \frac{\varepsilon_0 \varepsilon_r V}{d} = \varepsilon' \varepsilon_0 E = \sigma_{vac} + \sigma_{pol} \quad (8.3)$$

$$C = \frac{\varepsilon_0 \varepsilon_r A}{d} \quad (8.4)$$

where ε_r is the relative dielectric permittivity of the dielectric and σ_{pol} is the excess charge per unit surface area over the charge density in vacuum conditions due to the presence of the dielectric, as shown in Fig. 8.1. σ_{pol} has the same value and the same unit as the electrical polarization P of the dielectric. The surface charge density on the metal plate is also defined as the electric displacement D, which can be written as:

$$\sigma_{die} = D = \varepsilon_0 E + P \quad (8.5)$$

When an AC electric field is applied on the dielectric material, the dielectric permittivity ε^* shows a complex form: [3]

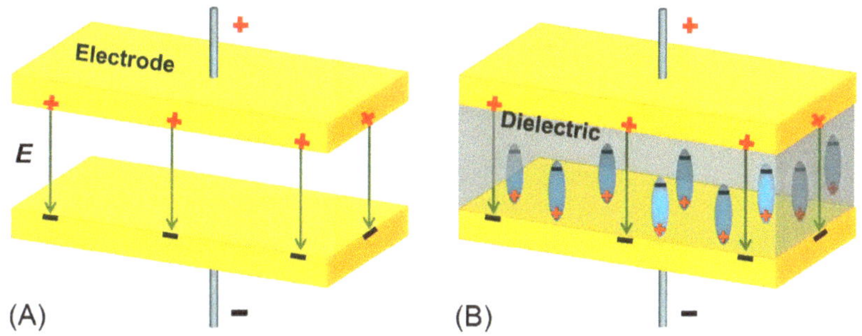

FIG. 8.1 (A) A parallel-plate capacitor of area *A* and separation *d* in a vacuum with an applied electric field. (B) A parallel-plate capacitor of area *A* and separation *d* within a dielectric with an applied electric field.

$$D = \varepsilon^* E = \varepsilon_r^* \varepsilon_0 E \tag{8.6}$$

$$\varepsilon_r^* = \frac{\varepsilon^*}{\varepsilon_0} = \varepsilon_r' - j\varepsilon_r'' \tag{8.7}$$

where ε_r^* is relative dielectric permittivity, which includes real part ε_r' (or dielectric permittivity) and imaginary part ε_r''. The phase difference between D and E is defined as the dielectric loss ($\tan\delta$):

$$\tan\delta = \frac{\varepsilon_r''}{\varepsilon_r'} \tag{8.8}$$

where δ reflects the phase difference 90-δ between the D and E.

8.1.1.2 Polarization Mechanisms

When an electric field is applied on the dielectric, a dipole moment is caused by the displacement of the electron cloud of an isolated neutral atom, which results in a polarization. Polarization is a frequency-dependent phenomenon, which may originate from different mechanisms. Generally, in a homogeneous material, the dielectric responses have four major mechanisms. The dielectric response of each polarization mechanism changes with frequency in different patterns. These polarizations can be subdivided into two regimes: the electronic and ionic polarization result in a resonant dielectric response, while the response of the orientational and space charge is relaxation polarization, as shown in Fig. 8.2 [1,3].

(1) *Electronic polarization*: The electric field causes the displacement of the negatively charged electron cloud from the inner positive nucleus. This polarization has a response time of $10^{-14} \sim 10^{-16} s$ and the polarization is independent of temperature. Electronic polarization occurs at most frequencies, and drops off at very high frequencies in all solids, liquids, and gases.

(2) *Ionic polarization*: Similar to electronic polarization, ionic charges can also respond to an applied electric field. The positive and negative charges

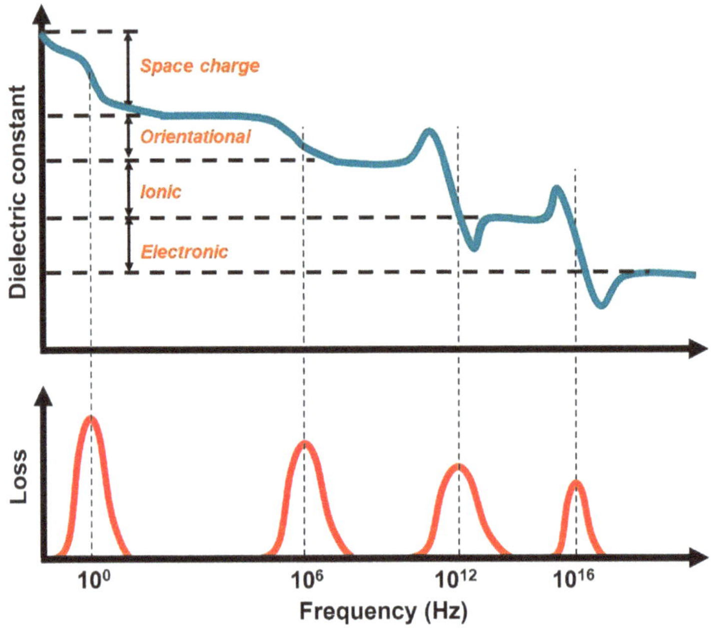

FIG. 8.2 Frequency dependence of the dielectric permittivity and loss for a dielectric with four different types of dielectric responses.

are displaced toward each other under an external electric field. This polarization has a response time of $10^{-12} \sim 10^{-13}\,s$ and the polarization is almost independent of temperature.

(3) *Orientational Polarization*: If dipoles exist in a material, the electric field E will generate a torque on each dipole, and the torque will typically cause dipoles to be aligned along the electric field direction. This polarization has a response time of $1 \sim 10^{-8}\,s$ and the polarization is strongly dependent on temperature. Due to orientation, polarization can occur at a very low frequency, and can affect the capacitive and insulative properties of materials, such as ceramics, and so forth.

(4) *Space charge polarization*: When the space charge appears in the dielectric, the electric field E would generate a force on the particle, which would separate the positive and negative charges. Therefore, some dipole moments are formed. This polarization has a response time that is larger than $10^{-5} \sim 1\,s$, and the polarization is strongly dependent on temperature.

8.1.1.3 Capacitors for Energy Storage

The capacitor can act as an energy storage device, or an electrical buffer, by blocking DC signals and storing surges of charges that would damage the electrical devices. For insulative applications, it is desirable to have a low dielectric

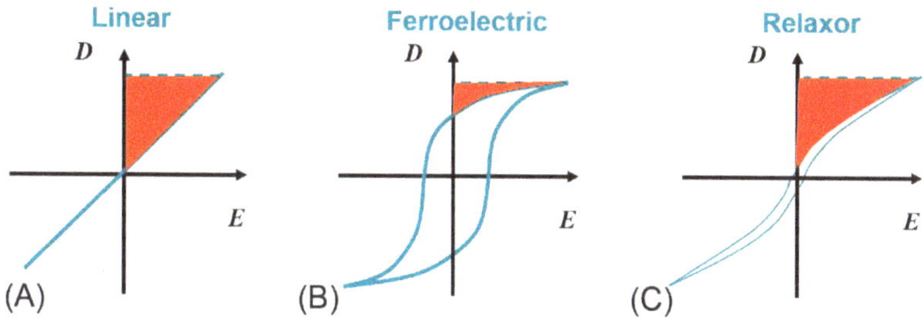

FIG. 8.3 Relationship of electric displacement vs. electric field: (A) linear, (B) normal ferroelectric, and (C) relaxor ferroelectric behaviors. The shaded areas represent the discharged energy density.

permittivity. However, for the capacitive purpose, a high relative dielectric permittivity is preferred, with low loss. In this chapter, dielectric material as a capacitor with high dielectric permittivity for energy storage is discussed.

The energy stored in a dielectric material under an electric field E can be expressed by the shadow area in Fig. 8.3, in which different relationships between E and D are presented: [4,5]

$$U_e = \int \vec{E} \cdot d\vec{D} \tag{8.9}$$

where U_e is energy storage density defined as the energy stored in a unit volume (J/m^3).

Based on Fig. 8.3 and Eq. (8.9), it can be concluded that a higher D and a higher E are very important in achieving a higher energy density. Additionally, the curvature of D versus E is also very critical to the U_e. For linear dielectrics, it is well known that the energy density of a dielectric material is proportional to the product of permittivity, and the square of the applied electric field, which can be expressed as Eq. (8.10). A higher dielectric strength, which would allow a high electric field to be applied and a higher permittivity, are highly desirable for the dielectrics used in energy storage devices.

$$U_e = \frac{1}{2}\varepsilon_r'\varepsilon_0\vec{E}^2 \tag{8.10}$$

In a word, for application, the dielectric properties of a material can be characterized by three very important parameters: dielectric permittivity ε, dielectric loss $\tan\delta$, and dielectric strength E_b applied on that material.

8.1.2 Dielectric Materials

According to various structures and dielectric properties, many dielectrics have been widely used in the industry for different applications [2,6]. By their structure, dielectric materials can be classified into two major categories: nonpolar materials and polar materials. By their composition, dielectrics can be classified into inorganic and organic materials [7].

8.1.2.1 Nonpolar Materials

A nonpolar material is a material with a molecule that does not have a permanent dipole moment [2]. The charges in a nonpolar molecule are equally distributed. When an electric field is applied on a nonpolar molecule, opposite polarities will occur, because the positive and the negative charges experience forces in opposite directions. The electron cloud of each nonpolar molecule will be displaced in the direction of the attraction, and the corresponding displacement will balance the internal electric force.

For both nonpolar inorganic and organic materials, they usually have a low dielectric permittivity compared with polar dielectric materials, such as silicon ($\varepsilon_r = 3.7$), diamonds ($\varepsilon_r = 5.5$–10), quartz ($\varepsilon_r = 4.4$), parfin ($\varepsilon_r = 1.9 \sim 2.5$), carbon tetrachloride ($\varepsilon_r = 2.0$), and so forth [2]. $(Ta_2O_5)_{0.92}$-$(TiO_2)_{0.08}$ has a dielectric permittivity value of 126 at room temperature [8], which is the highest dielectric permittivity among the nonpolar materials, until now. Nonpolar organic materials have been widely studied as dielectrics, owing to their easy processability, flexibility, high dielectric strength, and low melting temperature in different applications, including epoxy ($\varepsilon_r = 4$), polyethylene (PE, $\varepsilon_r = 2.2 \sim 2.4$), polystyrene (PS, $\varepsilon_r = 2.8 \sim 4.5$), poly(methyl methacrylate) (PMMA, $\varepsilon_r = 4 \sim 5$), polytetrafluoroethylene (PTFE, $\varepsilon_r = 2.1$), biaxially-oriented polypropylene (BOPP, $\varepsilon_r = 2.2$), and so forth [7–13].

8.1.2.2 Polar Materials

A polar material possesses a permanent dipole moment that is related to its molecule. If a material contains polar molecules, these polar molecules are generally in random orientations without external electric fields. When an electric field is applied, the material will be polarized by orienting the dipole moments of polar molecules. In some polar materials, the dipoles have strong interactions with each other and align along certain directions to form a spontaneous polarization (P_s) without an external electric field. These materials are called pyroelectric materials. If the P_s in a pyroelectric material can be switched by the external electric field, the material is called a ferroelectric (FE) material. For both polar inorganic and organic materials, especially FE materials, they usually exhibit a high dielectric permittivity, compared with nonpolar dielectric materials, because the electric field can switch the P_s.

The widely used polar inorganic materials are ferroelectric ceramics and crystals, which usually have high dielectric permittivity (high-k) at 1 kHz ($\sim 10^3$), such as $BaTiO_3$ (BT, $\varepsilon_r = 1500$), $Pb(Mg_{1/3}Nb_{2/3})O_3$ (PMN, $\varepsilon_r = 5500$), and $Pb(Zr_{0.52}Ti_{0.48})O_3$ (PZT, $\varepsilon_r = 1300$) [6]. Furthermore, their dielectric permittivity is strongly dependent on temperature. The ferroelectric materials have a critical temperature (i.e., Curie temperature). The material is ferroelectric when the temperature is lower than the Curie temperature, while the material is paraelectric (PE) when the temperature is higher than the Curie temperature. At temperatures higher than this phase transition temperature, the dielectric permittivity is dependent on the temperature by the Curie-Weiss law as [6]:

$$\varepsilon_r' = \frac{C}{T - T_c} \tag{8.11}$$

where T_c is the Curie temperature and C is the Curie constant.

Different from the well-known "normal" ferroelectrics, some ferroelectric ceramics/crystals have a weak temperature dependence on the dielectric permittivity around the Curie temperature, which are called relaxor ferroelectrics. The corresponding maximum value of ε_r' and ε_r'' appears at different temperatures, and exhibits a frequency dependent behavior. The phase change cannot be associated with a FE-PE phase transition, and the Curie-Weiss law is not fulfilled for temperatures around the temperature that is related to the maximum value of dielectric permittivity [9]. In recent years, many typical relaxor ferroelectrics have received special attention because of their extraordinary dielectric properties, including $PbMg_{1/3}Nb_{2/3}O_3$-$PbTiO_3$ (PMN-PT), $PbZn_{1/3}Nb_{2/3}O_3$-$PbTiO_3$ (PZN-PT), $Pb_{1-x}La_x(Zr_{1-y}Ti_y)_{1-x/4}O_3$ (PLZT), and some members of the relaxor Aurivillius family $[Bi_2O_2]^{2+}[A_{n-1}B_nO_{3n+1}]^{2-}$ [9–11].

Besides ferroelectrics, some inorganic non-ferroelectric materials have been developed that also show very high dielectric permittivity, including $CaCu_3Ti_4O_{12}$ (CCTO), $Bi_{2/3}Cu_3Ti_4O_{12}$, $La_{2x}Sr_xNiO_4$, and (Nb+In) co-doped TiO_2 rutile [12–14]. CCTO has gained the most attention among these materials because it does not have a piezoelectric effect, and its high dielectric permittivity is almost independent of the temperature over a broad temperature range at low frequencies. The study on CCTO provides a new research direction to develop polymer-based composites with high dielectric performance [15].

For polar organic materials, ferroelectric polymer poly(vinylidene fluoride) (PVDF) with relatively high dielectric permittivity (~11) is the most studied polar polymer. The monomer unit of PVDF is —CH_2—CF_2—, which contains a high electronegative fluorine atom with a van der Waals radius of 1.5 Å, and a hydrogen atom with radius of 1.2 Å. The calculated dipole moment of a monomer unit is about 7.06×10^{-30} C m [2,16]. Four different crystalline phases are observed in PVDF, including α (in a trans-gauche conformation: TGTG), β (in an all trans-planar zigzag conformation: TTTT), γ (TTTGTTTG'), and δ phase (a polar version of the α phase), as shown in Fig. 8.4 [5]. In these four conformations, only the α phase is nonpolar, and the β, γ, and δ phase are ferroelectric. The α phase can be obtained by cooling the melt and drying below at a normal rate. The β phase can be realized by mechanical stretching or high field poling. The δ phase can be formed by poling the α phase sample under an electric field at room temperature. The γ phase is often achieved by solution casting below 100°C, or annealing/crystallization at high temperatures. In recent years, much attention has been paid to PVDF-based polymers, such as

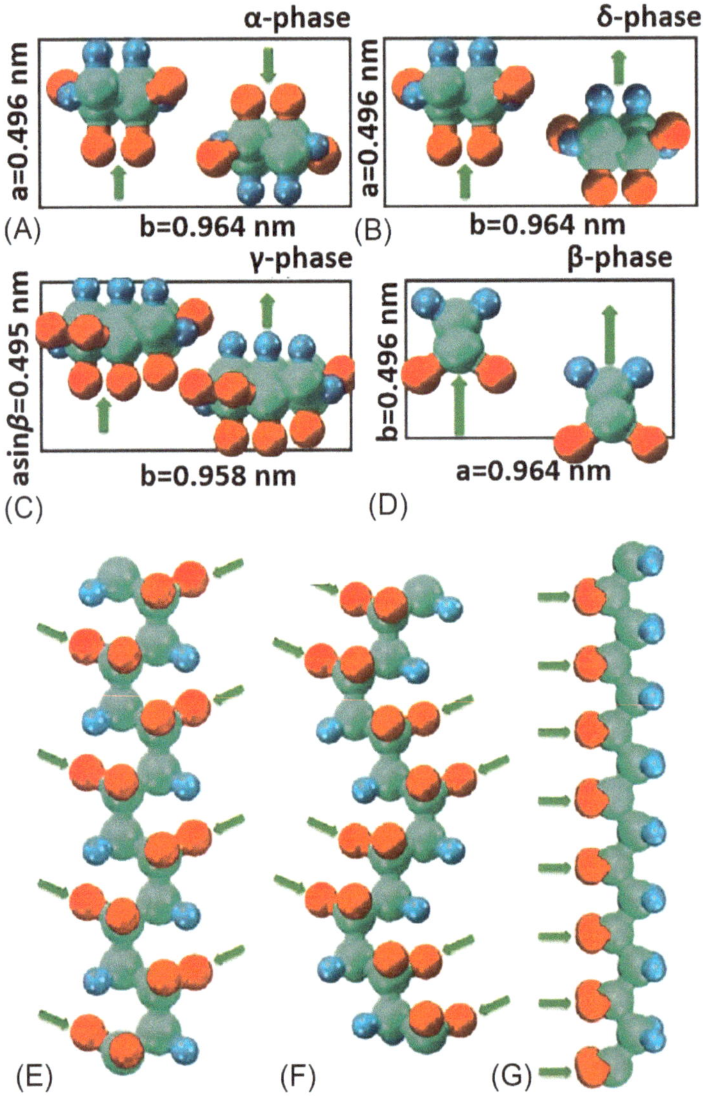

FIG. 8.4 Unit cells of (A) α, (B) δ, (C) γ, and (D) β forms of PVDF crystals viewed along the c-axes and schematic chain conformations for (E) TGTG′ (α/δ), (F) TTTGTTTG′ (γ), and (G) all-trans (β) rotational sequences. *Red*, *cyan*, and *blue* spheres represent F, C, and H atoms. The projections of dipole directions are indicated by green arrows. *(Reproduced with permission from L. Zhu, Q. Wang, Novel ferroelectric polymers for high energy density and low loss dielectrics, Macromolecules 45 (7) (2012) 2937–2954 (Copyright 2012 American Chemical Society).)*

poly(vinylidene fluoride-trifluoroethylene) [P(VDF-TrFE)], poly(vinylidene fluoride-chlorotrifluoroethylene) [P(VDF-CTFE)], poly(vinylidene fluoride-hexafluoropropylene) [P(VDF-HFP)] copolymers, and poly(vinylidene fluoride-trifluoroethylene-chlorofluoroethylene) [P(VDF-TrFE-CFE)], poly (vinylidene fluoride-trifluoroethylene-chlorotrifluoroethylene) [P(VDF-TrFE-CTFE)] terpolymers. They usually exhibit a high dielectric permittivity, which is higher than 10, or even as high as 60, due to different mole ratios of different parts, and different preparation methods [17,18].

8.1.3 Dielectric Composites

Considering solid dielectrics, inorganic and organic dielectric materials have been widely used in current electronic devices and systems. Inorganic dielectrics (i.e., ceramics) have a high dielectric permittivity and relatively low dielectric loss. They can work under high temperatures. However, they are brittle, and their breakdown field is relatively low. They also require a high processing temperature. Polymers have relatively low dielectric permittivity (usually smaller than 5) and exhibit a high breakdown field. They have excellent mechanical properties, and require a low process temperature. The comparison between inorganic and organic dielectric materials is listed in Table 8.1.

With the development of next-generation dielectric capacitors, it is essential to develop materials with high dielectric permittivity, low loss, high breakdown fields, low processing temperatures, easy fabrication, and low cost for advanced electronic devices and systems. Composites have been widely studied to fully utilize the advantages of different materials and combine their unique properties [15,19]. A composite is a material mixed from two or more constituent materials with significantly different physical or chemical properties from the individual

TABLE 8.1 Properties Comparison Between Inorganic and Organic Dielectric Materials

Properties	Inorganic Materials (Ceramics)	Organic Materials (Polymers)
Dielectric permittivity (ε)	High (10^3–10^4)	Low (<5)
Dielectric loss (tan δ)	Small	Relatively high
Dielectric breakdown strength (E_b)	Low (<5 MV/m)	High (>100 MV/m)
Processing temperature	High	Low
Mechanical properties	Brittle	Flexible
Fabrication	Hard	Easy

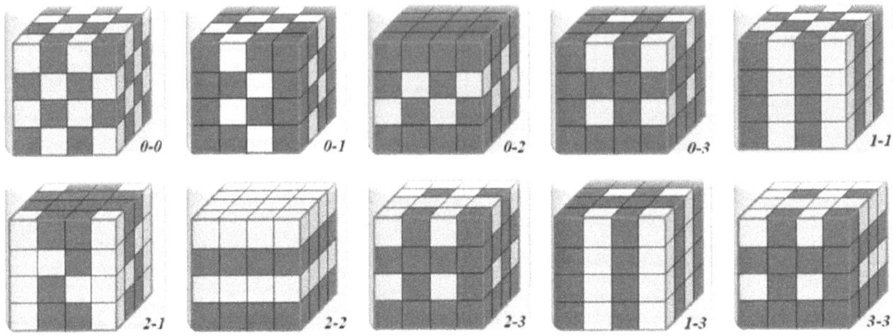

FIG. 8.5 Ten connectivity patterns in a diphasic composite. *(Reproduced with permission from R. E. Newnham, D.P. Skinner, L.E. Cross, Connectivity and piezoelectric-pyroelectric composites, Mater. Res. Bull. 13 (5) (1978) 525–536 (Copyright 1978 Elsevier Ltd.).)*

components. In the final composite, individual components remain separate and distinct from each other. Usually, the composite exhibits better, stronger, lighter, or lower-cost characterizations. In dielectric physical properties, the composite can be the sum, combination, and product of the dielectric properties of its constituents based on the physical properties and connectivity. The connectivity was first classified by Newnham and Cross [20]. There are ten different connectivity patterns in diphasic materials, which are 0-0, 1-0, 2-0, 3-0, 1-1, 2-1, 3-1, 2-2, 3-2, and 3-3, as shown in Fig. 8.5. The number 0/1/2/3 used here is to represent the dimension in continuity for each component in the composite [21,22]. For example, a 0-0 composite means it is a mixture of isolated particles (0-dimention) for both components. A 2-2 composite is a layer-by-layer structure in which each component is continuous in two dimensions. A 0-3 composite is defined as isolated particles embedded in a continuously 3-dimentional media.

Among these ten patterns of composites, the polymer-based 0-3 composite, which uses a dielectric polymer as a matrix and other materials as isolate fillers, has attracted more attention in recent years because this type is easy to prepare and retains the flexibility of the polymer matrix. According to the filler used, there are two main kinds of polymer-based 0-3 composites, as shown in Fig. 8.6 [15,23–25]. One is dielectric-dielectric composites in which the fillers are dielectric materials, such as ferroelectric ceramics, metal oxides, dielectric polymers, and so forth. The other is conductor-dielectric composites, where the fillers are conducting materials, including metal, carbon materials, conducting polymers, and so forth. For both types of composites, the polymer matrix strongly determines the dielectric performance of the composites [25]. The frequency and temperature dependence of dielectric properties and the breakdown field are mainly dependent on the polymer matrix. Many polymers have been selected as a matrix according to their glass transition temperature, melting temperature, flexibility, dielectric strength, and processability, including the nonpolar and polar polymers mentioned in the previous section [15].

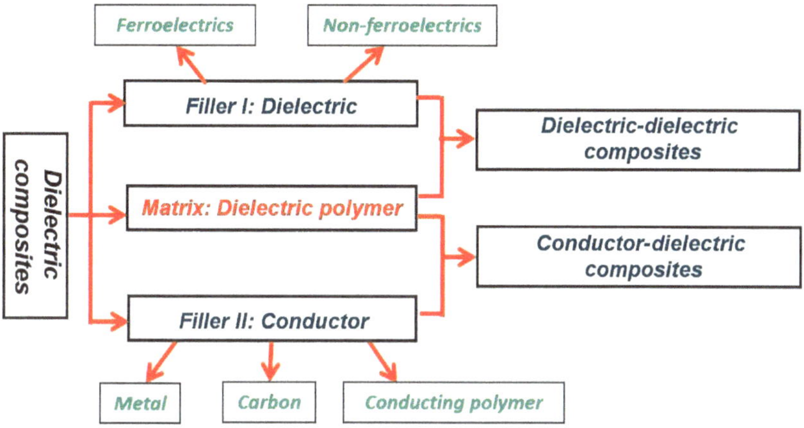

FIG. 8.6 Schematic of two types of polymer-based 0-3 dielectric composites.

8.2 DIELECTRIC-POLYMER NANOCOMPOSITES WITH HIGH PERMITTIVITY

For dielectric-dielectric composites, the dielectric permittivity increases with the filler content monotonously, and is also dependent on the volume fraction (φ) of the filler [26]. For this consideration, it is very important to find dielectric fillers with a high dielectric permittivity. Different dielectric fillers may result in different relationships between the dielectric properties of the composite. Rayleigh first studied the 0-3 dielectric composites in 1890s, and assumed that dielectric spheres in a rectangular order are embedded into another dielectric [27]. After that, many models and equations have been proposed to describe the relationship between the dielectric response and the content of the fillers; and other important factors have also been considered, including the size and shape of the filler particles [15,26,28]. Therefore, many fillers with high dielectric permittivity (i.e., ferroelectric ceramics) have been widely used in the development of polymer-based composites [24]. In this section, nanocomposites with different types of dielectric fillers are discussed.

8.2.1 Ferroelectric Ceramics as Fillers

As introduced in Section 8.1.2.2, ceramic-polymer nanocomposites have attracted significant interest recently due to the high-k ferroelectric ceramics that have been introduced [24]. The combination of ceramics with a high-dielectric-strength polymer matrix may result in high dielectric performance materials. In the past 20 years, many ceramic-polymer nanocomposites have been studied, such as BT, PMN, PMN-PT, PZT, and so forth [4,24,29,30]. Compared with micro-sized fillers, the nano-sized ceramic particles can prevent partial discharge degradation, increase voltage endurance, improve space-charge formation, reduce the thickness of composites, and maintain the flexibility. Most nanoparticles of ceramics have been provided by companies, while some have been prepared by solid state reactions, chemical methods, or even having their size reduced by high-energy-ball-milling [15]. For electrical energy storage concerns, the high dielectric permittivity of composites and the high electrical strength are both very important. Due to the processing method, there is always a problem with the surface issue between these inorganic particles and fluorinated polymers. The agglomeration of the ceramic introduced may cause electron conduction for a high dielectric loss, and undesirable porosity for dielectric failure at a much lower electrical field. In recent studies, the coupling agent and surface functionalization have been widely used for filler surface modification. The most used coupling agent is γ-aminopropyltriethoxy silane [KH550: $H_2NCH_2CH_2Si(OC_2H_5)_3$]. The —$NH_2$ of KH550 would react with the molecule of PVDF due to the possible formation of an F—H bond. Therefore, KH550 can act as a bridge-linked action to bind BT with PVDF, which is shown in Fig. 8.7 as a schematic chemical reaction process [31]. The dielectric permittivity of BT/PVDF composites with 40 vol% BT modified by 1.0 wt% KH550 can be improved by about 20%, in comparison with the composite without KH550 at a frequency from 1 to 100 kHz.

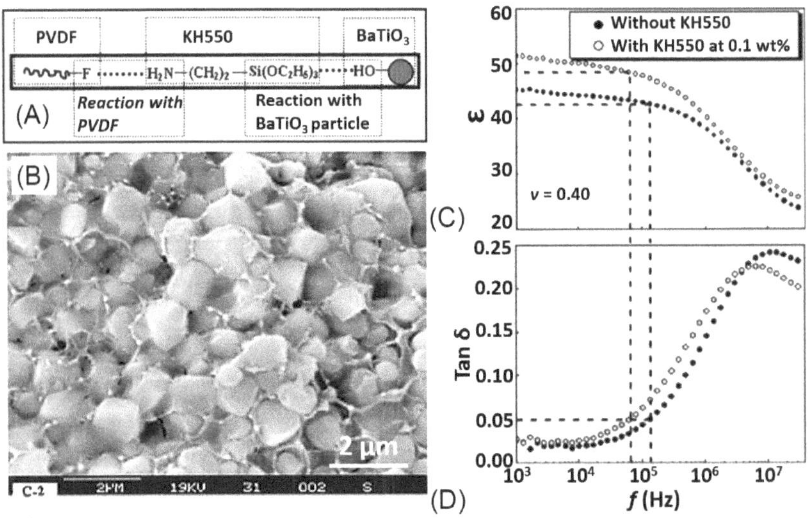

FIG. 8.7 (A) Schematic image of chemical reaction process of KH550 with both surfaces of BT and PVDF, (B) Morphologies of fractured surfaces of BT/PVDF composites 40.0 vol% BT with 1.0 wt% of KH550, (C) the dielectric permittivity and (D) dielectric loss of the BT/PVDF composites 40.0 vol% BT with 1.0 wt% of KH550. *(Reprinted with permission from Z.-M. Dang, H.-Y. Wang, H.-P. Xu, Influence of silane coupling agent on morphology and dielectric property in $BaTiO_3$/polyvinylidene fluoride composites, Appl. Phys. Lett. 89 (11) (2006) 112902 (Copyright 2006 American Institute of Physics).)*

The selected dielectric properties of ceramic-polymer nanocomposites at room temperature are listed in Table 8.2. BT is widely used as filler because it is relatively inexpensive and easy to prepare in different sizes, from micro-sized to nano-sized. Almadhoun et al. reported that nanocomposites containing hydroxylated BT nanofillers exhibited significantly lower leakage of current densities concurrently with the enhancement in the effective permittivity and dielectric breakdown strength [35]. Tang et al. developed a new synthesis method for the growth of high-aspect-ratio BT nanowires with a high yield. The BT/P(VDF-TrFE-CFE) nanocomposites showed high energy density (10.48 J/cm^3) using surface-functionalized high aspect ratio BT nanowires [38]. Song et al. studied the effect of the polymer matrix filler's shape and surface modification on the dielectric properties of the nanocomposite [43]. The improved breakdown strength and enhanced ferroelectric properties are attributed to the combined effect of the surface modification by dopamine, the large aspect ratio of the BT nanofibers, and the improved crystallinity of the polymer nanocomposites. A similar study was also performed by Tang and coworkers using PZT nanowires and nanorods [39]. The composites containing nanowires have higher dielectric permittivity and lower dielectric loss than those with nanorods with the same volume fraction. As a result, the high aspect ratio PZT nanowires showed a 77.8% increase in energy density over the lower-aspect-ratio PZT nanorods, under an electric field of 15 MV/m. PMN-PT exhibits a much higher dielectric permittivity around room temperature than the BT, and does not show a phase transition. Therefore, PMN-PT is also a good candidate as a filler for the development of high-dielectric-permittivity composites. Bai and coworkers reported composites using PMN-PT with a diameter of 500 nm as a filler [42]. The dielectric permittivity can be higher than 200 at 1 kHz, and the loss is smaller than 0.05. The authors also predicted that the maximum stored energy of this composite could be more than 15 J/cm^3 if composites were prepared in a class 1000 clean bench without the air bulbs, dust, and residual solvent.

8.2.2 Other High-*k* Ceramics as Fillers

From the discussion in Section 8.2.1, ferroelectric ceramics such as BT, PZT, and PMN-PT are the preferred choice as the dielectric material in capacitors. However, high remnant polarization and saturation of the polarization may have limitations in enhancing energy density. Recently, the use of $Ba_xSr_{1-x}TiO_3$ (BST) has proved to provide a high dielectric permittivity while eliminating the remnant polarization, and ultimately the efficiency of the capacitor [44–49]. In the BST solid solution, when the Ba mole fraction decreases below 0.7 ($x < 0.7$), the phase of BST transferred from the ferroelectric to the paraelectric phase. Xia et al. reported dielectric and energy storage properties of BST/P (VDF-CTFE) composites using KH550 as a coupling agent to improve the compatibility and the quenching method to improve the breakdown field [45]. The

TABLE 8.2 Dielectric Properties of Ceramic-Polymer Nanocomposites at Room Temperature

Polymer Matrix	Ceramics	Size (nm)	Thickness (μm)	Volume Content	ε_r (at 1 kHz)	tan δ (at 1 kHz)	Ref.
Polycarbonate	BT	30–50	3.89	50%	20	<0.01	[32]
P(VDF-HFP)	BT	30–50	3.84	50%	37	<0.07	[32]
P(VDF-CTFE)91/9	BT	50–70	25	23%	25	<0.05	[33]
P(VDF-TrFE-CTFE)	BT	50–70	25	30%	65	<0.07	[33]
P(VDF-HFP)	BT	30–50	–	50%	32	<0.05	[34]
P(VDF-TrFE)70/30	BT	<100	0.18	20%	24	<0.05	[35]
PVDF	BT	100	10-30	20%	20	<0.03	[36]
Polystyrene	BT	63	1.84	33%	45	0.06	[37]
P(VDF-TrFE-CTFE)	BT	–	–	17.5%	70	0.09	[38]
PVDF	PZT	160	100–200	50%	43	<0.05	[39]
Polyurethane	PZT	–	–	30%	24	0.7	[40]
P(VDF-TrFE) 70/30	PMN-35PT	–	30	40%	50	<0.1	[41]
P(VDF-TrFE) 50/50	PMN-PT	500	20	50%	250	<0.05	[42]

dielectric permittivity can reach 82 at 100 Hz with 30 wt% BST. The discharge energy density of 6.5 J/cm^3 at 250 MV/m were obtained with 10 wt% BST. Tang et al. reported a new synthesis process to grow high aspect ratio nanowires with high yield. The PVDF nanocomposites with $Ba_{0.2}Sr_{0.8}TiO_3$ in a cubic phase exhibited an ultra-high energy density of 14.86 J/cm^3 at 450 MV/m, and provided microsecond discharge times due to the high aspect ratio fillers functionalized with amine groups, as shown in Fig. 8.8 [46]. The use of a pyroelelctric phase of BST was the main reason for the high charge-discharge efficiency of the nanocomposite. Furthermore, high aspect ratio nanowires also promote the dielectric performance of nanocomposites.

The study on high-k ceramic $CaCu_3Ti_4O_{12}$ (CCTO) with a perovskite-related structure and very different dielectric properties from ferroelectric ceramics provided a new research direction for preparing high-k composites. A dielectric permittivity of 20,000 for CCTO ceramics and 300,000 for CCTO single crystals at 1 kHz and at room temperature has been reported [50]. The colossal dielectric response of CCTO has been widely studied due to many possible extrinsic and intrinsic mechanisms, including internal barrier layer

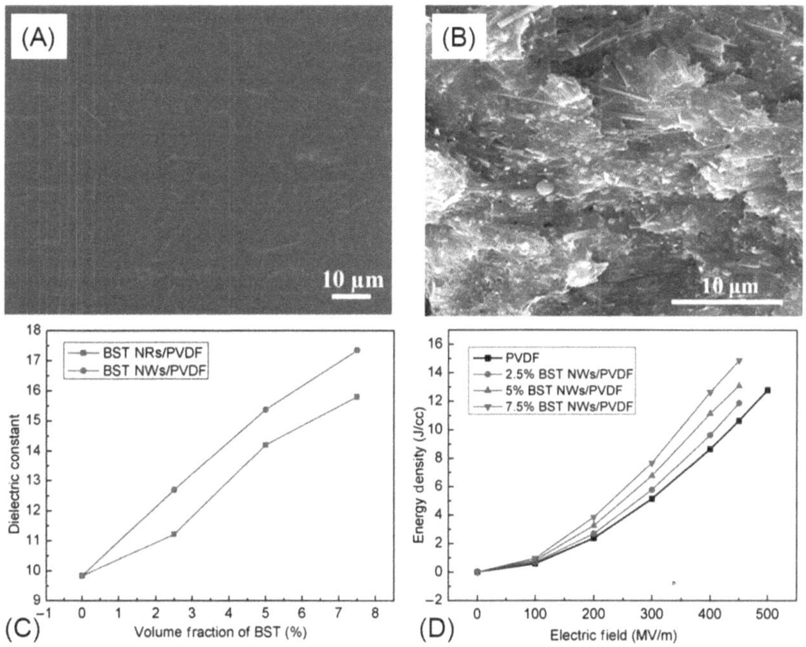

FIG. 8.8 (A) Top surface and (B) cross-section of 5% $Ba_{0.2}Sr_{0.8}$TiO3/PVDF nanocomposites. (C) Unipolar electric *D-E* loops for nanocomposites with different BST nanowires volume fractions, (D) energy density of the nanocomposite with different volume fraction as a function of the electric field calculated from *D-E* loops. *(Reprinted with permission from H. Tang, H.A. Sodano, Ultra high energy density nanocomposite capacitors with fast discharge using Ba0.2Sr0.8TiO3 nanowires, Nano Lett. 13 (4) (2013) 1373–1379 (Copyright 2013 American Chemical Society).)*

capacitances and surface barrier layer capacitances [51–53]. Besides the giant dielectric permittivity (>10^4), CCTO ceramics also possess other important advantages for composites, which do not have piezoelectric effects, and show a weak temperature dependence over a broad temperature range. Due to these reasons, there has been great interest in the investigation of CCTO-polymer composites [15,54]. Cheng's research group first systematically studied the CCTO-polymer composites with micro-sized and nano-sized fillers. Arbatti et al. reported a dielectric permittivity as high as 610 at 100 Hz at room temperature in CCTO/P(VDF-TrFE) using micro-sized fillers with a volume fraction of 50% [55]. The following work illustrated influences of process conditions on the microstructure and dielectric properties of CCTO-polymer composites. Using solution-cast and hot-pressing processes, the dielectric permittivity can reach 750 with a loss of around 0.2 at 100 Hz at room temperature, and can exceed 1000 at high temperatures [56].

Compared with micro-sized fillers, composites with nano-sized CCTO particles exhibited lower dielectric permittivity, but showed much lower loss and more flexibility [57]. Using the same hot pressing and solution casting methods, Zhang et al. synthesized nano-sized CCTO/P(VDF-TrFE) composites with uniform microstructures in a 6-layer configuration. Dielectric permittivity was 62, with a loss of 0.05 for nanocomposites with a 50 vol% CCTO at room temperature at 1 kHz [58]. Around the phase transition temperature (65–70°C) of the copolymer, a dielectric permittivity of 150 with a low loss (<0.1) was obtained in this nanocomposite, which was caused by the ferroelectric-paraelectric phase transition of P(VDF-TrFE). The differences between composites with micro-sized and nano-sized fillers are the amount of interfacial layers and the dielectric properties of the CCTO filler. The composite with nano-sized CCTO would have more interfacial layers in the composite because of the higher surface area of nanoparticles. Shan et al. used another polymer P(VDF-CTFE) 88/12 mol% as a matrix and also obtained nanocomposites with high dielectric permittivity and a small dielectric loss [59]. Different from P(VDF-TrFE), dielectric permittivity of this polymer and composites were almost independent of temperature from 25°C to 125°C, which could be useful for many applications, such as electrical power, and capacitors in a broad working temperature range. Dang et al. prepared a CCTO/polyimide composite with a dielectric permittivity of around 49 at 100 Hz, which is 14 times higher than the value of the pure PI matrix [60]. The preparation of composites for each group is very different, which may cause different results. In these works, CCTO ceramics were synthesized by solid-state reactions from metal oxides at high temperatures, and reduced to small particles by ball milling. The size distribution and surface morphology of particles were not ideal for dielectric nanocomposites.

Recently, other chemical methods have been proposed to obtain nanoparticles in large scale. Yang et al. studied the effect of particle size on the dielectric properties of CCTO/PVDF composites [61]. In Yang's report, samples with high content of CCTO (>30 vol%) could not be easily prepared, because it is difficult to mix the two components. The very high dielectric permittivity ($>10^6$) of nano-sized CCTO obtained with very high loss ($\sim$50) indicates the composites are conductor-like materials. Ehrhardt and coworkers prepared CCTO nanoparticles by two different soft-chemistry methods to yield nano-sized scale particles for an enlarged interfacial area [62]. The good chemical compatibility between the nonpolar chain of the surfactant on CCTO nanoparticles and the polymer matrix has been achieved, which was evidenced by the suspension of surface functionalized fine-ground powders, which were unchanged even after 4 weeks. Tang et al. first performed large-scale synthesis of CCTO nanowires by a two step-hydrothermal reaction, and then prepared nanocomposites with PVDF as a matrix [63]. It indicated that the aspect ratio of the filler can greatly influence the dielectric properties of the composites with high aspect ratio nanowires. As shown in Fig. 8.9A–C, the dielectric permittivity reached as high as 68, which was 7 times higher than the PVDF matrix (9.8) with loss as low as 0.081.

8.2.3 Metal Oxides

Due to the interface between the polymer matrix and ceramics, it may be not suitable, because dielectric permittivity of filler is hundreds, and even thousands of times that of polymers matrix. The very large difference of dielectric permittivity between filler and matrix may cause a highly inhomogeneous electric field, and thus a significantly reduced effective breakdown strength of the composite, resulting in very little of the electric energy stored. Metal oxide with high dielectric permittivity is a good choice for improving dielectric properties of nanocomposites considering the improvement of energy density. Zhang's group proposed that the dielectric permittivities of fillers ($\varepsilon_{TiO2}=47$ and $\varepsilon_{ZrO2}=21$) are close to that of the polymer matrix ($\varepsilon_{matrix}=42\sim50$), and can avoid enhanced local field effects and improve the energy storage [64,65]. For example, TiO_2 rod-shape nanoparticles (20×70 nm) were used as the filler, which has a relatively uniform distribution in P(VDF-TrFE-CTFE) [64]. The dielectric permittivity of nanocomposites are above 40 with a low loss smaller than 0.1. The energy density is 6.9 J/cm^3 at 200 MV/m for the composites with 10 vol% TiO_2. Another work reported a nanocomposite with ZrO_2 nanoparticles as the fillers in a different terpolymer matrix P(VDF-TrFE-CFE) 59.4/33.4/7.2 ($\varepsilon_r\approx50$) [65]. The high energy density of 10.5 J/cm^3 can be achieved under a lower field (300 MV/m). Tang et al. fabricated a high energy density nanocomposite capacitor by incorporating high aspect ratio functionalized TiO_2 nanowires into a PVDF matrix [66]. These nanocomposites exhibited energy density as high as 12.4 J/cm^3 at 450 MV/m, as shown in Fig. 8.10A–C. These studies open a new route toward dielectric nanocomposites with high electrical energy density by selecting a combination of polymer matrices and nanoparticles with balanced dielectric properties.

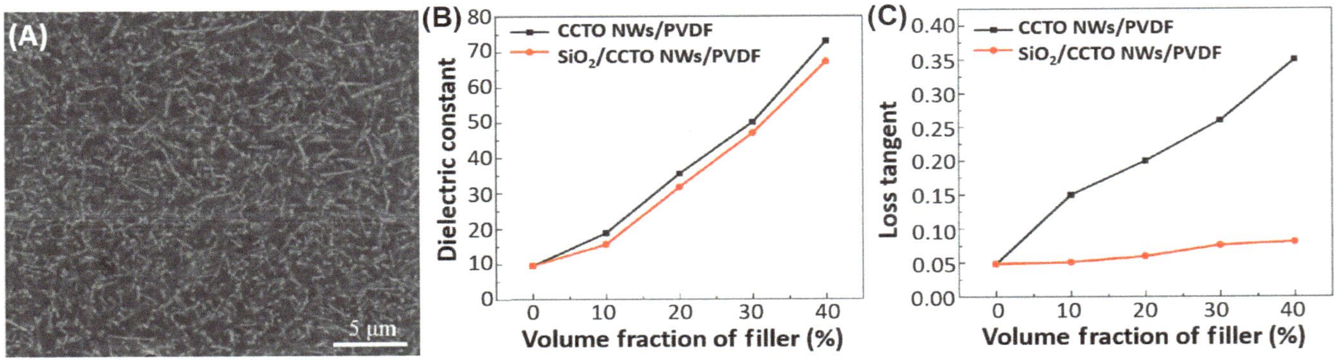

FIG. 8.9 (A) SEM image of a 30 vol% CCTO nanowires in PVDF nanocomposite. (B) Comparison of measured dielectric permittivity (at 1 kHz). (C) Loss tangent of nanocomposites with different fillers: CCTO nanowires and SiO_2/CCTO nanowires. *(Reproduced with permission from H. Tang, Z. Zhou, C.C. Bowland, H.A. Sodano, Synthesis of calcium copper titanate ($CaCu_3Ti_4O_{12}$) nanowires with insulating SiO_2 barrier for low loss high dielectric constant nanocomposites, Nano Energy 17 (Suppl C) (2015) 302–307 (Copyright 2015 Elsevier Ltd.).)*

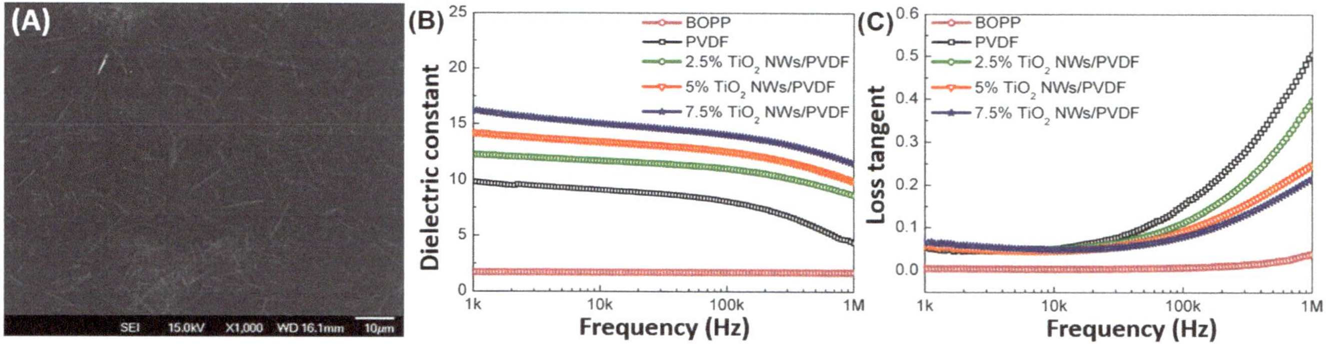

FIG. 8.10 (A) SEM image of nanocomposites with 5 vol% TiO_2 nanowires. (B) Dielectric permittivity of different TiO_2 nanowires' volume fractions in PVDF, and commercial biaxially oriented polypropylene from 1 kHz to 1 MHz. (C) Energy density of the nanocomposite with different volume fraction as a function of electric field calculated from D-E loops, also compared with nanocomposite with 10% TiO_2 nanopaticles in P(VDF-TrFE-CTFE). *(Reproduced with permission from H. Tang, H.A. Sodano, High energy density nanocomposite capacitors using non-ferroelectric nanowires, Appl. Phys. Lett. 102(6) (2013) 063901 (Copyright 2013 American Institute of Physics).)*

8.3 CONDUCTOR-POLYMER NANOCOMPOSITES WITH HIGH PERMITTIVITY

In spite of the significant developments on the polymer-based nanocomposites with dielectric fillers, another important class of fillers is from conductive materials. In previous reports, conductive fillers, such as metal particles/nanowires, metal nanowires, carbon materials, and conductive polymers are used as fillers to enhance the dielectric permittivity of the polymeric nanocomposites [67–73]. These polymer-based nanocomposites with conductive fillers are alternatives to the nanocomposites with dielectric fillers, and could also act as dielectrics for high-density energy storage capacitors.

The dielectric properties of a conductor-dielectric composite usually exhibit a percolation phenomenon [15,74]. When the amount of conducting filler is very small, the polymer matrix is still a dielectric, and its dielectric permittivity increases with an increase in the filler content. When the filler content is close to a certain value (i.e., percolation threshold φ_c), the dielectric permittivity of the composite increases dramatically, and reaches hundreds of times the value of the polymer matrix itself. When the filler content is higher than the φ_c, the conducting fillers form conductive channels/networks that connect the electrodes on both sides, and the composite becomes a conductor [75]. The φ_c is very critical for a conductor-dielectric composite, which is strongly dependent on the shape and size of the conducting fillers and their distribution in the polymer matrix [26]. Researchers also found that the φ_c is also strongly dependent on the fabrication methods of nanocomposites. Therefore, different types of conducting fillers with various shapes (nanospheres, nanotubes, nanobars, nanowires, nanoplates) have been studied [76].

8.3.1 Metal

Metal nanoparticles and metal nanowires are used as conductive fillers to improve dielectric permittivity. Silver is a metal with excellent electric conductivity, as high as 6.3×10^7 S/m at 20°C. Its nanoparticles and nanowires are widely used to embed into polymer matrices to improve the dielectric permittivity of nanocomposites. Qi et al. dispersed silver nanoparticles (with a diameter of around 40 nm) in an epoxy matrix, and found that a high dielectric permittivity of about 300 could be achieved near the percolation threshold of 23 vol% of silver [77]. In addition, the silver/epoxy composite also showed a dielectric stability in a temperature range from 23°C to 135°C. The possible reason for such thermal stability could be the in situ preparation and uniform distribution of silver nanoparticles in the matrix, which led to more resistance to the temperature-induced structure change [24]. Huang et al. investigated the dielectric properties and thermal conductivity of the PVDF nanocomposites with an addition of silver nanoparticles (diameter around 100 nm) [78]. The obtained nanocomposites possessed both high dielectric permittivity and high thermal conductivity, but without percolation behavior in the silver content of 0–20 vol%. Dang et al. incorporated silver nanoparticles into a polyimide resin by an in situ polymerization process [79]. The addition of silver, on the one hand, improved the thermal stability of the composites, and on the other hand, significantly enhanced the dielectric properties. With the addition of 12.5 vol% of silver nanoparticles, the composites can achieve a high dielectric permittivity of 400 at 1 kHz. Silver nanowires are one-dimensional materials with high aspect ratios. They are reported as fillers to improve the dielectric properties of polymers, such as PVDF [80], epoxy [81], polyurethane [82], and cellulose [83]. Zheng et al. embedded silver nanowires into PVDF by a simple casting method, and the nanocomposite film presented a high dielectric permittivity of 800 at 100 Hz, which remained at 379 at 1 kHz when the amount of silver nanowires was near the percolation amount of 20 vol% [80]. In addition, the nanocomposites could still keep a low dielectric loss of 0.15, which enables potential applications in advanced electronics [80]. In Chen's report, 16 vol% addition of silver nanowires led to the dielectric permittivity increasing to 100, one order of magnitude higher than the composites with silver nanowire amounts below the percolation threshold [81]. Mi et al. presented a super-flexible silver nanowire/polyurethane nanocomposite with excellent mechanical, thermal, and dielectric properties (Fig. 8.11) [82]. The increased amount of

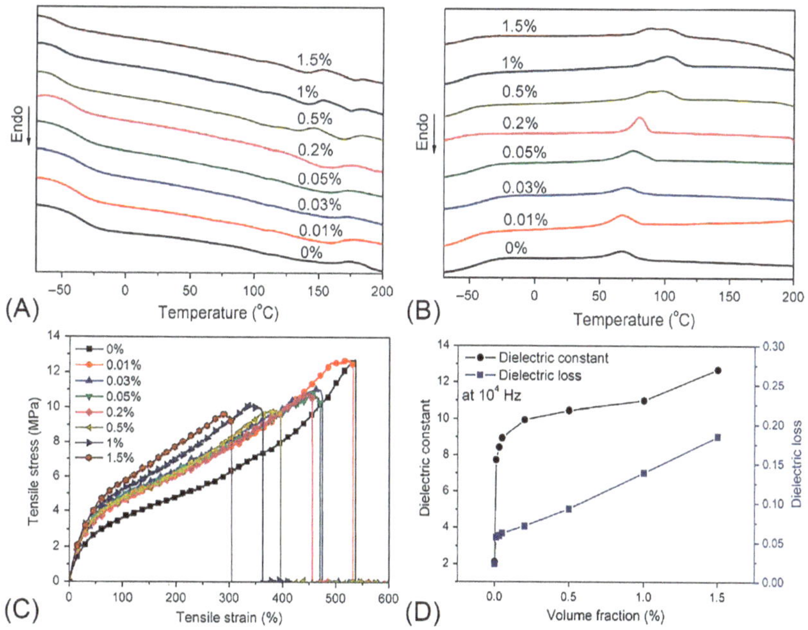

FIG. 8.11 Differential scanning calorimetry (DSC) measurements during (A) second heating process and (B) cooling process. (C) The typical stress-strain curves. (D) Dielectric properties at 10 kHz of silver nanowires/polyurethane nanocomposites at different filler volume fractions. *(Reproduced with permission from H.-Y. Mi, Z. Li, L.-S. Turng, Y. Sun, S. Gong, Silver nanowire/thermoplastic polyurethane elastomer nanocomposites: thermal, mechanical, and dielectric properties, Mater. Design (1980-2015) 56 (Suppl C) (2014) 398–404 (Copyright 2014 Elsevier Ltd.).)*

silver nanowires increased crystallization temperatures and enhanced both Young's modulus and the storage modulus of the nanocomposite films, but it could still have a high elongation at a break >300%. In addition, the addition of 1.5 vol% of silver nanowires caused the dielectric permittivity to be 9 times more than that of the pure polyurethane at 1 kHz, while the dielectric loss could still remain at a low level (<0.2). These silver nanowire/polyurethane nanocomposites would be ideal candidates for flexible electronic materials.

In another study, Inui et al. presented a flexible cellulose nanocomposite with silver fillers [83]. They compared the effect of silver nanoparticles (with a diameter of around 100 nm) and nanowires (with a diameter of around 100 nm, and an aspect ratio of about 100) as conductive fillers on the dielectric properties of the composites. The results showed that the addition of a small amount of silver nanowires (2.54 vol%) led to an extremely high dielectric permittivity of 726.5 at 1.1 GHz. As a comparison, with the much higher additional amount of 15.14 vol% silver nanoparticles, the dielectric permittivity of the composites was only 24. The possible reasons for these quite different dielectric permittivities from silver nanoparticles and nanowires could be explained by the percolation theory. Previous studies revealed that the dielectric permittivity of the metal-polymer composites significantly increased near the percolation threshold [24,84]. The silver nanowires' aspect ratio could be about 100, while the aspect ratio of silver nanoparticles could be 1. Therefore, silver nanowires can form percolation networks at a much lower content than the silver nanoparticles, which means the composites with silver nanowires would have a much lower percolation threshold.

Besides silvers, Ni was also adopted as a filler to improve the dielectric permittivity of nanocomposites. In 2003, Dang et al. fabricated Ni/PVDF composites via a simple blending and hot-molding process [67]. They demonstrated that the composites can reach a high dielectric permittivity of 400 near the percolation threshold of 17 vol% and possess a weak frequency and temperature dependence. Later, the same group further investigated the temperature dependence of Ni/PVDF composites [85]. There were two obvious regions from −10° C to 40°C (β region) and 100°C to 150°C (α region), which could be attributed to the polar effect of PVDF. In the first temperature region (β region), PVDF experienced a relaxation near the glass transition temperature of PVDF. In this region, the movement of dipolar groups in the amorphous PVDF was not enough to catch up with the frequency change of the electric field, and therefore led to low dielectric permittivity. In the second region (α region), the relatively high temperature provided sufficient energy to the PVDF molecules, and greatly changed the molecular motions and dipole direction along the applied electrical field. Therefore, a higher dielectric permittivity in this region could be observed. Another important work reported by Cheng's group studied the high percolation threshold in nanocomposites with Ni nanoparticles [86–88]. By using solution casting and hot pressing, two nanocomposite systems, Ni/P (VDF-TrFE) and Ni/P(VDF-CTFE) were prepared, as shown in Fig. 8.12.

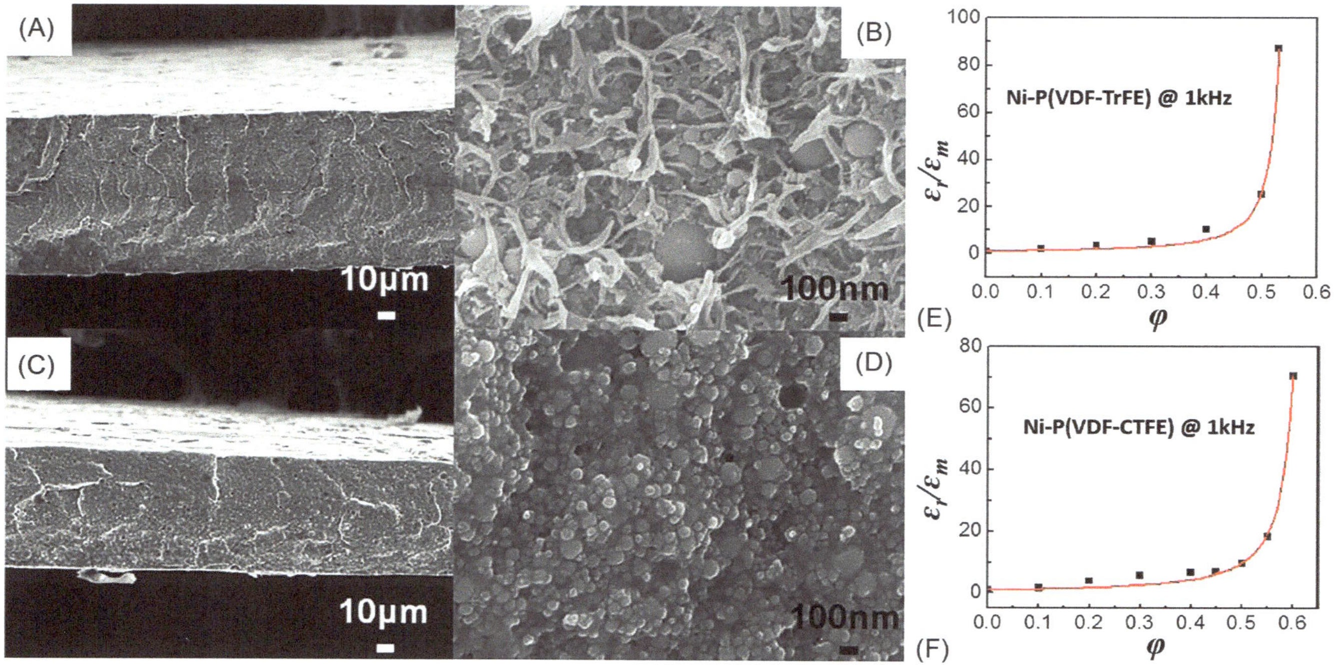

FIG. 8.12 SEM of composites with 50 vol% of Ni nanoparticles: (A), (B) Ni-P(VDF-TrFE); (C), (D) Ni-P(VDF-CTFE). The composition dependence of the $\varepsilon_r/\varepsilon_m$ at 1 kHz vs. φ for the nanocomposites: (E) Ni-P(VDF-TrFE) and (F) Ni-P(VDF-CTFE). *(Reproduced with permission from L. Zhang, W. Wang, X. Wang, P. Bass, Z.-Y. Cheng, Metal-polymer nanocomposites with high percolation threshold and high dielectric constant, Appl. Phys. Lett. 103 (23) (2013) 232903 (Copyright 2013 AIP Publishing LLC.).)*

Owing to the uniform microstructure, these nanocomposites had a high and similar percolation threshold (>55 vol%) and exhibited a high dielectric permittivity (>1000), which is more than 60 times that of the polymer matrix [86]. The nanocomposites with Ni content less than 50 vol% still showed a good insulation behavior. This work indicated that the percolation threshold can be pushed to high volume content by the uniform distribution of conducting fillers in the polymer matrix [89].

In addition, copper nanowires and stainless steel fibers were also reported as conductive metal fillers to improve the dielectric properties of composites. The copper nanowires/PVDF nanocomposites showed ascending dielectric permittivity, but increasing dielectric loss while increasing the amount of copper nanowires [90]. The large amount of nanocapacitor structures, and the conductive networks, could be attributed to the trend of dielectric permittivity and dielectric loss, respectively. Via a simple blending and hot pressing route, the stainless steel fibers were embedded into the PVDF matrix, and the resultant composite showed a percolation threshold at 9.4 vol%, which was much smaller than the other two-phase metal particle/polymer composites [84]. The stainless steel fiber/PVDF composites exhibited a dielectric permittivity of 427 at 50 Hz when the filler content was 10 vol%, which can be explained by the percolation theory.

8.3.2 Carbon Materials

8.3.2.1 Carbon Black

Carbon black is one kind of paracrystalline carbon powder material. Because of the very fine particle size in the range of several to tens of nanometers, carbon black has a high specific surface area [91]. It is usually used as filler for reinforcements, electric conductive agents, and also in inks as a color pigment. In 1982, Chung et al. studied the electrical conductivity and dielectric properties of carbon black/polyvinyl chloride composites in a broad frequency [92]. They demonstrated that the percolation theory was suitable for describing the trend of the dielectric permittivity with the filler loading level. When the filler amount approached the percolation threshold, the dielectric permittivity increased. But when the filler amount was above the percolation threshold, there were fully connected conductive channels, which led to the dramatic decrease of dielectric permittivity of the composites. Similar results were also reported by Broseau et al. in 1996 [93]. Later Calberg et al. introduced carbon black in a co-continuous two-phase PS/PMMA blend and found that the electrical and dielectric properties were dependent on the independent percolation threshold from the two phases of PS and PMMA, and the third phase of carbon black [94]. In 2004, Xu et al. adopted a high-gain, high-risk strategy to fabricate carbon black/epoxy composites as the embedded capacitor dielectrics [95]. They evaluated the effect of six different types of carbon blacks on the dielectric properties and found that four of them possessed a high dielectric permittivity (>1000) with a proper filler loading amount. Because of the interfacial polarization, this carbon black/epoxy composite also showed high frequency dependence.

8.3.2.2 Graphite

Graphite is well-known for its layered and planar structure. Because of its good electrical conductivity, graphite can be used as filler to fabricate polymer-based high permittivity composites. Miao et al. prepared graphite/PS nanocomposites by in situ polymerization, and investigated the structure and properties of the composites [96]. The graphite layers were distributed in the PS matrix with a thickness of less than 100 nm, and they formed a strong interfacial interaction between the PS and graphite. A percolation threshold of less than 8.2 wt% was observed, and the highest dielectric permittivity of 136 was achieved when the filler content was 6.5 wt%. He et al. exfoliated graphite flakes into graphite nanoplates and fabricated graphite nanoplates/PVDF nanocomposites by a simple solution-cast and hot-pressing process [97]. The homogenous dispersion of the fillers greatly enhanced the dielectric permittivity with a very low percolation threshold. When the filler amount was near the percolation threshold of 1.01 vol%, the dielectric permittivity could be up to 200 and 2700 at 1000 and 100 Hz, which was 20 and 270 times larger than the PVDF matrix, respectively. As expected, further increasing the filler amount above the percolation threshold led to the significant increase of dielectric loss of 229 at 1000 Hz, which could be due to the conductive channels in the composites. Similarly, Min et al. developed graphite nanoplatelets/epoxy nanocomposites through a typical interface design, as shown in Fig. 8.13 [98]. The nanocomposites possessed a high dielectric permittivity (>230) and a high thermal conductivity of 0.54 W/mK. The high aspect ratio of graphite nanoplatelets and the oxygen functional group on the basal planes of nanoplatelets were ascribed to the high performance of the nanocomposites. In another report, a one-step hydrothermal reaction was used to modify graphite powders with hydroxyl groups, and these functionalized graphite powders were applied to prepare PVDF-based nanocomposites [99]. The nanocomposites showed an ultrahigh dielectric permittivity of 5100, with a dielectric loss of around 3.0 when approaching the percolation threshold at 1 kHz.

8.3.2.3 Graphene

Graphene is an individual two-dimensional sheet, which can be exfoliated from graphite. It has outstanding properties, such as being ultra-light, mechanically strong, ultra-thin, electrically conductive, and immensely tough. Until now, graphene has found broad applications in the fields of energy, membranes, biomedicine, composites, coatings, sensors, and electronics. Because of interfacial polarization between the graphene fillers and the matrix, graphene was applied

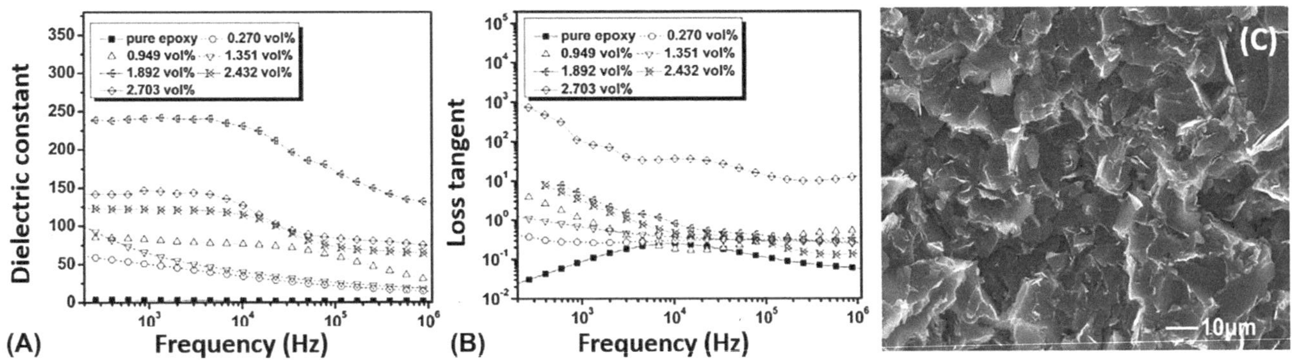

FIG. 8.13 Dependence of (A) the dielectric constant and (B) dissipation factor of the GNP/epoxy composites on frequency at room temperature. (C) SEM of GNP/epoxy composite (the content of GNPs is 1.892 vol%). *(Reproduced with permission from C. Min, D. Yu, J. Cao, G. Wang, L. Feng, A graphite nanoplatelet/epoxy composite with high dielectric constant and high thermal conductivity, Carbon 55 (2013) 116–125 (Copyright 2012 Elsevier Ltd.).)*

as a conductive filler to prepare polymer-based nanocomposites with high permittivity. Liu et al. adopted in situ polymerization to prepare high dielectric permittivity graphene/polyimide nanocomposites [100]. The dielectric permittivity was greatly improved by increasing the amount of graphene. Similarly, Devi and Kuma applied in situ chemical reduction of graphene oxide (rGO) using polypyrrole nanotubes (PPyNTs) as spacers to prepare rGO/PPyNTs nanocomposites [101]. In the composites, the PPyNTs were homogeneously dispersed onto the rGO, which led to significant improvement in the crystallinity and thermal stability. Because of the interfacial polarization at the interfaces of rGO and PPyNTs, the dielectric permittivity was also greatly improved. In order to achieve high flexibility, together with good dielectric performance, Fan et al. produced graphene/PVDF nanocomposites with a multi-layered structure [102]. The nanocomposites showed an extremely low percolation threshold of 0.0018 volume fraction of graphene, which was the lowest value among the reported PVDF-based nanocomposites. When the graphene volume fraction was 0.0018, the nanocomposites presented a high dielectric permittivity of 340 at 100 Hz. When further increasing the amount above the percolation threshold, the dielectric permittivity was further increased to a high value of 7940 at 100 Hz, but the dielectric loss was also remarkably increased. Wang et al. reported an interesting work by using a three-dimensional aerogel as a template to produce graphene/poly(vinyl alcohol) nanocomposites [103]. Because the barriers from poly(vinyl alcohol) could avoid the direct contacts between the conductive skeletons, the graphene aerogel/poly(vinyl alcohol) nanocomposites showed a significant enhancement on dielectric properties that the dielectric permittivity could be up to 1059, while the dielectric loss was as low as 0.08. In another report, Zhang et al. grafted different amounts of PS onto the reduced graphene oxide (rGO-PS) through an emulsion polymerization, and introduced rGO-PS into a PS matrix to prepare PS/rGO-PS nanocomposites [104]. The grafting of PS onto rGo greatly improved the interfacial polarization between rGO and PS, and therefore enhanced the dielectric properties. As shown in Fig. 8.14, when the rGO amount was fixed at 5 wt%, the nanocomposites with the highest amount of rGO-PS (PS/rGO-PS3) possessed a maximum dielectric permittivity of 394.6 at 100 Hz, which is 18, 2.2, and 2.9 times more than the values of PS/rGO, PS/rGO-PS1, and PS/rGO-PS2, respectively. In addition, PS/rGO-PS3 nanocomposites still maintained a relatively low dielectric loss of 0.45, which is even smaller than the value of PS/rGO (dielectric loss 1.74). This work provided an effective strategy to improve the dielectric performance through designing the interface structure of conductive fillers.

8.3.2.4 Carbon Nanofibers

Compared with carbon black and graphite powders, carbon nanofibers (CNFs) possess a high aspect ratio, and can achieve percolative levels in the nanocomposites at much smaller concentrations. Xu et al. prepared short electrospun

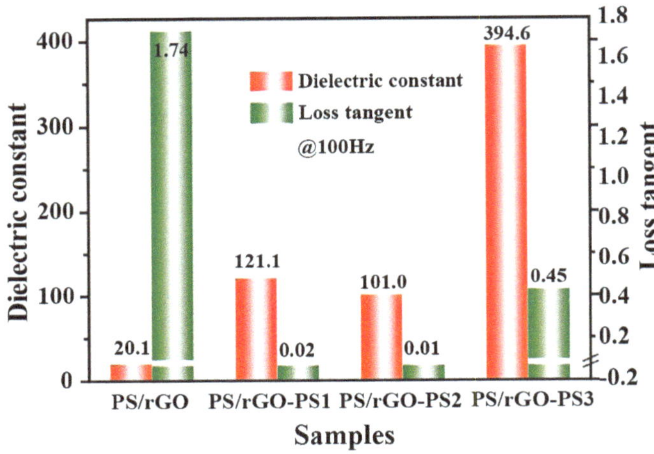

FIG. 8.14 Summary of the dielectric properties PS/rGO and PS/rGO-PS nanocomposites when the loading of fillers is fixed at 5 wt% at 100 Hz. *(Reproduced with permission from T. Zhang, W. Huang, N. Zhang, T. Huang, J. Yang, Y. Wang, Grafting of polystyrene onto reduced graphene oxide by emulsion polymerization for dielectric polymer composites: high dielectric constant and low dielectric loss tuned by varied grafting amount of polystyrene, Eur. Polym. J. 94 (2017) 196–207 (Copyright 2017 Elsevier Ltd.).)*

carbon nanofibers (SCNFs) and further fabricated SCNF/polyimide (SCNF/PI) nanocomposites by single spin-casting and thermal treatment [70]. The nanocomposites showed improved mechanical properties and dielectric performance. When the amount of SCNFs neared the percolation threshold of 4 wt %, the dielectric permittivity of the composites could be 60.79 at 100 Hz, while the dielectric loss was smaller than 0.5. Sui et al. blended CNFs with semi-crystalline polypropylene (PP) by a twin-screw extrusion [105]. The obtained PP nanocomposites with 5 wt% CNFs showed a very high dielectric permittivity of 600 under low frequency and 200 up to 4 kHz. In another report, Sun et al. designed a multi-layered structure with CNF and PVDF, and the CNF/PVDF nanocomposites achieved enhanced dielectric permittivity and low dielectric loss [106]. Similar CNF/PVDF dielectric nanocomposites were presented by Dang et al., whose research showed a low percolation threshold of 0.074 [107]. Besides of the preceding polymer matrix, CNFs were also blended with polyethylene [108] and an acrylic-polyurethane matrix [109]. Percolation theory and a microcapacitor model were used to explain the changes of dielectric properties altering the amount of CNFs.

8.3.2.5 Carbon Nanotubes

Carbon nanotubes (CNTs) are one-dimensional hollow nanomaterials. They have a high aspect length-to-diameter ratio, excellent mechanical properties with a tensile strength of 11–63 GPa, and a modulus of 270–950 GPa [110]. They also show superior electrical conductivity in that their electric resistance could be as low as 5×10^{-6} Ω/cm [111]. So CNTs are good candidates as fillers for producing polymer-based nanocomposites with improved properties. Xu et al. dispersed multi-walled CNTs (MWCNTs) into polyamic acid for electrospinning, and the obtained nanocomposite nanofibers were processed by thermal imidization and hot-pressing to produce the PI/MWCNTs nanocomposite films [112]. With this method, the MWCNTs could be homogeneously distributed in the PI matrix, which was beneficial to maintaining the mechanical and thermal properties [112]. In addition, the nanocomposites also presented high dielectric permittivity in the range of 147–161 at 1 kHz when the amount of MWCNTs was near the percolation threshold of around 12–14 vol%. By further increasing the amount of MWCNTs to 20 vol%, the highest dielectric permittivity of 217 at 1 kHz was achieved [112]. A similar preparation method with a combination of electrospinning and hot-pressing to produce CNT/polysulfone nanocomposites with improved dielectric properties was reported by Liu et al. [113]. In this report, in order to increase the dispersity of CNTs in a polysulfone solution, the CNTs were modified by jeffamines T403, and later formed a well-dispersed structural array nanophase in the polysulfone matrix. Because of the special core/shell structures, the nanocomposites exhibited a stable high dielectric permittivity, a low dielectric loss, and a large energy density [113]. CNTs were reported to be added into PVDF to improve the dielectric performance. Wang and Dang investigated dielectric properties of the untreated CNTs/PVDF nanocomposites [69]. They found the percolation threshold of the nanocomposite was only 1.61 vol%, and the highest dielectric permittivity was as high as 300 when the amount of CNTs was about 2.0 vol%. They believed that the large aspect ratio and the high conductivity of the CNTs were the reasons for the low percolation threshold. In another report, Li et al. compared the pristine, carboxylic and ester functionalized CNTs on the dielectric properties of the CNT/PVDF nanocomposites, and they found that all the three nanocomposites possessed an approximately equal percolation threshold of about 3.8 vol% [114]. The largest dielectric permittivity of 3600 at 1 kHz was achieved from the carboxylic functionalized CNT/PVDF nanocomposites with an amount of CNTs at about 8 vol%. Recently, Costa et al. pre-treated the CNTs by a corona, and then prepared CNT/PVDF nanocomposites [115]. When the amount of CNTs was 0.0419 volume fraction, the nanocomposites showed a high dielectric permittivity of near 3000, and a dielectric loss of about 2.8. Chen et al. prepared CNTs/PMMA nanocomposites by using surface-initiated atom transfer radical polymerization [116]. The CNT/PMMA nanocomposites showed tunable dielectric properties by controlling the PMMA shell on the CNT surface. The results indicated a higher percolation threshold from the thick PMMA shell, and the nanocomposites with 7 wt% PMMA@CNT possessed a high dielectric permittivity of 60, and a low dielectric loss of 0.26 at 1 kHz.

8.3.3 Conductive Polymers

Unlike the conductive metal and carbon materials, conductive polymers as fillers for the polymer-based nanocomposites have advantages of high polymerization yields, environmental stability, electrical conductivity, a simple blending process, good compatibility with the polymer matrix, and low production cost [117–119]. Conductive polymer fillers are good candidates for the preparation of all-organic, high dielectric permittivity, light, flexible, and inexpensive electronics. However, due to the insolubility in solvents, and their inability to be melted, most of the conductive polymer fillers were processed into nanoparticles/dispersion for application in nanocomposites. Until now, there have been few reports regarding the high dielectric, all-organic nanocomposites with conductive polymer fillers.

Among these studies, polyaniline (PANI) has been considered as one of the most promising conductive polymers. In 2003, Zhang and Huang incorporated conductive PANI particles into a P(VDF-TrFE-CTFE) terpolymer matrix, and the resultant allorganic polymer composites could achieve a superior dielectric permittivity (>1000) at 1 kHz, which was much higher than the value of the matrix (>50) [120]. Lu et al. applied in situ polymerization to disperse PANI into an epoxy matrix, and investigated the effect of the PANI loading amount on the dielectric properties of the composites [117]. The obtained PANI/epoxy composites possessed a high dielectric permittivity up to 3000, and the dielectric loss of 0.5 at 10 kHz. Yuan et al. dispersed the PANI particles into PVDF at the presence of solvent N,N-dimethylformamide (DMF) by grinding, and then obtained the PANI/PVDF composites by a simple film casting [121]. The allorganic composites with a PANI volume fraction of 0.05 (slightly larger than the percolation threshold of 0.042) possessed a dielectric permittivity as high as 385 at 1 kHz, with a breakdown strength of 60 MV/m, and energy storage density of 6.1 J/cm^3. In this report, the percolation theory and microcapacitor modeling were used to explain the trend of the filler amount dependent on dielectric properties. As shown in Fig. 8.15A, when the loading amount of PANI fillers was very small, the fillers were distributed well in the matrix, and only limited microcapacitors were formed, which led to a small dielectric permittivity (<20). As shown in Fig. 8.15B and C, when the amount of PANI approached the percolation threshold, more and more microcapacitors were formed, and the dielectric permittivity was also dramatically increased to 385. Further increasing the content of filler would lead to the connection of long conductive chains of polyaniline, which would cause the large leakage current and also result in the drop of dielectric permittivity to 247. Besides PANI, other conducting polymers, such as polypyrrole (PPy), have also been selected as fillers for all-organic nanocomposites [122]. The PPy with the morphology of nanoclips avoided the formation of long conductive nano-channels due to the tangle of long fibers, resulting in a low percolation threshold (<8%). The nanocomposite exhibited a dielectric permittivity of more than 1000 at room temperature, and about 2000 at temperatures higher than 98°C.

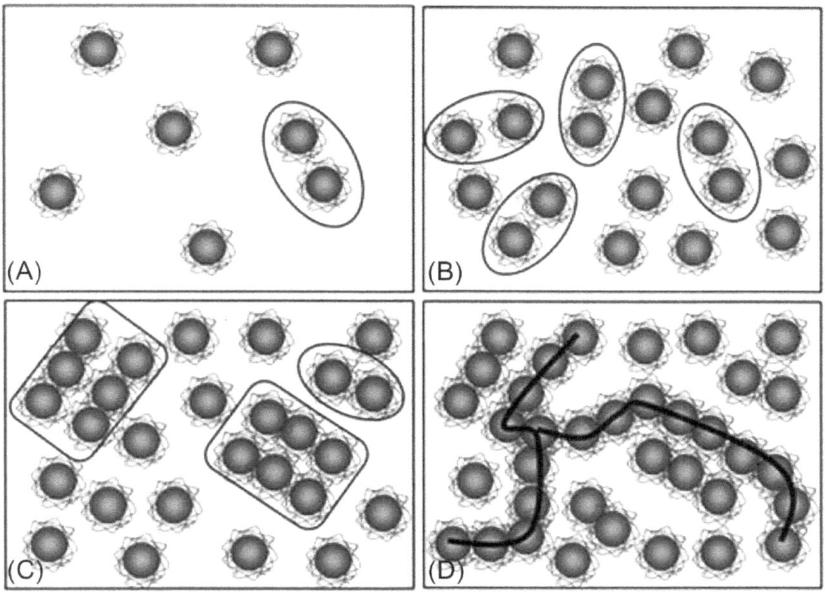

FIG. 8.15 Schematic images of the microstructure of PANI/PVDF films with (A) $0<f\leq 0.037$, (B) $0.037<f\leq 0.042$, (C) $0.042<f\leq 0.050$ and (D) $0.050<f\leq 0.060$ respectively. The microcapacitors were demonstrated by the ellipse, and the solid line showed the long conductive chains of polyaniline. *(Reproduced with permission from J.-K. Yuan, Z.-M. Dang, S.-H. Yao, J.-W. Zha, T. Zhou, S.-T. Li, J. Bai, Fabrication and dielectric properties of advanced high permittivity polyaniline/poly (vinylidene fluoride) nanohybrid films with high energy storage density, J. Mater. Chem. 20 (12) (2010) 2441–2447 (Copyright 2010 The Royal Society of Chemistry).)*

8.4 HIGH PERMITTIVITY POLYMER-BASED NANOCOMPOSITES WITH HYBRID FILLERS

In Sections 8.2 and 8.3, fillers with single components are used to prepare high permittivity polymer-based nanocomposites. However, there are some disadvantages when using single-component fillers. For example, dielectric fillers make it difficult to achieve improvements on mechanical and dielectric properties, while conductive fillers have problems of homogeneous dispersion in the polymer matrix. Therefore, it is necessary to develop hybrid fillers with different combinations to improve the dielectric performance of polymer-based nanocomposites.

8.4.1 Hybrid Fillers to Improve the Dispersion

Aggregation and large additions of fillers are always problems for the CNT/polymer high permittivity nanocomposites. Recently, Liao et al. adopted a solvent-exchange strategy to improve the dispersion of CNTs in a PI matrix by introducing the third phase of graphene [123]. The hybrid CNT/graphene fillers could construct a three-dimensional network, and be kept for a long time. Via an in situ polymerization and thermal imidization, the obtained CNT/graphene/PI nanocomposites showed a high dielectric permittivity of 124.9 at 100 Hz, which was 4000% higher than that of a neat PI matrix. As shown in Fig. 8.16A, two frequency regimes were observed, which means there were two different dielectric relaxation processes in the nanocomposites. Fig. 8.16B clearly presented that the CNT/graphene/PI nanocomposites from the solvent-exchange method had much larger dielectric permittivity than the CNT/PI nanocomposites, and CNT/graphen/PI from a simple blending method, which could be attributed to the much better dispersion of fillers from the sample solvent-exchange method. In addition, Fig. 8.16C compared the filler amount of dependent dielectric permittivity of different PI-based nanocomposite materials. The results indicated that the CNT/graphene/PI from the solvent-exchange possessed an important area, where high dielectric permittivity and low filler amount were achieved.

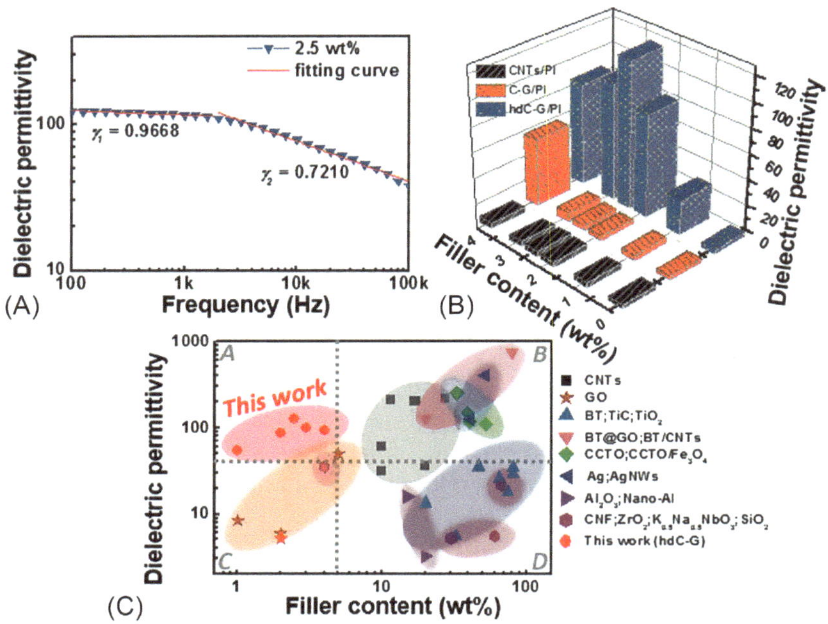

FIG. 8.16 Dielectric permittivity versus frequency of composite with 2.5 wt% and fitting curves (A); dependence of dielectric permittivity of the three kinds of PI based composite films on the weight percent of fillers (B); and dielectric permittivity (1 kHz) of PI composites in this work and other previous studies with different fillers (C). *(Reproduced with permission from X. Liao, W. Ye, L. Chen, S. Jiang, G. Wang, L. Zhang, H. Hou, Flexible hdC-G reinforced polyimide composites with high dielectric permittivity, Compos. Part A 101 (2017) 50–58 (Copyright 2017 Elsevier Ltd.).)*

To improve the dispersion of fillers, PANI, PPy, and polydopamine (PDA) were coated on the fillers. Yang et al. coated CNTs with polypyrrol by an inverse microemulsion polymerization [124]. The organic coating shell of PPy on the one hand ensured the good dispersion of CNTs in the PS matrix, and on the other hand, screened the movement to reduce the current leakage. Therefore, the obtained nanocomposites showed a high dielectric permittivity of about 44, and a quite small dielectric loss of <0.07. A similar strategy was also adopted by Xu et al. [125]. In their report, PANI was coated on CNTs by in situ polymerization to improve the dispersibility in the PU matrix, and the obtained composites showed good mechanical properties and dielectric properties. In another report, PANI was decorated on the reduced graphene oxide by in situ polymerization [126]. The PANI/graphene hybrid filler led to a higher dielectric permittivity and lower dielectric loss of the nanocomposites than the composites with only graphene as fillers. Dopamine could be easily polymerized under a basic solution and deposited on different substrates, including organic and inorganic materials [127]. Zhang et al. [128], Zhou et al. [129], and Zhu et al. [130] decorated PDA on silver nanoparticles, silver nanowires, and CNTs, respectively. The obtained hybrid fillers were blended with PVDF and PVA to form nanocomposites. Because of the improved distribution of fillers in the polymer matrix, all these nanocomposites presented a high dielectric permittivity accompanied with a low dielectric loss.

8.4.2 Dielectric and Conductive Hybrid Fillers

Besides the decoration of fillers by an additional polymer shell, another strategy by conbination of high dielectric permittivity inorganic fillers and conductive fillers is used to fabricate polymer-based high dielectric permittivity nanocoposites. Table 8.3 summarized the dielectric permittivity and dielectric loss of the polymer-based nanocomposites with hybrid dielectric and conductive fillers. BT and TiO_2 are two kinds of ferroelectric ceramics with high dielectric permittivity; while graphene, silver, and CNT are the conventional conductive fillers for the polymer-based nanocomposites. As shown in Table 8.3, many kinds of combinations could be used as hybrid fillers for different polymer matrice. The group of BT/graphene/PI nanocomposites with the addition of 17 vol% hybrid fillers showed a high dielectric permittivity of 31 with an extremely low dielectric loss of 0.0378 [131], while the conbination of BT/graphene/PI nanocmoposites with 50 vol% hybrid fillers presented a super high dielectric permittivity of 1062, but a relatively high dielectric loss of 0.23 [132]. When using PP as a polymer matric, the hybrid BT/silver and BT/CNT resulted in a dielectric permittivity of 160 and 28, at 1000 and 100 Hz, respectively [133,134]. At the same frequency of 100 Hz and using the same polymer matrix, the hybrid fillers, and the addition of 21.44 vol% BT /graphite led to a dielectric permittivity of 51.8 [135], while addition of 10.9 vol% TiO_2/

TABLE 8.3 Dielectric Properties of Polymer-Based Nanocomposites with Hybrid Dielectric and Conductive Fillers

Dielectric Filler	Conductive Filler	Filler Amount	Polymer	f (Hz)	ε_r	tanδ	Ref.
BT	Graphene	17 vol%	PI	100	31	0.038	[131]
BT	CNT	50 vol%	PI	100	1062	0.23	[132]
BT	silver	56.8 vol%	PP	1k	160	0.11	[133]
BT	CNT	42 wt%	PP	100	28	0.07	[134]
BT	Graphite	21.44 vol%	PS	100	51.8	0.249	[135]
TiO_2	Graphene	10.9 vol%	PS	100	1741	0.39	[136]
BT	Graphene	8 wt%	PVDF	1k	56.3	0.057	[137]
TiO_2	silver	35 vol%	P(VDF-HFP)	100	33.4	0.023	[138]
BT	MWCNT	37.1 vol%	P(VDF-HFP)	1k	71.6	0.04	[139]

graphene led to a much higher value of 1741 [136]. The possible reason for this significant difference could be attributed to the more homogenous dispersion of TiO_2/graphene than BT/graphite. In addition, the conbinations of BT/graphene and TiO_2/silver were reported as fillers to achieve high permittivity polymer-based nanocomposites. BT/graphene/PVDF and TiO_2/silver/P(VDF-HFP) nanocomposites exhibited dielectric permittivity/dielectric loss of 56.3/0.057 at 1 kHz and 33.4/0.023 at 100 Hz, respectively.

8.5 CONCLUDING REMARKS

In summary, polymer-based dielectric nanocomposites have been widely developed, and will continue to attract more research attention in the future to meet the requirements of the next-generation electronic devices and systems. According to the electric properties of fillers, two different kinds of nanocomposites have been investigated. Although it is not difficult to obtain high dielectric perimittivity, the critical factor of energy storage for polymer-based dielectric nanocomposites must be considered, to enhance dielectric permittivity and maintain high dielectric strength, or reduce dielectric loss simultaneously. The improvements on different types of nanocomposites may be a little different. For dielectric-polymer composites, higher filler concentrations may lead to low breakdown strength of the composites due to the inorganic fillers. The optimization of the interface between the organic matrix and inorganic fillers by surface functionalization of filler nanoparticles is critical for excellent compatibility between the fillers and the polymer matrix. The nature of the interfaces and the mechanism of the dielectric response of interfaces are still not well understood. The contribution from the interfaces can be exploited to improve the breakdown strength or enhancement of the polarization response. A systematic study of the chemical structures of polymers and fillers, and size and morphology of fillers versus dielectric performance at high electric fields has to be addressed in future studies. For conductor-polymer composites, there are some physical issues that are unclear, such as the dielectric response near the percolation threshold, the contribution of high dielectric permittivity is dominated by the dielectric relaxation process or electric conduction, and the tunneling effect in the adjacent conducting particles. Furthermore, the percolation threshold φ_c is the most important parameter, and many approaches have been proposed to control it. However, understanding the relationship between φ_c and the microstructural features of composites is still challenging. For nanocomposites with hybrid fillers, the optimization of the volume fraction of both fillers in the matrix strongly determines the dielectric properties of the final composites. The quantitative study on the charge migration and accumulation in the polymer matrix and on the surface of two different fillers is needed.

ACKNOWLEDGMENTS

This work was supported by the Priority Academic Program Development of Jiangsu Higher Education Institutions (PAPD), National Basic Research Program of China (Grant No. 2015CB654602), and 111 Project under Grant No. B14040 and the Fundamental Research Funds for the Central Universities at Xi'an Jiaotong University.

REFERENCES

[1] A.V. Hippel, Dielectric Materials and Applications, Technology Press of MIT, Cambridge, Boston, 1954.
[2] K.C. Kao, Dielectric Phenomena in Solids, Elsevier Academic Press, San Diego, CA, 2004.
[3] C.J. Dias, D.K. Das-Gupta, Ferroelectric Polymer and Ceramic-Polymer Composites, Trans Tech Publications Ltd., Switzerland, 1994
[4] Z.-M. Dang, J.-K. Yuan, S.-H. Yao, R.-J. Liao, Flexible nanodielectric materials with high permittivity for power energy storage, Adv. Mater. 25 (44) (2013) 6334–6365.
[5] L. Zhu, Q. Wang, Novel ferroelectric polymers for high energy density and low loss dielectrics, Macromolecules 45 (7) (2012) 2937–2954.
[6] M.W. Barsoum, Fundamentals of Ceramics, Institute of Physics Publishing, Bristol and Philadephia, 1997.
[7] H.S. Nalwa, Handbook of Low and High Dielectric Constant Materials and Their Applications, Academic Press, London, 1999.
[8] R.F. Cava, W.F. Peck Jr, J.J. Krajewski, Enhancement of the dielectric constant of Ta2O5 through substitution with TiO2, Nature 377 (1995) 215.
[9] A. Peláiz-Barranco, F. Calderón-Piñar, O. García-Zaldívar, Y. González-Abreu, Relaxor behaviour in ferroelectric ceramics, in: A.P. Barranco (Ed.), Advances in Ferroelectrics, InTech, Rijeka, 2012. Ch. 5.
[10] Z.Y. Cheng, R.S. Katiyar, X. Yao, A.S. Bhalla, Temperature dependence of the dielectric constant of relaxor ferroelectrics, Phys. Rev. B 57 (14) (1998) 8166–8177.
[11] R.Z. Hou, X.M. Chen, Y.W. Zeng, Diffuse ferroelectric phase transition and relaxor behaviors in Ba-based bismuth layer-structured compounds and La-substituted SrBi4Ti4O15, J. Am. Ceram. Soc. 89 (9) (2006) 2839–2844.
[12] W. Hu, Y. Liu, R.L. Withers, T.J. Frankcombe, L. Norén, A. Snashall, M. Kitchin, P. Smith, B. Gong, H. Chen, J. Schiemer, F. Brink, J. Wong-Leung, Electron-pinned defect-dipoles for high-performance colossal permittivity materials, Nat. Mater. 12 (2013) 821.
[13] M.A. Subramanian, D. Li, N. Duan, B.A. Reisner, A.W. Sleight, High dielectric constant in ACu3Ti4O12 and ACu3Ti3FeO12 phases, J. Solid State Chem. 151 (2) (2000) 323–325.
[14] A.P. Ramirez, M.A. Subramanian, M. Gardel, G. Blumberg, D. Li, T. Vogt, S.M. Shapiro, Giant dielectric constant response in a copper-titanate, Solid State Commun. 115 (5) (2000) 217–220.
[15] L. Zhang, Z.Y. Cheng, Development of polymer-based 0-3 composites with high dielectric constant, J. Adv. Dielectr. 01 (04) (2011) 389–406.
[16] H.S. Nalwa, Ferroelectric Polymers: Chemistry, Physics,and Applications, Marcel Dekker, Inc., New York, Basel, Hong Kong, 1995
[17] Z. Zhang, T.C.M. Chung, Study of VDF/TrFE/CTFE terpolymers for high pulsed capacitor with high energy density and low energy loss, Macromolecules 40 (4) (2007) 783–785.
[18] Z.-Y. Cheng, Q.M. Zhang, F.B. Bateman, Dielectric relaxation behavior and its relation to microstructure in relaxor ferroelectric polymers: high-energy electron irradiated poly(vinylidene fluoride–trifluoroethylene) copolymers, J. Appl. Phys. 92 (11) (2002) 6749–6755.

[19] V.K. Thakur, M.R. Kessler, Polymer nanocomposites: new advanced dielectric materials for energy storage applications, in: Advanced Energy Materials, John Wiley & Sons, Inc., 2014, pp. 207–257.
[20] R.E. Newnham, D.P. Skinner, L.E. Cross, Connectivity and piezoelectric-pyroelectric composites, Mater. Res. Bull. 13 (5) (1978) 525–536.
[21] D.P. Skinner, R.E. Newnham, L.E. Cross, Flexible composite transducers, Mater. Res. Bull. 13 (6) (1978) 599–607.
[22] N. Jayasundere, B.V. Smith, J.R. Dunn, Piezoelectric constant for binary piezoelectric 0-3 connectivity composites and the effect of mixed connectivity, J. Appl. Phys. 76 (5) (1994) 2993–2998.
[23] Y. Shen, X. Zhang, Y. Lin, C.-W. Nan, Polymer nanocomposites dielectrics for energy applications, in: Z. Lin, Y. Yang, A. Zhang (Eds.), Polymer-Engineered Nanostructures for Advanced Energy Applications, Springer International Publishing, Cham, 2017, pp. 511–534.
[24] Z.-M. Dang, J.-K. Yuan, J.-W. Zha, T. Zhou, S.-T. Li, G.-H. Hu, Fundamentals, processes and applications of high-permittivity polymer–matrix composites, Prog. Mater. Sci. 57 (4) (2012) 660–723.
[25] Z.-M. Dang, Polymer nanocomposites with high permittivity, in: S.-C. Tjong (Ed.), Nanocrystalline Materials, second ed., Elsevier, Oxford, 2014, pp. 305–333.
[26] L. Zhang, P. Bass, Z.Y. Cheng, Physical aspects of 0-3 dielectric composites, J. Adv. Dielectr. 05 (02) (2015).
[27] L. Rayleigh, LVI. On the influence of obstacles arranged in rectangular order upon the properties of a medium, Philos. Mag. Ser. 34 (211) (1892) 481–502.
[28] D.K. Hale, The physical properties of composite materials, J. Mater. Sci. 11 (11) (1976) 2105–2141.
[29] X. Huang, P. Jiang, Core–shell structured high-k polymer nanocomposites for energy storage and dielectric applications, Adv. Mater. 27 (3) (2015) 546–554.
[30] V.K.T. Prateek, R.K. Gupta, Recent progress on ferroelectric polymer-based nanocomposites for high energy density capacitors: synthesis, dielectric properties, and future aspects, Chem. Rev. 116 (7) (2016) 4260–4317.
[31] Z.-M. Dang, H.-Y. Wang, H.-P. Xu, Influence of silane coupling agent on morphology and dielectric property in BaTiO3/polyvinylidene fluoride composites, Appl. Phys. Lett. 89 (11) (2006).
[32] P. Kim, S.C. Jones, P.J. Hotchkiss, J.N. Haddock, B. Kippelen, S.R. Marder, J.W. Perry, Phosphonic acid-modified barium titanate polymer nanocomposites with high permittivity and dielectric strength, Adv. Mater. 19 (7) (2007) 1001–1005.
[33] J. Li, J. Claude, L.E. Norena-Franco, S.I. Seok, Q. Wang, Electrical energy storage in ferroelectric polymer nanocomposites containing surface-functionalized BaTiO3 nanoparticles, Chem. Mater. 20 (20) (2008) 6304–6306.
[34] P. Kim, N.M. Doss, J.P. Tillotson, P.J. Hotchkiss, M.-J. Pan, S.R. Marder, J. Li, J.P. Calame, J.W. Perry, High energy density nanocomposites based on surface-modified BaTiO3 and a ferroelectric polymer, ACS Nano 3 (9) (2009) 2581–2592.
[35] M.N. Almadhoun, U.S. Bhansali, H.N. Alshareef, Nanocomposites of ferroelectric polymers with surface-hydroxylated BaTiO3 nanoparticles for energy storage applications, J. Mater. Chem. 22 (22) (2012) 11196–11200.
[36] K. Yu, H. Wang, Y. Zhou, Y. Bai, Y. Niu, Enhanced dielectric properties of BaTiO3/poly (vinylidene fluoride) nanocomposites for energy storage applications, J. Appl. Phys. 113 (3) (2013).

[37] H.M. Jung, J.-H. Kang, S.Y. Yang, J.C. Won, Y.S. Kim, Barium titanate nanoparticles with diblock copolymer shielding layers for high-energy density nanocomposites, Chem. Mater. 22 (2) (2010) 450–456.

[38] H. Tang, Y. Lin, H.A. Sodano, Synthesis of high aspect ratio BaTiO3 nanowires for high energy density nanocomposite capacitors, Adv. Energy Mater. 3 (4) (2013) 451–456.

[39] T. Haixiong, L. Yirong, A. Clark, A.S. Henry, Nanocomposites with increased energy density through high aspect ratio PZT nanowires, Nanotechnology 22 (1) (2011).

[40] K.S. Lam, Y.W. Wong, L.S. Tai, Y.M. Poon, F.G. Shin, Dielectric and pyroelectric properties of lead zirconate titanate/polyurethane composites, J. Appl. Phys. 96 (7) (2004) 3896–3899.

[41] K.H. Lam, H.L.W. Chan, H.S. Luo, Q.R. Yin, Z.W. Yin, C.L. Choy, Dielectric properties of 65PMN–35PT/P(VDF–TrFE) 0–3 composites, Microelectron. Eng. 66 (1) (2003) 792–797.

[42] Y. Bai, Z.Y. Cheng, V. Bharti, H.S. Xu, Q.M. Zhang, High-dielectric-constant ceramic-powder polymer composites, Appl. Phys. Lett. 76 (25) (2000) 3804–3806.

[43] Y. Song, Y. Shen, H. Liu, Y. Lin, M. Li, C.-W. Nan, Improving the dielectric constants and breakdown strength of polymer composites: effects of the shape of the BaTiO3 nanoinclusions, surface modification and polymer matrix, J. Mater. Chem. 22 (32) (2012) 16491–16498.

[44] P. Wu, M. Zhang, H. Wang, H. Tang, P. Bass, L. Zhang, Effect of coupling agents on the dielectric properties and energy storage of Ba0.5Sr0.5TiO3/P(VDF-CTFE) nanocomposites, AIP Adv. 7 (7) (2017).

[45] W. Xia, Z. Xu, F. Wen, Z. Zhang, Electrical energy density and dielectric properties of poly (vinylidene fluoride-chlorotrifluoroethylene)/BaSrTiO3 nanocomposites, Ceram. Int. 38 (2) (2012) 1071–1075.

[46] H. Tang, H.A. Sodano, Ultra high energy density nanocomposite capacitors with fast discharge using Ba0.2Sr0.8TiO3 nanowires, Nano Lett. 13 (4) (2013) 1373–1379.

[47] Y. Song, Y. Shen, P. Hu, Y. Lin, M. Li, C.W. Nan, Significant enhancement in energy density of polymer composites induced by dopamine-modified Ba0.6Sr0.4TiO3 nanofibers, Appl. Phys. Lett. 101 (15) (2012).

[48] L. Zhang, P. Wu, Y. Li, Z.Y. Cheng, J.C. Brewer, Preparation process and dielectric properties of Ba0.5Sr0.5TiO3–P(VDF–CTFE) nanocomposites, Compos. Part B 56 (2014) 284–289.

[49] P. Wu, L. Zhang, X. Shan, Microstructure and dielectric response of BaSrTiO3/P(VDF-CTFE) nanocomposites, Mater. Lett. 159 (2015) 72–75.

[50] C.C. Homes, T. Vogt, S.M. Shapiro, S. Wakimoto, A.P. Ramirez, Optical response of high-dielectric-constant perovskite-related oxide, Science 293 (5530) (2001) 673.

[51] Z. Guozhong, Z. Jialiang, Z. Peng, W. Jinfeng, W. Chunlei, Grain boundary effect on the dielectric properties of CaCu 3 Ti 4 O 12 ceramics, J. Phys. D. Appl. Phys. 38 (11) (2005) 1824.

[52] M.A. Subramanian, A.W. Sleight, ACu3Ti4O12 and ACu3Ru4O12 perovskites: high dielectric constants and valence degeneracy, Solid State Sci. 4 (3) (2002) 347–351.

[53] S.F. Shao, J.L. Zhang, P. Zheng, W.L. Zhong, C.L. Wang, Microstructure and electrical properties of CaCu3Ti4O12 ceramics, J. Appl. Phys. 99 (8) (2006).

[54] Y. Shen, A. Gu, G. Liang, L. Yuan, High performance CaCu3Ti4O12/cyanate ester composites with excellent dielectric properties and thermal resistance, Compos. A: Appl. Sci. Manuf. 41 (11) (2010) 1668–1676.

[55] M. Arbatti, X. Shan, Z.Y. Cheng, Ceramic–polymer composites with high dielectric constant, Adv. Mater. 19 (10) (2007) 1369–1372.

[56] L. Zhang, X. Shan, P. Bass, Y. Tong, T.D. Rolin, C.W. Hill, J.C. Brewer, D.S. Tucker, Z.Y. Cheng, Process and microstructure to achieve ultra-high dielectric constant in ceramic-polymer composites, Sci. Rep. 6 (2016).

[57] L. Zhang, X. Shan, P.W. Wu, J.S. Song, Z.-Y. Cheng, Microstructure and dielectric properties of CCTO-P(VDF-TrFE) nanocomposites, Ferroelectrics 405 (1) (2010) 92–97.
[58] L. Zhang, X. Shan, P. Wu, Z.Y. Cheng, Dielectric characteristics of CaCu3Ti4O12/P(VDF-TrFE) nanocomposites, Appl. Phys. A 107 (3) (2012) 597–602.
[59] X. Shan, L. Zhang, X. Yang, Z.Y. Cheng, Dielectric composites with a high and temperature-independent dielectric constant, J. Adv. Ceram. 1 (4) (2012) 310–316.
[60] Z.-M. Dang, T. Zhou, S.-H. Yao, J.-K. Yuan, J.-W. Zha, H.-T. Song, J.-Y. Li, Q. Chen, W.-T. Yang, J. Bai, Advanced calcium copper titanate/polyimide functional hybrid films with high dielectric permittivity, Adv. Mater. 21 (20) (2009) 2077–2082.
[61] W. Yang, S. Yu, R. Sun, R. Du, Nano- and microsize effect of CCTO fillers on the dielectric behavior of CCTO/PVDF composites, Acta Mater. 59 (14) (2011) 5593–5602.
[62] C. Ehrhardt, C. Fettkenhauer, J. Glenneberg, W. Munchgesang, H.S. Leipner, M. Diestelhorst, S. Lemm, H. Beige, S.G. Ebbinghaus, A solution-based approach to composite dielectric films of surface functionalized CaCu3Ti4O12 and P(VDF-HFP), J. Mater. Chem. A 2 (7) (2014) 2266–2274.
[63] H. Tang, Z. Zhou, C.C. Bowland, H.A. Sodano, Synthesis of calcium copper titanate (CaCu3Ti4O12) nanowires with insulating SiO2 barrier for low loss high dielectric constant nanocomposites, Nano Energy 17 (Suppl C) (2015) 302–307.
[64] J. Li, S.I. Seok, B. Chu, F. Dogan, Q. Zhang, Q. Wang, Nanocomposites of ferroelectric polymers with TiO2 nanoparticles exhibiting significantly enhanced electrical energy density, Adv. Mater. 21 (2) (2009) 217–221.
[65] B. Chu, M. Lin, B. Neese, X. Zhou, Q. Chen, Q.M. Zhang, Large enhancement in polarization response and energy density of poly(vinylidene fluoride-trifluoroethylene-chlorofluoroethylene) by interface effect in nanocomposites, Appl. Phys. Lett. 91 (12) (2007).
[66] H. Tang, H.A. Sodano, High energy density nanocomposite capacitors using non-ferroelectric nanowires, Appl. Phys. Lett. 102 (6) (2013).
[67] Z.M. Dang, Y.H. Lin, C.W. Nan, Novel ferroelectric polymer composites with high dielectric constants, Adv. Mater. 15 (19) (2003) 1625–1629.
[68] R. Sengupta, M. Bhattacharya, S. Bandyopadhyay, A.K. Bhowmick, A review on the mechanical and electrical properties of graphite and modified graphite reinforced polymer composites, Prog. Polym. Sci. 36 (5) (2011) 638–670.
[69] L. Wang, Z.-M. Dang, Carbon nanotube composites with high dielectric constant at low percolation threshold, Appl. Phys. Lett. 87 (4) (2005).
[70] W. Xu, Y. Feng, Y. Ding, S. Jiang, H. Fang, H. Hou, Short electrospun carbon nanofiber reinforced polyimide composite with high dielectric permittivity, Mater. Lett. 161 (2015) 431–434.
[71] L.J. Romasanta, M. Hernández, M.A. López-Manchado, R. Verdejo, Functionalised graphene sheets as effective high dielectric constant fillers, Nanoscale Res. Lett. 6 (1) (2011) 508.
[72] C. Huang, Q. Zhang, Enhanced dielectric and electromechanical responses in high dielectric constant all-polymer percolative composites, Advanced Functional Materials14(5) (2004) 501–506.
[73] K. Deshmukh, M.B. Ahamed, S.K. Pasha, R.R. Deshmukh, P.R. Bhagat, Highly dispersible graphene oxide reinforced polypyrrole/polyvinyl alcohol blend nanocomposites with high dielectric constant and low dielectric loss, RSC Adv. 5 (76) (2015) 61933–61945.
[74] C.W. Nan, Y. Shen, J. Ma, Physical properties of composites near percolation, Annu. Rev. Mater. Res. 40 (1) (2010) 131–151.
[75] L. Zhang, P. Bass, Z.-M. Dang, Z.Y. Cheng, Characterization of percolation behavior in conductor–dielectric 0-3 composites, J. Adv. Dielectr. 04 (04) (2014).

[76] Z.-M. Dang, M.-S. Zheng, J.-W. Zha, 1D/2D carbon nanomaterial-polymer dielectric composites with high permittivity for power energy storage applications, Small 12 (13) (2016) 1688–1701.
[77] L. Qi, B.I. Lee, S. Chen, W.D. Samuels, G.J. Exarhos, High-dielectric-constant silver–epoxy composites as embedded dielectrics, Adv. Mater. 17 (14) (2005) 1777–1781.
[78] X. Huang, P. Jiang, L. Xie, Ferroelectric polymer/silver nanocomposites with high dielectric constant and high thermal conductivity, Appl. Phys. Lett. 95 (24) (2009).
[79] Z.-M. Dang, B. Peng, D. Xie, S.-H. Yao, M.-J. Jiang, J. Bai, High dielectric permittivity silver/polyimide composite films with excellent thermal stability, Appl. Phys. Lett. 92 (11) (2008).
[80] W. Zheng, X. Lu, W. Wang, Z. Wang, M. Song, Y. Wang, C. Wang, Fabrication of novel Ag nanowires/poly(vinylidene fluoride) nanocomposite film with high dielectric constant, Phys. Status Solidi A 207 (8) (2010) 1870–1873.
[81] Y.G. Chen, S.J. Wang, High dielectric constant of silver nanowires-epoxy composites, Mater. Sci. Forum, 2016, p. 9 Trans Tech Publications Ltd.
[82] H.-Y. Mi, Z. Li, L.-S. Turng, Y. Sun, S. Gong, Silver nanowire/thermoplastic polyurethane elastomer nanocomposites: thermal, mechanical, and dielectric properties, Mater. Design (1980-2015) 56 (Suppl C) (2014) 398–404.
[83] T. Inui, H. Koga, M. Nogi, N. Komoda, K. Suganuma, High-dielectric paper composite consisting of cellulose nanofiber and silver nanowire, 14th IEEE International Conference on Nanotechnology, 2014, pp. 470–473.
[84] Y.-J. Li, M. Xu, J.-Q. Feng, Z.-M. Dang, Dielectric behavior of a metal-polymer composite with low percolation threshold, Appl. Phys. Lett. 89 (7) (2006).
[85] H.-P. Xu, Z.-M. Dang, N.-C. Bing, Y.-H. Wu, D.-D. Yang, Temperature dependence of electric and dielectric behaviors of Ni/polyvinylidene fluoride composites, J. Appl. Phys. 107 (3) (2010).
[86] L. Zhang, W. Wang, X. Wang, P. Bass, Z.-Y. Cheng, Metal-polymer nanocomposites with high percolation threshold and high dielectric constant, Appl. Phys. Lett. 103 (23) (2013).
[87] L. Zhang, P. Bass, Z.-Y. Cheng, Revisiting the percolation phenomena in dielectric composites with conducting fillers, Appl. Phys. Lett. 105 (4) (2014).
[88] L. Zhang, X. Wang, Z.Y. Cheng, A case study of conductor-dielectric 0-3 composites using Ni-P(VDF-CTFE) nanocomposites, J. Adv. Phys. 4 (4) (2015) 362–369.
[89] L. Zhang, P. Bass, G. Wang, Y. Tong, Z. Xu, Z.Y. Cheng, Dielectric response and percolation behavior of Ni–P(VDF–TrFE) nanocomposites, J. Adv. Dielectr. (2017).
[90] A.B. da Silva, M. Arjmand, U. Sundararaj, R.E.S. Bretas, Novel composites of copper nanowire/PVDF with superior dielectric properties, Polymer 55 (1) (2014) 226–234.
[91] R. Alcántara, J.M. Jiménez-Mateos, P. Lavela, J.L. Tirado, Carbon black: a promising electrode material for sodium-ion batteries, Electrochem. Commun. 3 (11) (2001) 639–642.
[92] K. Chung, A. Sabo, A. Pica, Electrical permittivity and conductivity of carbon black-polyvinyl chloride composites, J. Appl. Phys. 53 (10) (1982) 6867–6879.
[93] C. Brosseau, F. Boulic, P. Queffelec, C. Bourbigot, Y. Le Mest, J. Loaec, A. Beroual, Dielectric and microstructure properties of polymer carbon black composites, J. Appl. Phys. 81 (2) (1997) 882–891.
[94] C. Calberg, S. Blacher, F. Gubbels, F. Brouers, R. Deltour, R. Jérôme, Electrical and dielectric properties of carbon black filled co-continuous two-phase polymer blends, J. Phys. D. Appl. Phys. 32 (13) (1999) 1517.
[95] J. Xu, M. Wong, C. Wong, Super high dielectric constant carbon black-filled polymer composites as integral capacitor dielectrics, Electronic Components and Technology Conference, 2004. Proceedings. 54th, IEEE, 2004, pp. 536–541.

[96] J. Sun, Q. Xue, L. Chu, Q. Guo, T. Lei, F. Xia, Z. Zhang, Ultra-high dielectric constant of poly (vinylidene fluoride) composites filled with hydroxyl modified graphite powders, Polym. Compos. 37 (2) (2016) 327–333.

[97] F. He, S. Lau, H.L. Chan, J. Fan, High dielectric permittivity and low percolation threshold in nanocomposites based on poly (vinylidene fluoride) and exfoliated graphite nanoplates, Adv. Mater. 21 (6) (2009) 710–715.

[98] C. Min, D. Yu, J. Cao, G. Wang, L. Feng, A graphite nanoplatelet/epoxy composite with high dielectric constant and high thermal conductivity, Carbon 55 (2013) 116–125.

[99] M. Xiao, L. Sun, J. Liu, Y. Li, K. Gong, Synthesis and properties of polystyrene/graphite nanocomposites, Polymer 43 (8) (2002) 2245–2248.

[100] H. Liu, P. Xu, H. Yao, W. Chen, J. Zhao, C. Kang, Z. Bian, L. Gao, H. Guo, Controllable reduction of graphene oxide and its application during the fabrication of high dielectric constant composites, Appl. Surf. Sci. 420 (2017) 390–398.

[101] M. Devi, A. Kumar, Structural, thermal and dielectric properties of in-situ reduced graphene oxide—polypyrrole nanotubes nanocomposites, Mater. Res. Bull. 97 (2018) 207–214.

[102] P. Fan, L. Wang, J. Yang, F. Chen, M. Zhong, Graphene/poly (vinylidene fluoride) composites with high dielectric constant and low percolation threshold, Nanotechnology 23 (36) (2012).

[103] Z. Wang, N.M. Han, Y. Wu, X. Liu, X. Shen, Q. Zheng, J.-K. Kim, Ultrahigh dielectric constant and low loss of highly-aligned graphene aerogel/poly(vinyl alcohol) composites with insulating barriers, Carbon 123 (Suppl C) (2017) 385–394.

[104] T. Zhang, W. Huang, N. Zhang, T. Huang, J. Yang, Y. Wang, Grafting of polystyrene onto reduced graphene oxide by emulsion polymerization for dielectric polymer composites: high dielectric constant and low dielectric loss tuned by varied grafting amount of polystyrene, Eur. Polym. J. 94 (2017) 196–207.

[105] G. Sui, S. Jana, W. Zhong, M. Fuqua, C. Ulven, Dielectric properties and conductivity of carbon nanofiber/semi-crystalline polymer composites, Acta Mater. 56 (10) (2008) 2381–2388.

[106] L. Sun, B. Li, Y. Zhao, G. Mitchell, W. Zhong, Structure-induced high dielectric constant and low loss of CNF/PVDF composites with heterogeneous CNF distribution, Nanotechnology 21 (30) (2010).

[107] Z.-M. Dang, J.-P. Wu, H.-P. Xu, S.-H. Yao, M.-J. Jiang, J. Bai, Dielectric properties of upright carbon fiber filled poly (vinylidene fluoride) composite with low percolation threshold and weak temperature dependence, Appl. Phys. Lett. 91 (7) (2007).

[108] Z. Dang, Y. Shen, L. Fan, N. Cai, C. Nan, S. Zhao, Dielectric properties of carbon fiber filled low-density polyethylene, J. Appl. Phys. 93 (9) (2003) 5543–5545.

[109] S. Pan, Z. Wang, M. Chen, Y. Liu, R. Fan, Enhanced permittivity in flexible carbon-fiber and acrylic-polyurethane composites, Mater. Lett. 205 (2017) 44–47.

[110] M.-F. Yu, O. Lourie, M.J. Dyer, K. Moloni, T.F. Kelly, R.S. Ruoff, Strength and breaking mechanism of multiwalled carbon nanotubes under tensile load, Science 287 (5453) (2000) 637–640.

[111] C. Schönenberger, A. Bachtold, C. Strunk, J.-P. Salvetat, L. Forró, Interference and interaction in multi-wall carbon nanotubes, Appl. Phys. A 69 (3) (1999) 283–295.

[112] W. Xu, Y. Ding, S. Jiang, J. Zhu, W. Ye, Y. Shen, H. Hou, Mechanical flexible PI/MWCNTs nanocomposites with high dielectric permittivity by electrospinning, Eur. Polym. J. 59 (2014) 129–135.

[113] H. Liu, Y. Shen, Y. Song, C.W. Nan, Y. Lin, X. Yang, Carbon nanotube array/polymer core/shell structured composites with high dielectric permittivity, low dielectric loss, and large energy density, Adv. Mater. 23 (43) (2011) 5104–5108.

[114] Q. Li, Q. Xue, L. Hao, X. Gao, Q. Zheng, Large dielectric constant of the chemically functionalized carbon nanotube/polymer composites, Compos. Sci. Technol. 68 (10–11) (2008) 2290–2296.
[115] P. Costa, J. Silva, S.L. Mendez, Strong increase of the dielectric response of carbon nanotube/poly (vinylidene fluoride) composites induced by carbon nanotube type and pre-treatment, Compos. Part B 93 (2016) 310–316.
[116] Z. Chen, L. Xie, X. Huang, S. Li, P. Jiang, Achieving large dielectric property improvement in polymer/carbon nanotube composites by engineering the nanotube surface via atom transfer radical polymerization, Carbon 95 (2015) 895–903.
[117] J. Lu, K.-S. Moon, B.-K. Kim, C. Wong, High dielectric constant polyaniline/epoxy composites via in situ polymerization for embedded capacitor applications, Polymer 48 (6) (2007) 1510–1516.
[118] C.-H. Ho, C.-D. Liu, C.-H. Hsieh, K.-H. Hsieh, S.-N. Lee, High dielectric constant polyaniline/poly (acrylic acid) composites prepared by in situ polymerization, Synth. Met. 158 (15) (2008) 630–637.
[119] P. Tsotra, O. Gryshchuk, K. Friedrich, Morphological studies of epoxy/polyaniline blends, Macromol. Chem. Phys. 206 (7) (2005) 787–793.
[120] C. Huang, Q. Zhang, J. Su, High-dielectric-constant all-polymer percolative composites, Appl. Phys. Lett. 82 (20) (2003) 3502–3504.
[121] J.-K. Yuan, Z.-M. Dang, S.-H. Yao, J.-W. Zha, T. Zhou, S.-T. Li, J. Bai, Fabrication and dielectric properties of advanced high permittivity polyaniline/poly (vinylidene fluoride) nanohybrid films with high energy storage density, J. Mater. Chem. 20 (12) (2010) 2441–2447.
[122] L. Zhang, Z. Liu, X. Lu, G. Yang, X. Zhang, Z.Y. Cheng, Nano-clip based composites with a low percolation threshold and high dielectric constant, Nano Energy 26 (2016) 550–557.
[123] X. Liao, W. Ye, L. Chen, S. Jiang, G. Wang, L. Zhang, H. Hou, Flexible hdC-G reinforced polyimide composites with high dielectric permittivity, Compos. Part A 101 (2017) 50–58.
[124] C. Yang, Y. Lin, C. Nan, Modified carbon nanotube composites with high dielectric constant, low dielectric loss and large energy density, Carbon 47 (4) (2009) 1096–1101.
[125] W. Xu, Y. Ding, Y. Yu, S. Jiang, L. Chen, H. Hou, Highly foldable PANi@ CNTs/PU dielectric composites toward thin-film capacitor application, Mater. Lett. 192 (2017) (2017) 25–28.
[126] M. Li, X. Huang, C. Wu, H. Xu, P. Jiang, T. Tanaka, Fabrication of two-dimensional hybrid sheets by decorating insulating PANI on reduced graphene oxide for polymer nanocomposites with low dielectric loss and high dielectric constant, J. Mater. Chem. 22 (44) (2012) 23477–23484.
[127] H. Lee, S.M. Dellatore, W.M. Miller, P.B. Messersmith, Mussel-inspired surface chemistry for multifunctional coatings, Science 318 (5849) (2007) 426–430.
[128] L. Zhang, S. Yuan, S. Chen, D. Wang, B.-Z. Han, Z.-M. Dang, Preparation and dielectric properties of core–shell structured Ag@ polydopamine/poly (vinylidene fluoride) composites, Compos. Sci. Technol. 110 (2015) 126–131.
[129] Y. Zhou, Z. Luo, X. Zhuang, F. Liu, Multilayer-structured high-performance nanocomposites based on a combination of silver nanoparticles and nanowires, Mater. Lett. 182 (2016) 323–327.
[130] J. Zhu, X. Ji, M. Yin, S. Guo, J. Shen, Poly (vinylidene fluoride) based percolative dielectrics with tunable coating of polydopamine on carbon nanotubes: toward high permittivity and low dielectric loss, Compos. Sci. Technol. 144 (2017) 79–88.
[131] J. Liu, G. Tian, S. Qi, Z. Wu, D. Wu, Enhanced dielectric permittivity of a flexible three-phase polyimide–graphene–BaTiO 3 composite material, Mater. Lett. 124 (2014) 117–119.

[132] W. Xu, Y. Ding, S. Jiang, L. Chen, X. Liao, H. Hou, Polyimide/BaTiO 3/MWCNTs three-phase nanocomposites fabricated by electrospinning with enhanced dielectric properties, Mater. Lett. 135 (2014) 158–161.

[133] S. Luo, S. Yu, R. Sun, C.-P. Wong, Nano Ag-deposited BaTiO3 hybrid particles as fillers for polymeric dielectric composites: toward high dielectric constant and suppressed loss, ACS Appl. Mater. Interfaces 6 (1) (2013) 176–182.

[134] C.-R. Yu, D.-M. Wu, Y. Liu, H. Qiao, Z.-Z. Yu, A. Dasari, X.-S. Du, Y.-W. Mai, Electrical and dielectric properties of polypropylene nanocomposites based on carbon nanotubes and barium titanate nanoparticles, Compos. Sci. Technol. 71 (15) (2011) 1706–1712.

[135] F.-A. He, K.-H. Lam, J.-T. Fan, L.-W. Chan, Novel syndiotactic polystyrene/BaTiO 3-graphite nanosheets three-phase composites with high dielectric permittivity, Polym. Test. 32 (5) (2013) 927–931.

[136] C. Wu, X. Huang, L. Xie, X. Wu, J. Yu, P. Jiang, Morphology-controllable graphene–TiO2 nanorod hybrid nanostructures for polymer composites with high dielectric performance, J. Mater. Chem. 21 (44) (2011) 17729–17736.

[137] Y.-J. Wan, P.-L. Zhu, S.-H. Yu, W.-H. Yang, R. Sun, C.-P. Wong, W.-H. Liao, Barium titanate coated and thermally reduced graphene oxide towards high dielectric constant and low loss of polymeric composites, Compos. Sci. Technol. 141 (2017) 48–55.

[138] N. Xu, X. Xiao, H. Yang, E. Yu, Q. Zhang, Enhanced dielectric constant and suppressed dielectric loss of ternary composites based on Ag-P (VDF-HFP) matrix and TiO 2 nanowires, Ceram. Int. 42 (10) (2016) 12475–12481.

[139] Y. Jin, N. Xia, R.A. Gerhardt, Enhanced dielectric properties of polymer matrix composites with BaTiO3 and MWCNT hybrid fillers using simple phase separation, Nano Energy 30 (Suppl C) (2016) 407–416.

Chapter 9

Proton-Conducting Materials Used as Polymer Electrolyte Membranes in Fuel Cells

Jia Wei
PCI Synthesis, Devens, MA, United States

Chapter Outline

9.1 INTRODUCTION

The gradual depletion of fossil resources is becoming a severe challenge, and the control of the release of carbon dioxide is becoming a practical issue across the Earth. Efforts to find reusable resources such as biomass, wind energy, solar energy, and fuel cells are under investigation in both academic and industrial organizations. Fuel cells are one of the most promising ways of solving the problem of transportation pollution in cities [1]. As a clean energy, fuel cells utilize hydrogen and oxygen as the energy source, and produce water as a byproduct, making them a suitable candidate for replacing the combustion engines used in cars and buses. Many different materials have been applied in the study of proton transport, including Nafion [1–3], carbon nanotubes [4,5], ionic liquids [6,7], graphene [8–10], and hydrocarbon polymer materials [11]. However, many of these materials suffer from mechanical instability at elevated temperatures. Thus, more and more research has been focusing on exploring polymer electrolyte membranes (PEM) that are suitable for high temperature proton transport with low humidity. By working at elevated

Polymer-Based Multifunctional Nanocomposites and Their Applications
https://doi.org/10.1016/B978-0-12-815067-2.00009-3

temperatures, some key issues can be managed, including water management, CO poisoning, cooling, and heat recovery. In addition, use of the expensive platinum in electrodes can be reduced, lowering the overall cost of fuel cells. Among various types of PEMs for fuel cells, heterocycles such as imidazole- and triazole-based PEMs have garnered more attention in recent years for polymer electrolyte membrane fuel cell (PEMFC) applications. These composites offer a good balance between charge carrier density and mechanical stability, making them good candidates for future PEMFC applications. This chapter mainly covers the current progress in using heterocycle-polymer systems for PEMFC development.

9.2 PROTON TRANSPORT MECHANISMS IN FUEL CELLS

The development of PEMs is critical for producing efficient fuel cells [12]. Fuel cells consist of a cathode, an anode, and conducting electrolytes (see Fig. 9.1). The cathode and anode are usually assembled using a polar plate, a gas diffusion layer, and platinum metal. Fuel cells work via the mechanism as shown in Fig. 9.1. First, a hydrogen molecule diffuses through the polar plate and a gas diffusion layer. Second, the hydrogen molecule is oxidized, and two protons generate two electrons. Last, the protons diffuse through the PEM and reach the active catalyst layer on the anode. At the anode, the recombination between the protons and electrons occurs via catalytic reduction of oxygen to form a water molecule as the side product. The fact that water (see Eq. 9.1) is produced as the

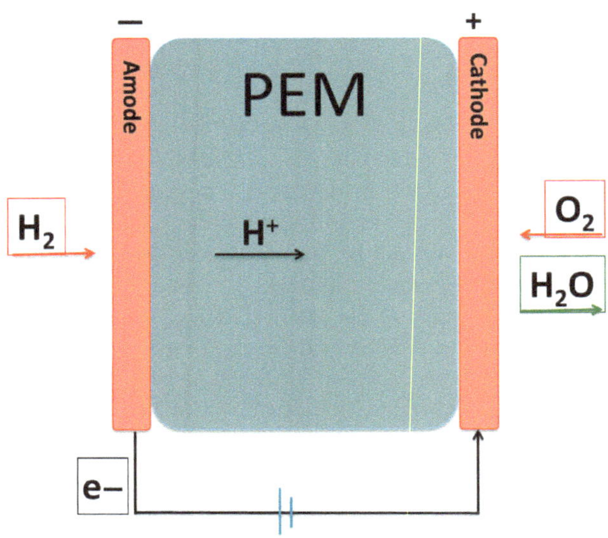

FIG. 9.1 The diagram of a simplified single fuel cell.

only byproduct makes this process efficient and environmentally friendly, and therefore, of great interest.

The chemistry reaction involved can be described as follows:

$$\begin{aligned} \text{Anode}: 2H_2 &\rightarrow 4H^+ + 4e^- \\ \text{Cathode}: O_2 + 4H^+ + 4e^- &\rightarrow 2H_2O \\ \text{Overall}: 2H_2 + O_2 &\rightarrow 2H_2O \end{aligned} \tag{9.1}$$

9.3 TYPES OF FUEL CELLS

Fuel cells are named based on the specific electrolytes used in the cells including: polymer electrolyte membrane fuel cells (PEMFC), alkaline fuel cells (AFC), direct methanol fuel cells (DMFC), phosphoric acid fuel cells (PAFC), molten carbonate fuel cells (MCFC), solid oxide fuel cells (SOFC), and microbial fuel cells (MFC) [13,14]. In addition to this naming system, fuel cells are generally distinguished based on their operating temperature: low temperature fuel cells and high temperature fuel cells.

Among the various types of fuel cells mentioned, PEMFC is of particular interest. In general, fuel cells can be operated at either low temperatures (normally with high relative humidity) or elevated temperatures (usually above 80°C at low relative humidity). PEMFCs follow a different proton transport mechanism at different operating temperatures. By operating at an elevated temperature, the fuel cell efficiency is increased, and a less expensive platinum metal is needed for the electrodes (reducing the cost). The cost of platinum metal used in the electrodes for fuel cells accounts for about one third of the overall cost of the fuel cells. Reducing the use of platinum is a widely accepted method for the purpose of commercializing fuel cell cars. For these reasons, the exploration of PEMFCs suitable for operation at high temperatures (above 80°C) is of significant importance.

Among the various types of fuel cells, anhydrous proton conducting related PEMFCs have attracted attention because of their unique features, such as: lower operating temperatures, rapid start-up, and their portability. PEMFCs can be further subcategorized based on the polymer composites, including:

- Aromatic polymer backbone-acid composites such as polybenzimidazole (PBI) [15,16], poly(arylene ether sulfone) (PSU) [17], Poly (ether ether ketone) (PEEK) [18], and poly(2,6-dimethyl-1,4-phenylene oxide) (PPO) [19]
- Polymer-heterocycle systems (triazole, benzyltriazole, imidazole, pyrazole)
- Metal organic frameworks (MOF) [20]

Among different polymer composites, heterocycle-polymer composites hold great potential due to their advantage of mixing both the charge carrier density and mechanical stability into one system. Heterocycles such as imidazole (pKa_1=7.18) [21,22] and 1H-1,2,3-triazole (pKa_1=1.17) [23] are known to show excellent proton conductivity; however, they lack thermal and mechanic stability at elevated temperatures. One solution to solve the mechanical instability problem is to immobilize the heterocycles with a polymer backbone. The resulting polymer-heterocycle composites, in general, show good mechanical and thermal stability, making them useful for a system targeting efficient proton transport without sacrificing the mechanical stability. The proton transfer in their systems turns out to be related to many factors, such as charge carrier density, mobility of the polymer, micro-phase separation, and other polymer properties. The study of the proton transport in the heterocycle-polymer system at anhydrous conditions is thus of great interest in fuel cell related research. This chapter covers the current progress of heterocycle-polymer systems in PEMFCs for anhydrous proton conduction. Additionally, Nafion will be briefly discussed as a proton transport material for its application in fuel cells.

9.4 MECHANISM OF PROTON CONDUCTING

Two proton transport mechanisms are widely accepted: [12] the vehicle mechanism [24] and Grotthuss mechanism [25]. With modern characterization techniques such as small angle X-ray scattering (SAXS) and NMR, widely accepted proton conducing mechanisms have been proposed, even though the exact mechanism of proton diffusion can be more complicated. In brief, the vehicle mechanism is usually applied in conditions where high humidity and low temperature are combined. Water molecules come together with the electrolyte membrane to form a proton-conducting channel (cluster), and water molecules act as a vehicle for transporting protons. One classic example is Nafion, which has been widely studied and applied in fuel cell applications. Nafion is a poly (tetrafluoroethylene)-based polymer with sulfonic acid groups arranged at intervals across the backbone (see Fig. 9.2). The fluorocarbon-based backbone and the sulfonic acid side chains possess drastically different polarities, leading to nanoscale phase separation and formation of water clusters [2,26]. More specifically, it has been widely accepted that the formation of nano-channels (or clusters) whose inner walls are lined with sulfonic acid groups account for the high conductivity of Nafion in PEM applications [1,3,11,27]. However, the proton conductivity starts to decrease when temperature increases (humidity decreases in this case too), as the water clusters are no longer connected at elevated

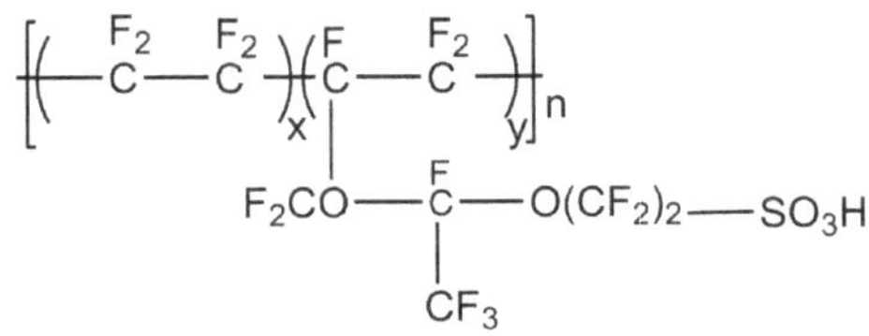

FIG. 9.2 The chemical structure of Nafion.

temperatures. This limits the application of Nafion in HTPEMFC [28]. Conditions involving low humidity and elevated temperatures are better modeled by the Grotthuss mechanism, where the proton transport is based on a so-called proton hopping process. Under this condition, protons are transported through long-range hydrogen bonding [25,29]. Most proton transport at elevated temperatures and low humidity generally follows this mechanism.

9.5 HIGH-TEMPERATURE POLYMER ELECTROLYTE MEMBRANES FUEL CELLS

Many types of high-temperature polymer electrolyte membranes fuel cells (HTPEMFC) have been studied with heterocycle-based polymeric materials, such as triazoles, imidazoles, pyrazoles, benzimidazoles, and benzyltriazoles. The major reason that these heterocycles are not used as electrolytes alone is due to the fact that the mechanical stability is poor and the heterocycles could not bear long-term use in fuel cells. For example, imidazole exhibits high proton conductivity (4.6×10^{-3} cm^{-1}, at 120°C) because of self-dissociation; however, the mechanical stability is poor, with a melting point of 89–91°C [30]. To increase the mechanical stability of these heterocycles, they can be attached to a polymer backbone offering sufficient mechanical strength. Different types of polymer backbones have been studied and reported on, including: aliphatic hydrocarbon-based polymers, aromatic hydrocarbon-based polymers, and polysiloxane-based polymers.

Despite the fact that the heterocycle-polymer systems show good proton conductivity, acid doping (e.g., phosphoric acid) these systems can further boost the proton conductivity. However, the addition of acid usually increases the complexity of the heterocycle-polymer system. To better understand the structure-property relationship in the heterocyle-polymer system, a clear understanding of the factors that affect proton transport is critical. As learned from the published work, the factors that affect proton transport are often complex.

9.6 CURRENT DEVELOPMENT OF HETEROCYCLE-POLYMER SYSTEMS FOR HTPEMFC

In 1998, the Maier group [18] reported the use of sulfonated polyetherketone/imidazole (pyrazole) systems for a proton conductivity study. Sulfonated polyether ketone was physically blended with imidazole at several ratios, and the resulting polymer composites were explored for their proton conductivities. At elevated temperatures (150°C), the proton conductivity obtained is comparable with the proton conductivity obtained at 50°C in hydrated form. In 2001, the same group [31] explored the use of imidazole terminated ethylene oxide as electrolytes (see Fig. 9.3). Proton conductivities up to 8×10^{-5} S/cm at 120°C was achieved in the absence of water. In addition to proton conductivity measurement, the ^{19}F diffusion coefficients by pulsed magnetic field gradient-NMR

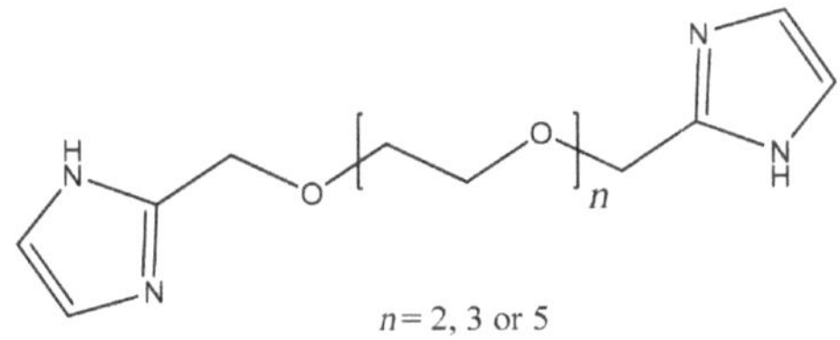

FIG. 9.3 The structure of imidazole bonded ethylene oxide.

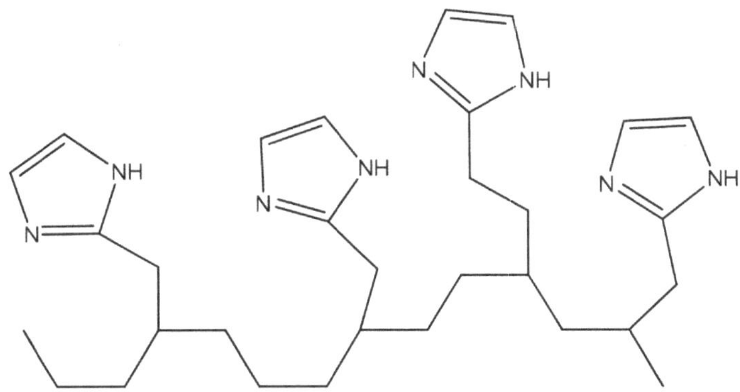

FIG. 9.4 Imidazole bonded polystyrene.

were used to study the diffusion of protons. The authors proposed that long-range hydrogen bonding is the main reason for the transportation of protons in this case, agreeing with the Grotthus mechanism.

In 2003, Herz et al. [32] reported two types of herterocycle-based polymeric systems: polystyrene with imidazole terminated flexible side chains (see Fig. 9.4), and benzylimidazole covalently bonded to a silicon oxide network via a flexible spacer (see Fig. 9.5). Proton conductivities of up to 7×10^{-4} were achieved at 200°C under anhydrous conditions. The two systems also maintained good mechanical stability.

In 2001, Wegner et al. [33] reported the use of polyurethane for the study of proton transport in this system (see Scheme 9.1). Because imidazole self-dissociates and shows good proton conductivity, the polymer composites synthesized showed that a conductivity of up to 1×10^{-3} S/cm was achieved at 150°C under anhydrous conditions. The heterocycle-polymer system showed good mechanical stability too.

In 2005, the Honma group [34] reported poly(vinylphosphonic acid)-heterocycle (PVPA-heterocycle: PVPA-Im, PVPA-MeIm, PVPA-Py) composites (see Fig. 9.6) and measured their proton conductivities in an anhydrous proton transport study. In this system, the strong acid polymer (PVPA) was mixed with an organic heterocycle such as imidazole, pyrazole or 1-methylimidazole, which

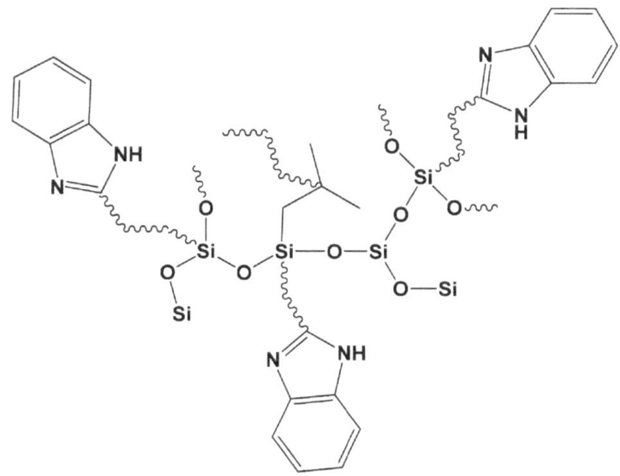

FIG. 9.5 Benzylimidaozle bonded polysiloxanes.

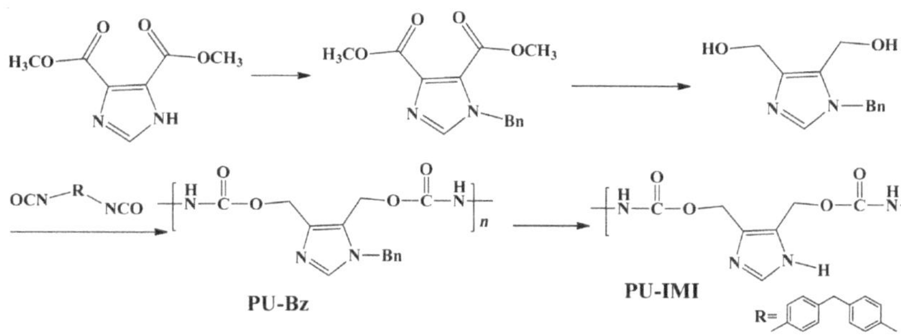

SCHEME 9.1 The synthesis of PU-IMI.

all bear high proton exchange capacities. PVPA-Im showed proton conductivity of 7×10^{-3} S/cm at 150°C under anhydrous conditions. A mechanism similar to the Grotthuss mechanism was proposed. The protonated and non-protonated nitrogen in the heterocycle act as donor and acceptor in the proton-conducting channel. The molecular cluster characteristics of imidazole are the main reason for the relatively high proton conductivity of the PVPA-Im, compared with the PVPA-MeIm or PVPA-Py.

In 2005, the Li group [35] reported the polymer electrolyte membranes using 1H-1,2,3-triazole attached to a polyethylene backbone (see Fig. 9.7). Another analog, poly(4-vinylimidazole), is also tested as a comparison. Under anhydrous conditions, the proton conductivity of poly(4-vinyl-1,2,3-triazole) is 10^6 times higher than that of poly(4-vinylimidazole) at elevated temperatures. It is concluded that the long-range hydrogen bonding through the triazole

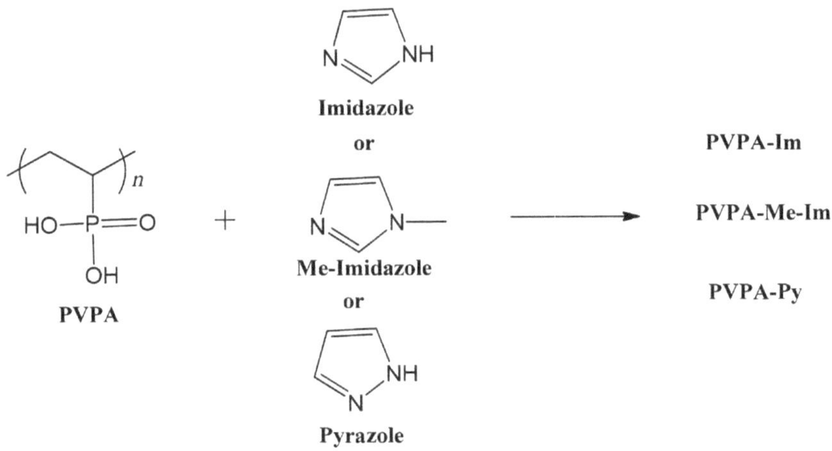

FIG. 9.6 Structure of PVPA, imidazole, me-imidazole, and pyrazole.

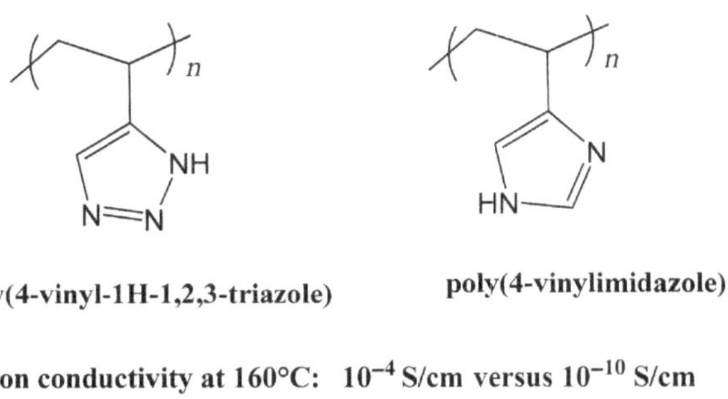

FIG. 9.7 Structure of poly(4-vinyl-1H-1,2,3-triazole) and poly(4-vinylimidazole).

network facilitates the faster proton transport in the poly(4-vinyl-1,2,3-triazole) material.

In 2009, the Coughlin group [36] reported on triazole-bonded polymethacrylate polymers (see Fig. 9.8) applied as polymer electrolytes. Polymethacrylates containing different amounts of 1H-1,2,3-triazoles were explored for proton conductivity. At a temperature range of 60–200°C, the triazole bonded polymethacrylates showed proton conductivity ranging from 10^{-9} to 10^{-4} S/cm. Proton conductivity increases as temperature increases with all three triazole-bonded polymethacrylate polymers (Table 9.1). At 160°C, the proton conductivities for Mono-Triazole-polymethacrylate, Bis-Triazole- polymethacrylate, and Tri-Triazole-polymethacrylate are all around 1×10^{-6} S/cm The proton conductivity is comparable at temperatures above 140°C. The increase in the mass fraction of triazole rings (see Table 9.1) did not lead to a significant increase in proton conductivity. This can be explained by the fact that

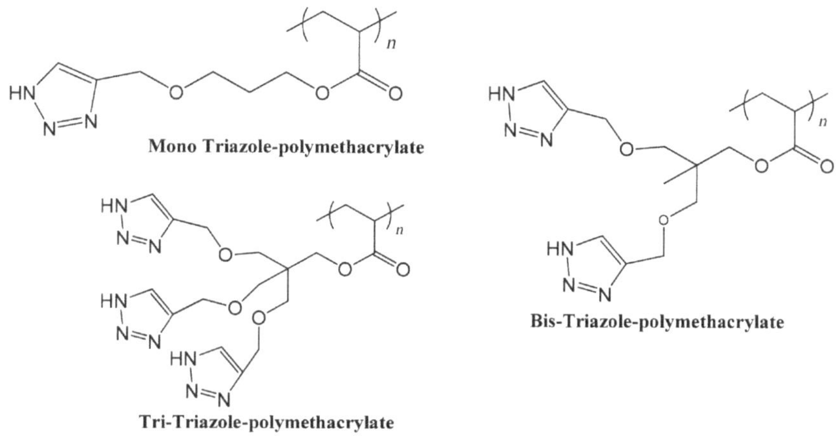

FIG. 9.8 Structures of mono-, bis- and tri-substituted polymethacrylates.

TABLE 9.1 Summary of the Properties of Triazole-polymethacryaltes

Polymer-triazole	T_g (°C)[a]	Wt% of Triazole	σ (S/cm) at 160°C
Mono-Triaozle-polymethacrylate	12	32	10^{-6}
Bis-Triaozle-polymethacrylate	44	40	10^{-6}
Tri-Triaozle-polymethacrylate	52	47	10^{-6}

[a]*T_g refers to glass transition temperature.*

the increase in the charge carrier density was offset by the increased glass transition temperature with increased triazole substitution.

Chen et al. [37] reported enhanced proton transport by forming self-assembling nano-channels in comb polymers. The comb polymers (Fig. 9.9; 10^{-3} S/cm at 180°C for Compound **A**) exhibited an increase in conductivity of 100 times at elevated temperatures as compared with its analogue without the decyl chain (Fig. 9.9; 10^{-5} S/cm at 180°C for Compound **B**). The decyl chain helped create an ordered structure in compound **A**, in which the benzyl-triazole groups are immiscible with the hydrophobic decyl chain, facilitating the formation of a localized triazole-conducting domain in a lamellae micro-phase (Fig. 9.10). The increased concentration of benzyltriazole in the conducting

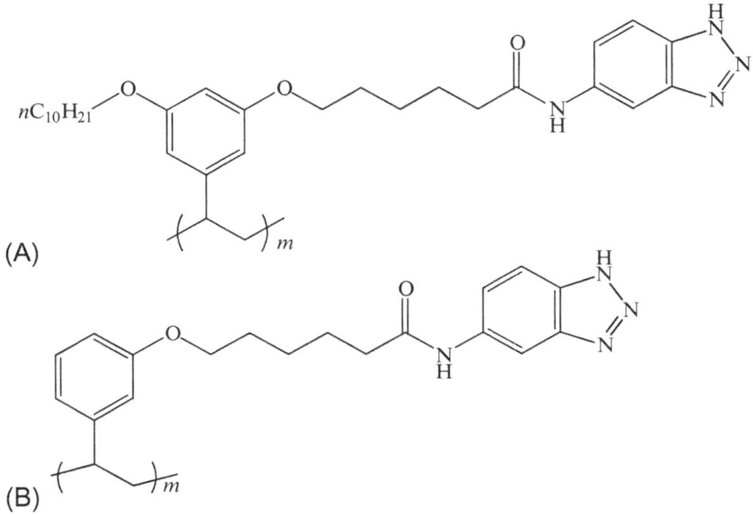

FIG. 9.9 Compound **A** (*up*) with decyl chain and **B** (*bottom*) without decyl chain.

FIG. 9.10 Lamellae structure for compound **A**. Black domain represents the benzyltriazole part; the *red* domain represents the decyl chain part.

domain led to increased proton conductivity, while the glass transition temperature ($T_g = 55°C$ and $67°C$ for compound **A** and **B** respectively) only varied moderately for the two compounds [37].

Wei et al. [38,39] reported the synthesis of triazole-derived polybutadiene homo-polymers (see Fig. 9.11), which show proton conductivity of 10^{-5} S/cm at 120°C. This serves as the first example of studying the proton conductivity with polybutadiene polymers. The use of the polybutadiene polymer backbone offers a relatively less rigid polymer backbone compared with their polynorbornene-based analogs. The two side chains with triazole rings attached

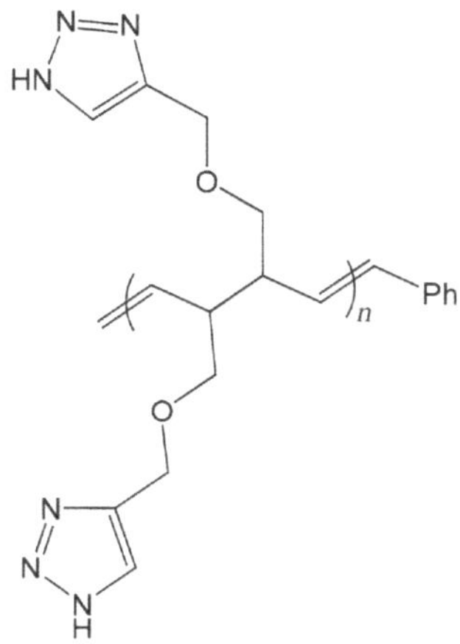

FIG. 9.11 Triazole-bonded polybutadiene.

are intended to increase the charge carrier density per unit. The polybutadiene-based diblock copolymers were also synthesized to further explore the proton conductivity of the block copolymers. The clear relationship between the proton transport and the micro-phase structure is still unclear, but this work offers a promising polymeric system to further explore the proton transport and the effect of the micro-environment.

Granados-Focil et al. [40] reported a group of triazole derived polysiloxane derivatives (see Fig. 9.12) that show proton conductivity in the range of 10^{-5} to 10^{-3} S/cm at 120–180°C in the case in which no acids were doped. The glass transition temperature and the volume fraction of the triazole functional groups were shown to heavily affect the proton conductivity. For comparison, analogs such as imidazole-polysiloxanes were used to explore proton conductivity, but

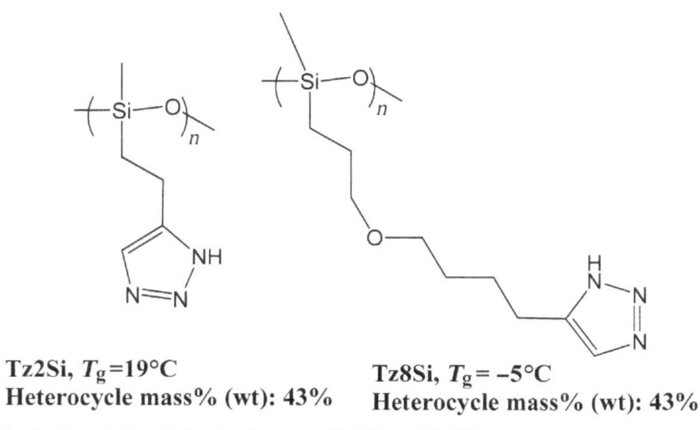

FIG. 9.12 Imidazole bonded polysiloxanes: Tz2Si and Tz8Si.

no significant difference in proton conductivity was seen with either imidazole or triazole. The imidazole-bonded systems tend to show slightly higher proton conductivity due to imidazoles forming molecular clusters via a hydrogen bonding network. In addition, the glass transition temperature of the polymer matrix and the mass fraction of the proton-conducting motifs tend to affect the proton conductivity of polymer materials. In the trizaole-bonded polysiloxane system, materials with a higher heterocycle mass fraction showed higher proton conductivity at elevated temperatures. For example, Tz2Si (43% heterocycle mass fraction % in weight) showed a proton conductivity of 1×10^{-4} S/cm at 150° C, while Tz8Si (28% heterocycle mass fraction % in weight) showed 1×10^{-5} S/cm at 160°C (Fig. 9.12).

Jannasch et al. [16] reported PEMs with benzimidazole units tethered to polysiloxanes (see Fig. 9.13). The relative thermally stable benzimidazole groups were attached to polysiloxane polymer backbones with a relatively low glass transition temperature. Without acid doping, the polymeric system showed a conductivity of 7×10^{-6} S/cm at 140°C with a polysiloxane polymer carrying 57 mol% benzimidazole under anhydrous conditions.

The Higuchi group [41] reported the Pt(II)-based supramolecular polymer and imidazole system, in which platinum-based polymer (polyPtC) with carboxylic acid groups was blended with imidazole to form imidazole-loaded polymer (polyPtC-Im) film (see Fig. 9.14). PolyPtC-Im prepared showed proton conductivity of 1.5×10^{-5} S/cm at 120°C under anhydrous conditions. This is six orders of magnitude higher than that of the polyPtC polymer. The highly ordered 1D imidazole channels (about 330 nm in diameter for polyPtC-Im) through the polymer chains were confirmed by SEM and FTIR, and a highly-ordered 1D channel was shown to lead to the improved proton conductivity. At elevated temperatures, a possible proton hopping mechanism, similar to the Grotthuss mechanism, was proposed. The rotation of the intercalated imidazoles and the vibration of their N—H bonds may facilitate the transport of protons through long-range hydrogen bonding.

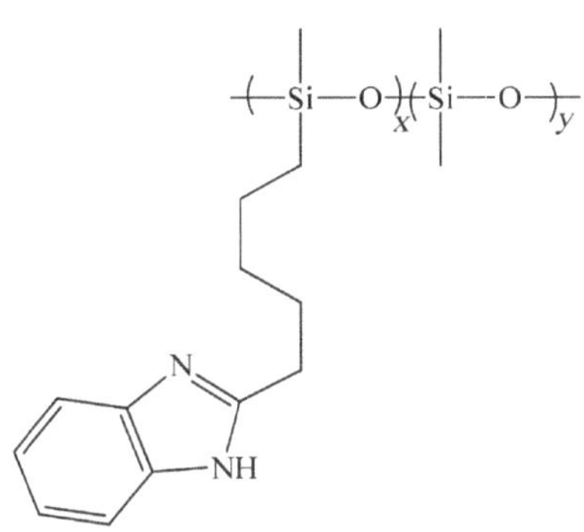

FIG. 9.13 Benzi-imidazole-attached polysiloxane block copolymers.

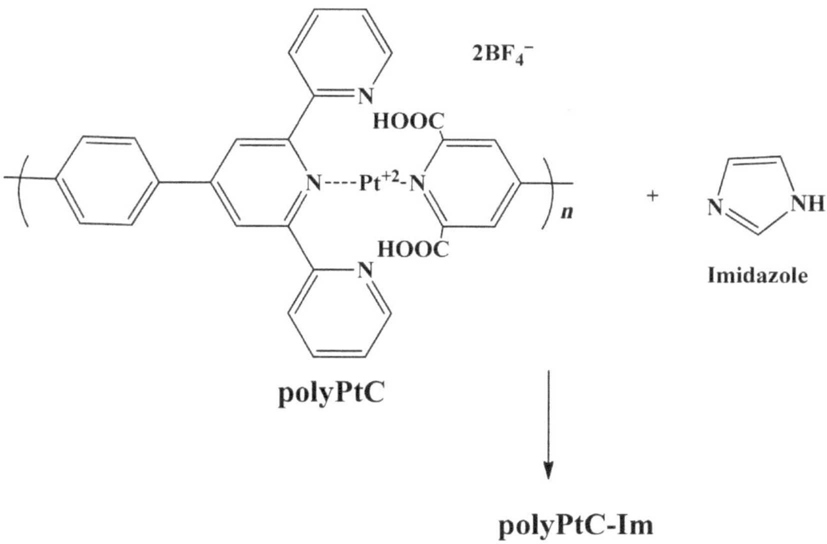

FIG. 9.14 The synthesis of polyPtC-Im.

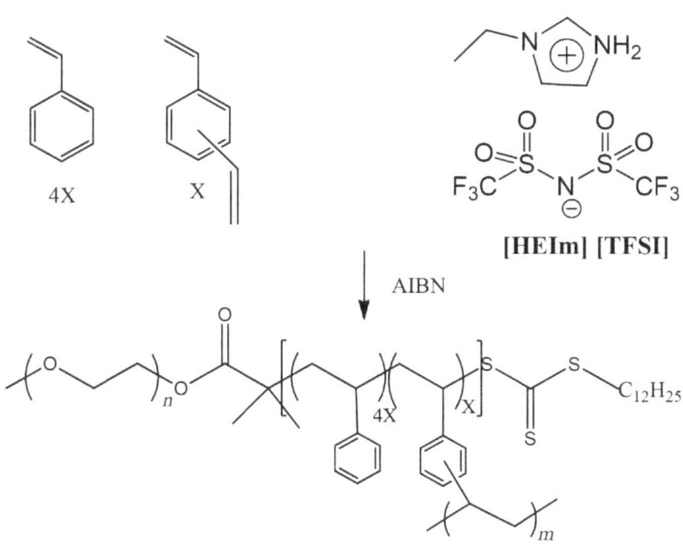

FIG. 9.15 The synthesis of polystyrene-ionic liquid composites.

In 2016, the Lodge group [42] reported the preparation of nanostructure polymer electrolyte membranes incorporating a protic ionic liquid ([HEIm] [TFSI]) (see Fig. 9.15). SEM and SAXS data showed that the PEM composite bears a bi-continuous morphology consisting of micro-phase separated domains (in other words, lack of long-range order). The increased charge carrier density in the conducing domain helps to promote proton transport. A proton conductivity of up to 1×10^{-2} S/cm at 180°C was obtained. The high proton conductivity is proposed to be a result of polymerization-induced micro-phase separation (PIMS).

9.7 CHALLENGES AND FUTURE PERSPECTIVES IN HTPEMFC

The application of proton transport in fuel cells continues to be a complex challenge. Knowledge of heterocyle-polymer systems is still limited, despite multiple studies exploring the relationships between the composites' microstructures and proton transport. Challenges still remain, as complex factors affect proton transport in PEMs. PEMs capable of operating at elevated temperatures are being explored as promising systems for fuel cell application. The universal factors affecting the proton transport in the PEMs are still not fully understood. One of the current challenges facing the field is exploring materials with high proton conductivity and good mechanical stability. In the case of heterocycle-polymer systems, proton conducting channels and the potential micro-phase properties are playing a critical role. Many questions remain at this stage regarding the polymer structure-property relationship. Thus, it is worth exploring different types of polymers, other than those that have been discussed in this chapter.

Regarding heterocycle-polymer systems, most studies have used polyhydrocarbons and polysiloxane for the heterocycle-polymer composites. Another category of polymers, polyphosphazenes, with $[N{=}PR_2]_n$ repeating units, are also potentially good candidates for PEMs in fuel cell applications. The synthesis of polyphosphazenes is well documented [43–45], and their application as polymer electrolytes has also been reported, despite the fact that the existing examples usually require external humidification [28]. Polyphosphazenes doped or functionalized with sulfonic acid [46,47] or phosphonic acid [48] for fuel cell applications have been proven to be stable under fuel cell operating conditions. Thus, further exploring polyphosphazenes for their use in PEMFCs is worthwhile, as they are very promising candidates, that are potentially able to fulfill all the requirements for water-free proton conducting membranes.

9.8 CONCLUSIONS

Novel PEMFCs require materials that exhibit good mechanical flexibility and maintain high proton conductivity at temperatures above 100°C. High temperature operation reduces the complexities in the control of moisture and pressure, and reduces the need for platinum metals. Development of materials able to fulfill all these critical requirements remains a critical challenge. The existing samples for heterocycle-polymer systems have applied different polymer backbones attached to, or blended with, different heterocycles. Many of these samples show excellent proton transport capabilities at elevated temperatures. With the use of modern analytical techniques such as TEM, SAXS, and SEM, proton transport behavior at the nano-scale can be explored. The proton conductivities in the heterocycle-polymer systems vary from one system to another because many factors affect the proton transport, including charge carrier density, glass transition temperature, formation of a long-range proton transport channel, and micro-phase separation. Further exploration of the structure-property in PEMs will be of great help in guiding the future design of PEMFCs.

REFERENCES

[1] O. Diat, G. Gebel, Nat. Mater. 7 (2008) 13.
[2] K.A. Mauritz, R.B. Moore, Chem. Rev. 104 (2004) 4535.
[3] K. Schmidt-Rohr, Q. Chen, Nat. Mater. 7 (2007) 75.
[4] R.H. Tunuguntla, F.I. Allen, K. Kim, A. Belliveau, A. Noy, Nat. Nanotechnol. 11 (2016) 639.
[5] Z. Cao, Y. Peng, T. Yan, S. Li, A. Li, G.A. Voth, J. Am. Chem. Soc. 132 (2010) 11395.
[6] C.M. Evans, G.E. Sanoja, B.C. Popere, R.A. Segalman, Macromolecules 49 (2016) 395.
[7] J. Stoimenovski, E.I. Izgorodina, D.R. MacFarlane, Phys. Chem. Chem. Phys. 12 (2010).
[8] Y. Liu, A. Goswami, F. Liu, D.L. Smith, P.P. Ruden, J. Appl. Phys. 116 (2014) 234301.
[9] J.L. Achtyl, R.R. Unocic, L. Xu, Y. Cai, M. Raju, W. Zhang, R.L. Sacci, I.V. Vlassiouk, P.F. Fulvio, P. Ganesh, D.J. Wesolowski, S. Dai, A.C.T. van Duin, M. Neurock, F.M. Geiger, Nat. Commun. 6 (2015) 6539.
[10] G. He, X. He, X. Wang, C. Chang, J. Zhao, Z. Li, H. Wu, Z. Jiang, Chem. Commun. 52 (2016) 2173.
[11] M. Rikukawa, K. Sanui, Prog. Polym. Sci. 25 (2000) 1463.
[12] K.-D. Kreuer, Chem. Mater. 8 (1996) 610.
[13] O.Z. Sharaf, M.F. Orhan, Renew. Sust. Energ. Rev. 32 (2014) 810.
[14] K. Pourzare, Y. Mansourpanah, S. Farhadi, Biofuel Res. J. 3 (2016) 496.
[15] M.A. Haque, A.B. Sulong, K.S. Loh, E.H. Majlan, T. Husaini, R.E. Rosli, Int. J. Hydrog. Energy 42 (2017) 9156.
[16] J.C. Persson, P. Jannasch, Macromolecules 38 (2005) 3283.
[17] K.T. Park, S.G. Kim, J.H. Chun, D.H. Jo, B.-H. Chun, W.I. Jang, G.B. Kang, S.H. Kim, K.B. Lee, Int. J. Hydrog. Energy 36 (2011).
[18] K.D. Kreuer, A. Fuchs, M. Ise, M. Spaeth, J. Maier, Electrochim. Acta 43 (1998) 1281.
[19] T. Xu, D. Wu, L. Wu, Prog. Polym. Sci. 33 (2008) 894.
[20] B. Wu, J. Pan, L. Ge, L. Wu, H. Wang, T. Xu, Sci. Rep. 4 (2014) 4334.
[21] H. Walbe, R.W. Isensee, J. Org. Chem. 21 (1956) 702.
[22] R.A. Gagliano, R.C. Knowlton, L.D. Byers, J. Org. Chem. 54 (1989) 5247.
[23] J. Catalan, R.M. Claramunt, J. Elguero, J. Laynez, M. Menendez, F. Anvia, J.H. Quian, M. Taagepera, R.W. Taft, J. Am. Chem. Soc. 110 (1988) 4105.
[24] K.-D. Kreuer, A. Rabenau, W. Weppner, Angew. Chem. Int. Ed. Eng. 21 (1982) 208.
[25] C.J.T. Grotthuss, Ann. Chim. 58 (1806) 54.
[26] L. Rubatat, A.L. Rollet, G. Gebel, O. Diat, Macromolecules 35 (2002) 4050.
[27] H. Böhme, D. Morf, E. Mundlos, Chem. Ber. 89 (1956) 2869.
[28] Q. Li, R. He, J.O. Jensen, N. Bjerrum, J. Chem. Mater. 15 (2003) 4896.
[29] Y. Ren, G.H. Chia, Z. Gao, Nano Today 8 (2013) 577.
[30] M.F.H. Schuster, W.H. Meyer, Annu. Rev. Mater. Res. 33 (2003) 233.

[31] M. Schuster, W.H. Meyer, G. Wegner, H.G. Herz, M. Ise, M. Schuster, K.D. Kreuer, J. Maier, Solid State Ionics 145 (2001) 85.

[32] H.G. Herz, K.D. Kreuer, J. Maier, G. Scharfenberger, M.F.H. Schuster, W.H. Meyer, Electrochim. Acta 48 (2003) 2165.

[33] C.B. Yoon, W.H. Meyer, G. Wegner, Synth. Met. 119 (2001) 465.

[34] M. Yamada, I. Honma, Polymer 46 (2005) 2986.

[35] Z. Zhou, S. Li, Y. Zhang, M. Liu, W. Li, J. Am. Chem. Soc. 127 (2005) 10824.

[36] S. Martwiset, O. Yavuzcetin, M. Thorn, C. Versek, M. Tuominen, E.B. Coughlin, J. Polym. Sci. A Polym. Chem. 47 (2009) 188.

[37] Y. Chen, M. Thorn, S. Christensen, C. Versek, A. Poe, R.C. Hayward, M.T. Tuominen, S. Thayumanavan, Nat. Chem. 2 (2010) 503.

[38] J. Wei, W. Trout, F.S. Granados, Prepr. Pap.-Am. Chem. Soc., Div. Energy Fuels Preprints 58 (2013) 132.

[39] J. Wei, W. Trout, Y.C. Simon, S. Granados-Focil, J. Polym. Sci. A Polym. Chem. 55 (2017) 1929.

[40] S. Granados-Focil, R.C. Woudenberg, O. Yavuzcetin, M.T. Tuominen, E.B. Coughlin, Macromolecules 40 (2007) 8708.

[41] C. Chakraborty, U. Rana, R.K. Pandey, S. Moriyama, M. Higuchi, ACS Appl. Mater. Interfaces 9 (2017).

[42] S.A. Chopade, S. So, M.A. Hillmyer, T.P. Lodge, ACS Appl. Mater. Interfaces 8 (2016) 6200.

[43] B. Wang, Macromolecules 38 (2005) 643.

[44] H.R. Allcock, C.A. Crane, C.T. Morrissey, M.A. Olshavsky, Inorg. Chem. 38 (1999) 280.

[45] B. Wang, E. Rivard, I. Manners, Inorg. Chem. 41 (2002) 1690.

[46] Q. Guo, P.N. Pintauro, H. Tang, S. O'Connor, J. Membr. Sci. 154 (1999) 175.

[47] H.R. Allcock, R.M. Wood, J. Polym. Sci. B Polym. Phys. 44 (2006) 2358.

[48] H.R. Allcock, M.A. Hofmann, C.M. Ambler, S.N. Lvov, X.Y. Zhou, E. Chalkova, J. Weston, J. Membr. Sci. 201 (2002) 47.

Chapter 10

Smart Adhesion Surfaces

Quan Xu and Weijun Li
State Key Laboratory of Heavy Oil Processing, China University of Petroleum, Beijing, China

Chapter Outline

10.1 INTRODUCTION

After incessant natural selection, biological systems have evolved into many diverse hierarchy structures with the versatility to adapt to a wide range of environmental changes, optimizing their performance [1–9]. Many biomimetic systems combine the properties of nature's creatures to achieve the synthesis of artificial smart surfaces [10–12]. Aristotle commented two thousand years ago "A gecko can run up and down a tree in any way, even with the head downwards." In other words, a gecko's feet can provide strong and reversible adhesion on numerous surfaces, in order to climb up and down [13–15]. In addition to the gecko's magic adhesion systems, mussels, sea stars, oysters, and barnacles can also affix themselves to rocks, and thereby enhance their chances for survival in the turbulent intertidal zone, which has attracted scientists' and engineers' attention [16,17].

Advanced characterization and modeling techniques reveal that these unique properties of natural materials rely on a sophisticated hierarchical structure, from the millimeter to the nanometer range [10,18]. Geckos are an appealing

Polymer-Based Multifunctional Nanocomposites and Their Applications
https://doi.org/10.1016/B978-0-12-815067-2.00010-X

creature, with the most efficacious adhesives in nature. In addition, geckos have complex hierarchical structures. Each toe pad has millions of small hairs named setae, which are approximately 110 μm in length and 4.2 μm in diameter. Each seta is further branched into 100–1000 smaller hairs named spatulae (approximately 0.2 μm in length and in width). When the gecko's spatulae come into contact with different surfaces via the accumulated van der Waals forces, geckos can generate sufficient force to lift themselves up to climb on sloping surfaces, even ceilings. This unique property is called the dry adhesion system [19,20].

Sessile organisms such as mussels can settle themselves down by hold fasting, which is known as byssus. The byssus is also called a "beard," with a bundle of dispersedly arrayed threads. The byssus consists of four parts, including the root, stem, threads (distal and proximal portions), and plaques. The spatulate adhesive plaques attach to the outside surfaces and connect with the end of the distal thread. The threads are procreated from the stem, which extends from the root. Both the threads and plaques contain the unusual 3,4-dihydroxyphenylalanine (DOPA) residue. These DOPA proteins play a key role in the performance of the threads and formation of the adhesive plaques, which is known as wet adhesion [21–23]. These nature-ideal structures are of great interest to scientists working to achieve better performance in bionic materials.

In addition to the studies dedicated to further understanding the mechanisms, extensive efforts are underway to develop methods to synthesize and replicate these unprecedented natural phenomena. The adhesion in nature can be divided into two categories, according to its mechanism. One is based on the multi-scale micro/nano structure of the reversible adhesion, such as gecko toes that can freely crawl on a smooth surface. The other one is bio-secreting high viscosity permanent adhesives, such as those that mussels use to adhere to different substrate surfaces underwater.

Learning from nature is an ongoing theme in the development of smart composite materials [24]. This chapter focuses on the adhesion mechanism of geckos and mussels, as well as their corresponding biomimetic polymer adhesives, which have been widely used in diverse adhesives, coatings, sealants, and anchoring equipment. A brief overview of other important applications of the inspired gecko feet and mussel byssi is also discussed.

10.2 REVERSIBLE ADHESION OF MULTISCALE MICRO/NANOSTRUCTURE

10.2.1 The Structure of Gecko Feet

Gecko can firmly adhere to a variety of substrate surfaces, or even hang upside down from smooth surfaces, while maintaining a clean foot [25–31]. This efficient, reversible, and self-cleaning adhesion mainly derives from the special surface morphology of skin on their toes, which results from van der Waals forces between the multi-scale microstructure and surface [14,25].

More than 1000 species of geckos with various morphologies and attachment pads have been applied in the bionic field [32,33] [34]. One of the largest, tokay geckos (Gekko gecko), is the most widely used in scientific research due to their availability, and excellent adhesion force [35,36]. The length of tokay geckos is approximately 300–400 and 200–300mm, for males and females, respectively. Their weight is up to 300g, and they have a unique blue or gray body with orange or red spots [36]. Fig. 10.1 shows the surface morphology of the skin on the gecko toes from macroscopic to nanoscopic. The skin includes lamellae, setae, branches, and spatulae with a sophisticated hierarchical structure [34]. As shown in Fig. 10.1B, each toe consists of about 20 lamellae, and each lamella has uniformly distributed setae arrays (Fig. 10.1D). A single seta is approximately 3–130μm in length and 4.2μm in diameter. Each seta further branches into 100–1000 even smaller structures known as spatulae, the length and diameter of which are, respectively, 200 and 200–500nm [37,38]. The apex of a spatulae is approximately 0.2 μm in width, 0.5 μm in length, and 0.01 μm in

FIG. 10.1 Structural hierarchy of the gecko adhesive system. (A) Ventral view of a tokay gecko climbing a vertical glass; (B) ventral view of the foot, showing a mesoscale array of seta-bearing adhesive lamellae (scansors); (C) cross-sectional micrograph through an isolated setal array; (D) microscale arrays of setae are arranged in a grid pattern on the ventral surface of each scansor. Each rhombus structure is the branched end of a group of four setae aggregated together in a quadrant; (E) nanoscale array of hundreds of spatular tips branching from a single seta; (F) and (G) the terminal spatular branches are widened and flattened at the top end. *(Image from K. Autumn, P.H. Niewiarowski, J.B. Puthoff, Annu. Rev. Ecol. Evol. Syst., 45 (2014) 445–470.)*

thickness, and is referred to as a spatula. Sufficient contact is generated on substrate surfaces with the spatulae because of these multiscale microstructures. Albeit a single seta can produce tiny adhesive force, 500 million spatulae can maximize van der Waals force and generate superlative adhesion [34,35,39,40].

10.2.2 Adhesion Model of Spatulae

For typical rough and rigid surfaces, contact generally occurs in a small effective area. Practically, the apparent contact area is approximately two to six factors of the order larger than actual contact area [41,42]. The introduction of surface energy is helpful for calculating the effect of division on contacts in a dry environment. For a hemisphere spatula with radius R adhered to a plane surface, the JKR (Johnson-Kendall-Roberts) theory [43] can be calculated according to $F_{JKR}=(3\pi RW)/2$, which can roughly estimate the adhesion force. The adhesion mechanic model is shown in Fig. 10.2A, which can be used to predict the actual value, roughly. The work of adhesion (W) is about 50 mJ m^{-2} and R is 100 nm, the predicted adhesive force is $F_{JKR}=23.6$ nN [25,44]. This value is approximately twice that of the atomic force's experimental value, using atomic force microscope techniques [45]. The Kendall peel model [46], $F_{peel}=Wb/(1-\cos\theta)$, where b is the width of a spatula, θ is the tape pulled angle, and the spatula's elastic energy is neglected, is also used to model the actual adhesion. Spatulae are considered as nano-scale adhesive tapes in this model [19,45,47]. As shown in Fig. 10.2B, this model contains a distinct

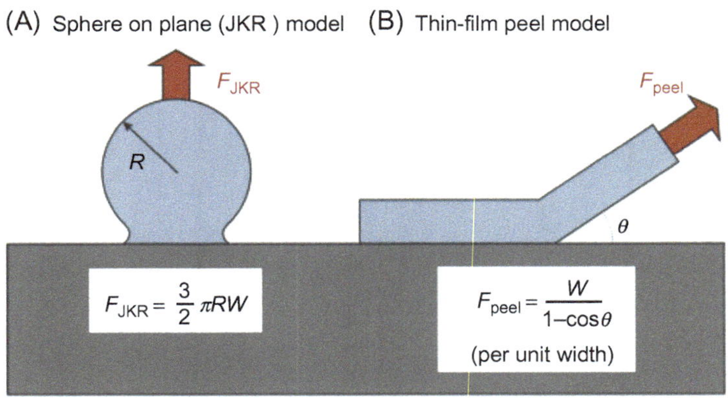

FIG. 10.2 Basic adhesion mechanics models. (A) The Johnson-Kendall-Roberts (JKR) model describes the force required to detach an adhesive sphere of radius R. (B) The thin-film peel model describes the force of the adhered film pulled at an angle θ, and will detach. Both models involve the work of adhesion (W). *(Image from K. Autumn, P.H. Niewiarowski, J.B. Puthoff, Annu. Rev. Ecol. Evol. Syst., 45 (2014) 445–470.)*

dependence on the delamination orientation. Practically, when $\theta = 90$ degree and $W = 50\,\mathrm{mJ\,m^{-2}}$ the peel model predicts $b = 200\,\mathrm{nm}$, the peel force is about 10 nN, which is approximately equal to the actual value. In addition, these peeling conditions are likely executed in the whole foot-sticking force measurement, in which the foot is dragged, rather than lifted [25,34,48].

10.2.3 van der Waals Forces and Capillary Forces

There are many types of intermolecular forces, including chemical interactions and forces between ions, polar molecules, and nominally nonpolar, but polarizable, substances. Unlike other forces, van der Waals forces exist extensively in various separation processes, from large (~50 nm), to atomic separation (~0.3 nm). It is generally believed that van der Waals forces provide the clinging force between the spatulae and the surfaces. Existing measurements and observations have proven van der Waals forces [14,25]. According to the van der Waals force mechanism of action, adhesion depends on the size and structure of spatulae, but has no relationship with surface chemistry. The relationship of the adhesion and tip size has also been confirmed by different prepared spatulae materials [14,49,50]. The capillary effect is another factor of intermolecular forces. Recent studies indicated that the higher humidity, the greater the sticking force of a gecko [50,51]; however, the adhesion decreased sharply in water. These conclusions confirmed the importance/necessity of the capillary effect [52]. The capillary effect can produce the Laplace pressure, which is in direct proportion to the tension of liquid. Another experiment concluded that the capillary effect is not necessary, but is influential in the gecko's sticking system [53].

10.2.4 Gecko-Inspired Polymer

Inspired by the gecko's feet, many attempts have been made to fabricate gecko-mimic adhesives by integrating the chemical, physical, biological, tribological, and mechanical principles. A great deal of different materials have been applied to duplicate the adhesives, including silicon [54–58], silicone rubbers [59–66], carbon nanotubes (CNTs) [67–70], polymathyl methacrylate (PMMA) [71], polyurethane (PU) [62,63,66], polyurethane acrylate (PUA) [55–57,72], polydimethylsioxane (PDMS) [73–75], and others [76–79]. The development of these fire-new functional systems has been summarized [80]. The dry adhesives are manufactured from some basic materials such as PDMS, PU, and PUA. A relatively efficient adhesive has been manufactured using polyimide by the e-beam lithography method (Fig. 10.3) [81]. These high aspect ratio nano polyimide hairs can generate relatively high normal adhesion (~3 N/m^2). However, major drawbacks of this method are slow speed and high costs. Hence, this material cannot be used for large-scale production and application. Subsequently, Fig. 10.4 shows two different PDMS pillars with

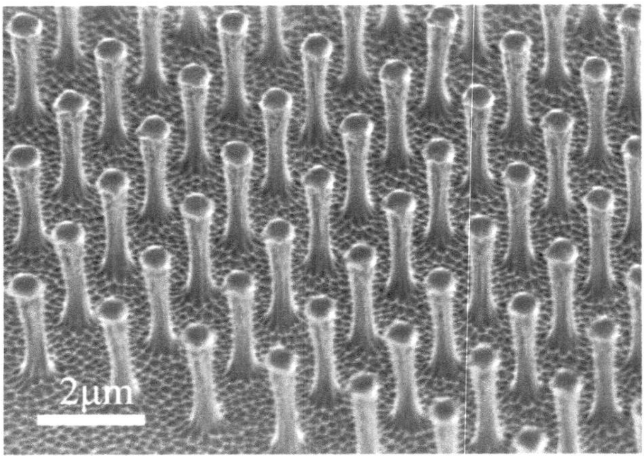

FIG. 10.3 The SEM of gecko-like synthetic adhesive, which was microfabricated with polyimide hairs. *(Image courtesy of A.K. Geim, S. Dubonos, I. Grigorieva, K. Novoselov, A. Zhukov, S.Y. Shapoval, Nat. Mater., 2 (2003) 461–463.)*

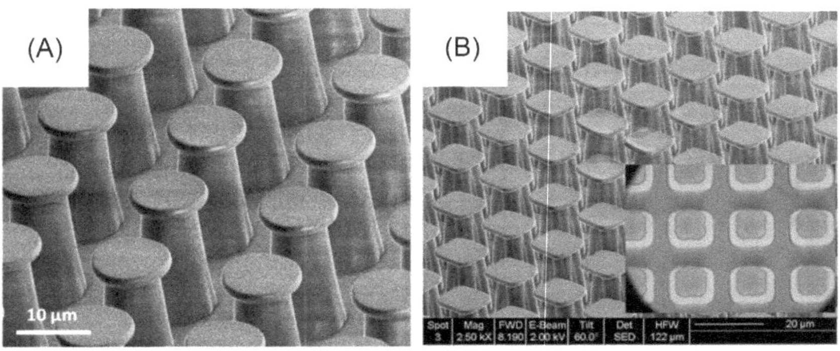

FIG. 10.4 Two images of PDMS dry adhesives with overhanging caps. (A) Isotropic adhesives with symmetric caps produced in photoresist molds. Image taken with permission from [74]. (B) Anisotropic adhesives with offsetting caps formed in a photoresist based mold. Image taken with permission from [74].

overhanging caps [74]. These gecko-inspired dry adhesives are directly duplicated from molds using silicon or photoresists. The importance of formation of such small caps is to greatly increase the adhesion, which can be applied for more directional behavior or rough surface compatibility. Qu et al. [68] reported a vertically aligned, multi-walled nanotube array (Fig. 10.5). The character of this adhesive is arrays with different heights, but without entangled layer of nanotubes. They discovered that large shear forces ($\sim 100\,N\,cm^{-2}$), but small normal forces ($\sim 10\,N\,cm^{-2}$), exist in this material when shear loading is absent.

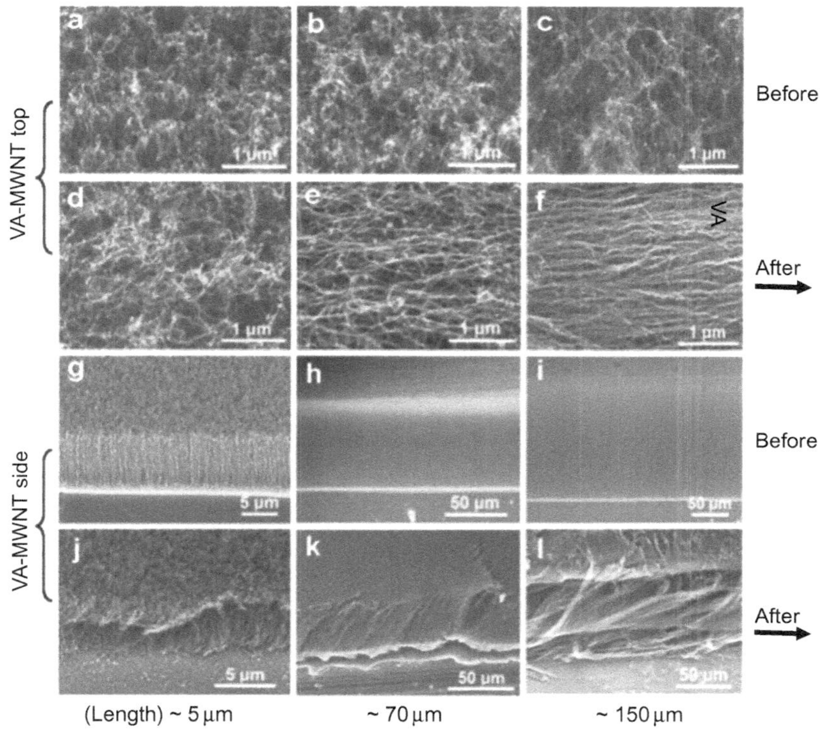

FIG. 10.5 SEM images of vertical aligned multi-wall nanotubes (VA-MWCNT) before ((A), (B), (C), (G), (H), (I)) and after ((D), (E), (F), (J), (K), (L)) being loaded in shear. For the longest VA-MWCNTs, the degree of alignment in the shear direction is the highest, demonstrating the permanent deformation that occurs. *(Image from L. Qu, L. Dai, M. Stone, Z. Xia, Z.L. Wang, Science, 322 (2008) 238–242.)*

Nevertheless, the fiber array will create permanent deformations with the increase of the pull-off force.

In addition to the significant adhesion, gecko feet also have super hydrophobicity and dry self-cleaning properties. Although the bioinspired dry self-cleaning surfaces have been studied in depth, the artificial dry self-cleaning surface is still in its infancy. The first gecko-mimicking adhesive with dry self-cleaning properties has been reported by Lee & Fearing [82] using a kind of high aspect ratio polypropylene array. This adhesive array exhibited about 30% recovery of cleaning particles when shear adhesion was tested for 30 steps on a relatively clean, dry, and rigid surface (Fig. 10.6A and B). Whereafter, polypropylene has been used to manufacture a gecko-inspired self-cleaning surface because of its hardness and tenderness. For example, Gillies et al. [83]

FIG. 10.6 SEM images of a polypropylene (PP) fibrillary adhesive and conventional pressure-sensitive adhesives. (A) A fibrillar adhesive contaminated with gold microspheres. (B) Fibrillar adhesive after 30 contacts on a clean glass substrate. *(Figure is taken from J. Lee, R.S. Fearing, Langmuir, 24 (2008) 10587–10591.)*

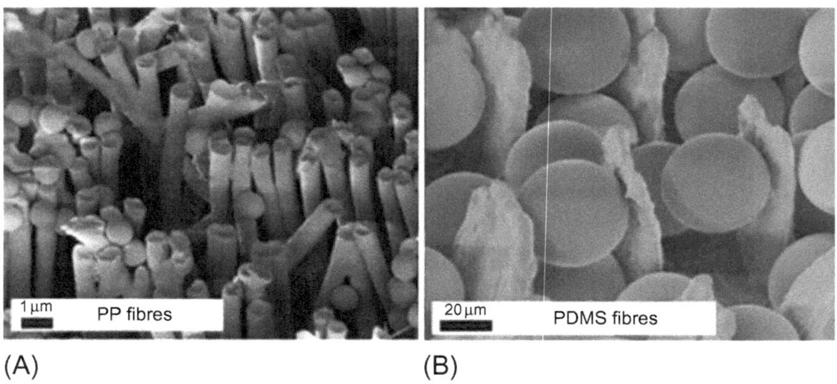

FIG. 10.7 SEM images of a polypropylene (PP) fibrillary adhesive and conventional pressure-sensitive adhesives. (A) PP fibers contaminated with 3210mm glass spheres, some of which can be seen deeply embedded (EM) between fibers. (B) Polydimethylsiloxane (PDMS) fibers fouled with 40250mm glass spheres after 40 recovery steps. Particles can be readily seen embedded between the fibers (EM). *(Image from A.G. Gillies, J. Puthoff, M.J. Cohen, K. Autumn, R.S. Fearing, ACS Appl. Mater. Interfaces, 5 (2013) 6081–6088.)*

obtained a fibrillar structure using polypropylene. A recovery of 96%–115% of shear adhesion has been found in hard fibers after being polluted by various size particles. However, the recovery of soft fibers was 55% after fouling with large particles. Because these large particles were easily embedded into the soft material (Fig. 10.7A and B), this restricted further cleaning. Furthermore, vertically aligned elastomer fibers have been designed to test self-cleaning properties by Mengüç et al. [84]. After a load-drag-unload contact test, they found that large granules can drop off during the drag procedure. When the actual particle size was less than the distance between two adjacent fibers, these particles may

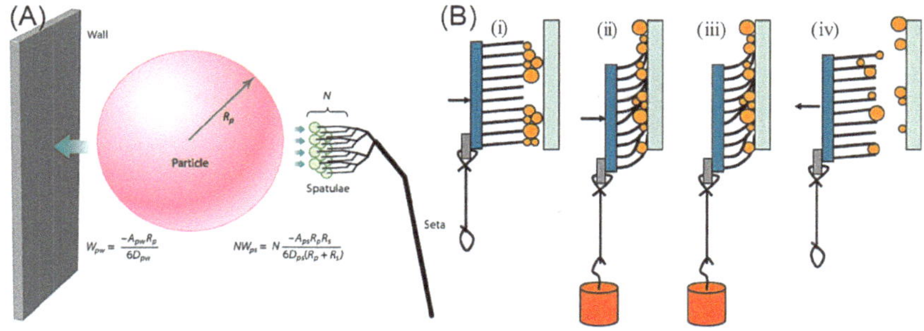

FIG. 10.8 (A) Model of interactions between N gecko spatulae of radius R_s, a spherical dirt particle of radius R_p, and a planar wall. (B) One cycle of simulated steps in contact with an initially clean glass slide: (i) applying normal compressive force, (ii) shear load added to the compressive load by a hanging weight, (iii) removing the compressive load (pure shear loading) and (iv) detaching the sample. *(Image from W.R. Hansen, K. Autumn, Proc. Natl. Acad. Sci. U. S. A., 102 (2005) 385–389 and J. Lee, R.S. Fearing, Langmuir, 24 (2008) 10587–10591.)*

embed into the fibers. When the particles were much tinier than the size of the fibers, they may be detained on the fiber surface, and become hard to remove.

The self-cleaning mechanism of gecko feet-inspired structures has also been widely discussed. This ability of gecko feet to self-clean was discovered by Hansen and Autumn in 2005. Hansen and Autumn found that the self-cleaning occurred when a particle-substrate clinging force exceeded the seta-particle clinging force (Fig. 10.8A) [19]. Nevertheless, this model does not give a complete explanation for some of the particles that exhibit a preference for binding to the toe pad over the substrate surface. To resolve this contradiction, Lee et al. [82] suggested that the earthquake of particles during shear adhesion is the key factor of the shedding of particles (Fig. 10.8B). Futhermore, Hu et al. [85] proposed the dynamic effect of gecko feet cleaning related to digital hyperextension. The self-cleaning of gecko spatulae arrays may be affected by the high-speed peel-off velocity (Fig. 10.9A). Subsequently, Xu et al. [86] tested a particle-substrate adhesive force and seta-particle clinging force during different peel-off velocities and shear adhesions. They found that the particle-wall adhesion is dependent on velocity, but the spatula-particle adhesion is not. It is this difference that gives gecko feet their superb dry self-cleaning properties. When the gecko steps, digital hyperextension can yield high peel-off and shear speed, which can drop off dust particles efficiently. Based on this understanding, a single gecko seta can achieve rapid particle micromanipulation. For example, a single seta can quickly pick up or drop particles to assemble complex patterns (Fig. 10.9B). The gecko-mimicking polymer materials have a very wide range of applications in micromanipulations, adhesives, self-cleaning equipment, and so forth.

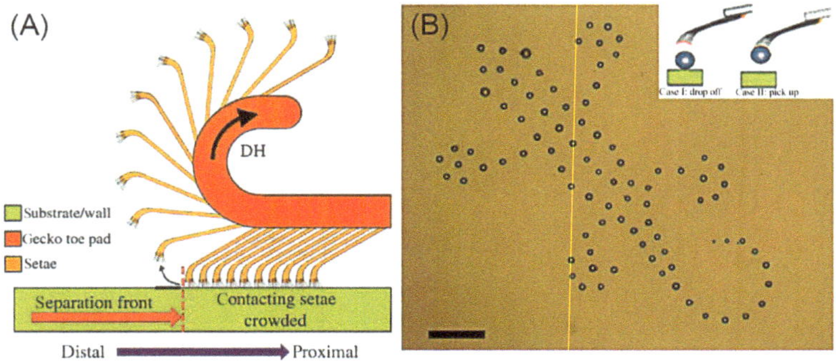

FIG. 10.9 (A) Schematics of gecko toe peeling caused by digital hyperextension. (B) Optical image of gecko-like log constructed using SiO_2 microparticles (size d ¼ 1–25 mm) by precisely assembling them on a glass slide with a biomimetic micromanipulator and using an atomic force microscope. *(Image from S. Hu, S. Lopez, P.H. Niewiarowski, Z. Xia, J. R. Soc. Interface, 9 (2012) 2781-2790 and Q. Xu, Y. Wan, T.S. Hu, T.X. Liu, D. Tao, P.H. Niewiarowski, Y. Tian, Y. Liu, L. Dai, Y. Yang, Nat. Commun., 6 (2015) 8949.)*

10.3 PERMANENT ADHESION OF STICKY POLYMERS

10.3.1 Compositions of Mussel Adhesion

10.3.1.1 The Holdfast of Mussel: Byssus

Mussels can secretly adhere to various surfaces (such as reefs, bottoms, and even animal crustaceans) to form permanent adhesion. This adhesion is extremely water-resistant and weather-proof and remains firmly attached to the substrate under billowing scour. These efficient bio-adhesives in moist conditions are mostly a multi-protein mixture, non-toxic and environment-friendly, which is synthesized in the presence of water [87,88]. Mussels dwell in shells, and are protected by shells, while their ectogenous holdfasts affix them to wet rocks. When they sweep over reefs by their feet and find more appropriate surfaces, they temporarily fix the body and produce a byssus of 3–10 per minute by ventral grooves in their feet that resemble reaction injection models [89]. Finally, radial byssus make mussels efficiently stick to various surfaces. The radiation range is about 5–6 cm (Fig. 10.10A). Byssus structures are comprised of four parts, including saptulate sticking plaques, threads (hard distal and compliant proximal parts), flexible roots, and supple stems. As shown in Fig. 10.10B, the plaques linking to the distal protein can attach to the substrate surface. The proximal proteins are partly protected by the shell, and the adiate from the stem, which combines with other tissues. The motivation of the byssus is precisely controlled by 12 various retractor muscles [90]. This byssus has promising potential applications in futuristic bionic devices.

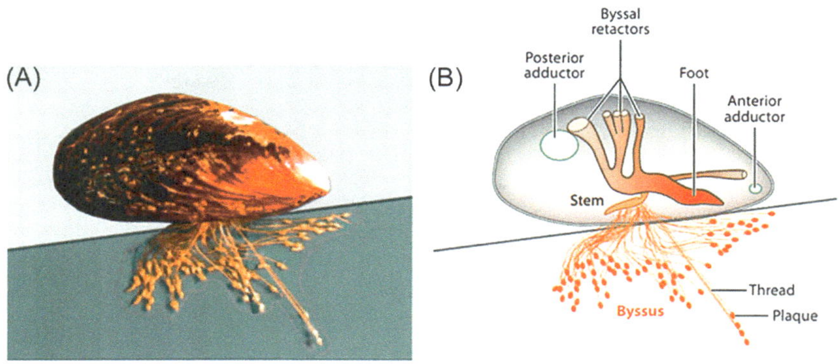

FIG. 10.10 Adhesion in the marine mussel *Mytilus californianus*. (A) Adult mussel exhibiting extensive byssus attached to a mica surface. (B) Schematic mussel on a half-shell. Each byssus is a bunch of threads tipped with adhesive plaques. The threads are joined at the stem, which inserts into the base of the foot. Byssal tension is controlled by 12 byssal retractor muscles, 6 per valve. The two adductor muscles open and close the valves. *(Image is taken from B.P. Lee, P.B. Messersmith, J.N. Israelachvili, J.H. Waite, Annu. Rev. Mater. Res., 41 (2011) 99–132)*

10.3.1.2 The Microstructure of Plaque

Mussel byssus plaques have an amazing foamy hierarchical structure (Fig. 10.11). The distal thread can connect to a plaque, which has a porous core matrix, and granular phase separation. Plaques exhibit ~40% porosity and conspicuity of the hole diameter gradient. These open pores can connect to each other through channels by a smooth wall (Fig. 10.11D). There are still many porous pillars on the interface that remain intact (Fig. 10.11E). The presence of these pillars indicates the possible connection between the plaque and the interface [90].

10.3.2 Byssus Protein Diversity

Although byssus has not been fully characterized, it has 20 known protein components, most of which are highly localized. Five of 7–8 different proteins are unique to plaque. Plaque proteins include mussel foot protein (mfp-1), prepolymerized proximal collagen (preCol-P), prepolymerized distal collagen (preCol-D), prepolymerized nongradient collagen (preCol-NG), and thread matrix protein (tmp-1). The main component of the cuticle consists of mfp-1, which plays a protective role on the outside. Other proteins ranging from mfp-2 to -6 are internal, and related to adhesion [90,91]. The key component of wet attachment is the post-translationally modified amino acid 3,4-dihydroxyphenyl-L-alanine (Dopa), which has moderate to strong cations, but in different sequences. Most of them are polymorphic families or result from alternative splicing.

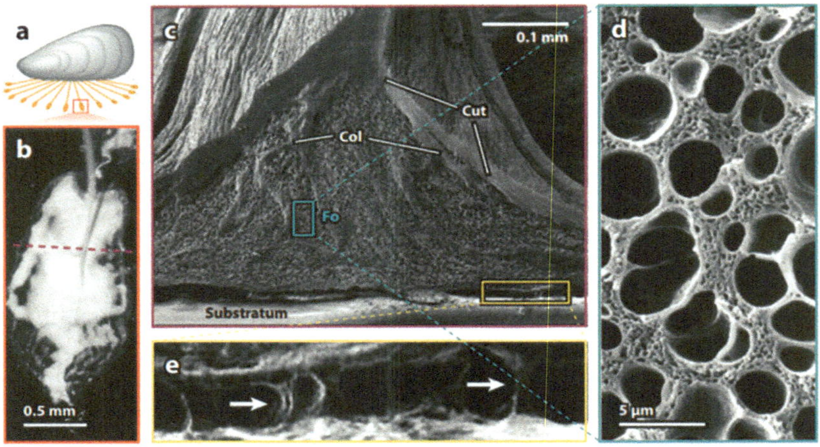

FIG. 10.11 The ultrastructure of a byssal adhesive plaque. (A) A byssated mussel shows the orientation of the sampled adhesive plaque (*red*). (B) Microscopic reflected light image of an attached plaque of *Mytilus californianus*. The underlying foam-like structure causes the light-scattering effect (white). The purple dashed line indicates the location of the frozen fracture. (C) SEM view of a freeze-fractured plaque illustrating the cuticle (Cut), the collagen fibers (Col) from the core, and the foam (Fo). (D) An enlargement of the foam-like structure shown in the blue box in panel C. (E) The interfacial area between the plaque and the substratum shown in the yellow box in panel C. Many pillars (arrows) remain intact, despite partial plaque delamination due to drying. *(Image from B.P. Lee, P.B. Messersmith, J.N. Israelachvili, J.H. Waite, Annu. Rev. Mater. Res., 41 (2011) 99–132.)*

10.3.2.1 Mfp-1

A molecular mass of Mfp-1 is about 108 kDa, which is a hydrophilic protein, and the first polyphenol protein to be mensurated and purified. The position of mfp-1 is byssus cuticles, oxidative cross-linking by polyphenols to form a stiff protective sleeve [92]. This protein contains 10%–15% dopa residues. The sequence and surface behavior of this protein have been overviewed [93,94]. In short, mfp-1 has an extensible and open structure in physiologic solutions, with evidence for the left helix. The oxidation easily occurs due to the existence of 15-mol% dopa, especially at or beyond neutral pH; however, oxidation also occurs in acidic conditions [95]. It is generally believed that the cuticle has a high hardness and intensity due to the chelation of iron (III) and dopa at a pH of 7.5 [95]. The most suitable ratio of Fe^{3+} has been exhaustively discussed [96,97], suggesting that the best chelation is interrelated with a tris-catecholato- iron (III) formation.

10.3.2.2 Mfp-2

The mfp-2 is the highest content component in byssal plaques, accounting for 25% of the total plaque protein. Its molecular weight is about 45 kDa. Mfp-2 consists of 11 tandem repeats of the epidermal growth factor domain, which is similar to the binding stabilized by three disulfide bonds, and is a common motif in extracellular proteins [98]. In addition, it contains a considerable number of secondary structures and relative resistance to a variety proteases, a key characteristic of integrity. Mfp-2 is a sticking protein with a stabilizing effect in the byssus, which is proven by the relatively high cysteine content (6 mol%), coupled with tandem and epidermal growth factors. The content of dopa is only 2 to 3 mol%, and dopa residues are mainly located in terminal areas of the protein. Protein sequences have been detected by Rzepecki et al. [99], and at least eight sequence variants have been detected in mfp-2.

10.3.2.3 Mfp-3

Up to now, Mefp-3 has been known as the smallest adhesion protein, with a molecular mass of about 5 to 7 kDa. It contains 20 to 25 mol% dopa, no repeats, and 4-hydroxy arginine and tryptophan residues. The most polymorphic in all the plaque proteins is Mfp-3. According to the data, as many as 35 different sequence variants exist, but these sequences cannot be expressed in each plaque [100]. Moreover, due to the post-translational modification of the protein tyrosine, there are many different variations. Arginine is converted to dopa and 4-hydroxyarginine, respectively. This conversion is particularly evident in the purified variants, which can be easily discerned by the number of sub-peaks detected by mass spectrometry [101,102]. In electrophoresis, these variants can be divided into two groups, including slow and fast moving isomers. The latter is rich in dopa, with typical content greater than 20 mol%, and the positive charge density is relatively high. For the former, the dopa content is 5–10 mol%, and the charge density is low [102]. The other amino acids in mfp-3 account for more than 10 mol%, including glycine, asparagine, tryptophan, and lysine. In-situ MALDI-TOF (matrix-assisted laser desorption ionization time-of-flight) mass spectrometry was used to analyze the footprint distribution in the plaque, and the results showed that mfp-3 is always particularly pronounced in the plaque.

10.3.2.4 Mfp-4

Mfp-4 is another of the larger plaque proteins, which is identified in the plaque with a molecular mass of 90 kDa, and completely measured gene sequences from *M. californianus*. A special histidine-rich decapeptide was found in tandem, which has a repeating 35 series [103]. Low levels of dopa (4 mol%) are limited to the N and C ends of the protein. Because mfp-4 has a tendency toward proximal plaque distribution, it is proposed that in situ validation may be performed by the interaction of the copper complex with the histidine-rich terminal of preCOLs [103], but this needs to be confirmed.

10.3.2.5 Mfp-5

Mfp-5 has the least polymorphism, in which only the single-nucleotide amino acid sequence was detected in *M. edulis* and the *M. californianus*. Intriguingly, these plaques contain two closely interrelated variants [104–106]. This protein contains 99 residues, of which 28 are consistent with dopa. Mfp-5 is rich in serine, glycine, and lysine, of which serine can be modified to phosphoserine. The charge density of mfp-5 is the highest among all plaque proteins; one for every three residues. Plaque footprints analyzed by in situ MALDI-TOF mass spectrometry indicated that mfp-5 is relatively small, with a molecular mass of 9.5 kDa, and it requires very high laser power. The desorption and ionization of footprinted proteins during MALDI detection strongly suggest the presence of protein. However, the most direct proof is missing, possibly because strong adsorption during MALDI limits desorption.

10.3.2.6 Mfp-6

Mfp-6 is a relatively small protein with a molecular mass of 11.6 kDa, which contains very high levels of lysine, tyrosine, and glycine, and a small amount of dopa, with a molecular mass of less than 2 mol% [105]. In addition, the cysteine has a molecular mass of 11 mol%, of which only a small portion is presented as disulfide cysteine. The high thiol content indicates that mfp-6 can utilize cysteine as a nucleophile of dopaquinone to form a cysteinyl-Dopa cross-link, which is one stable way to couple to other proteins in the plaque [105]. 5-*S*-cysteinyl-Dopa (about 1 mol%) can be detected in *M. californianus* plaques, which is crucial for maintaining the redox balance during foot filament formation; however, this is not the only function for mfp-6. Mfp-6 can effectively reduce dopaquinone to dopa to maintain a redox balance [107].

10.3.2.7 PreCOLs

PreCOL represents a class of natural block copolymers for fusion of collagens; these proteins can be self-assembled into a helical fiber core. There are three different types of preCOLs according to differentiating letter distribution: preCOL-D, preCOL-P, and preCOL-NG correspond to distal (D), proximal (P), and non-aliquoting [108–110], respectively. The collagen domain, or block, is centrally located, and is repeated by [Glycine-amino acid 2-amino acid 3]$_n$ in other collagens, which represents half of the total mass of each strand in the preCOL trimer. The preCol-P has a molecular weight of 95 kDa and a very strong extensibility of 160% [111]. The preCol-P comprises seven domains, including a representative amino and carboxy terminus, a large central collagen domain, elastin-like domains on both sides, and small histidine domains flanking it. PreCol-D is harder than preCol-P. Similar to preCol-P, preCol-D (molecular weight 97 kDa) also contains seven separate domains, namely, a representative amino and carboxy terminus, a central collagen region larger than preCol-P, a collagen domain located in the collagen domain, and histidine-rich domains flanking the silk fibroin domain [112]. PreCol-NG (molecular weight 76kDa) is a type of intermediate between PreCol-D and PreCol-P. It consists of a three-part unique structure comprised of a central collagen domain, acidic patches, and a histidine-rich end, which is similar to preCol-D and preCol-P [110]. The glycine- and asparagine-matrix proteins can separate preCOLs into a linear shape in the ventral groove of the mussel foot [89]. It is believed that the covalent metal-binding of the terminal histidine locks the preCOLs assembly into cohesion [108], and is capable of conducting the interaction between preCOLs and mfp-4 [103]. PreCOLs are critical to helix toughness, but may not involve surface attachment plaque.

10.3.3 Location and Interactions

The researchers achieved induced byssus plaque formation by injecting a 0.55-M potassium chloride (KCl) solution into the bottom of the toe ganglion organ of the mussel [113]. The result suggests that the formation of plaque begins within 1 min injection. In the experiment, the first detected protein was mfp-3, followed by mfp-6. After 30 s, Mfp-5 was also detected by MALDI-TOF; however, this particular protein requires higher laser power. Specific protein organization and locations are shown in Fig. 10.12. The interaction between various proteins is specifically described as follows. For example, mfp-2 momentarily binds to mfp-5, rather than mfp-3 [114]. In fact, mfp-3 seems to replace

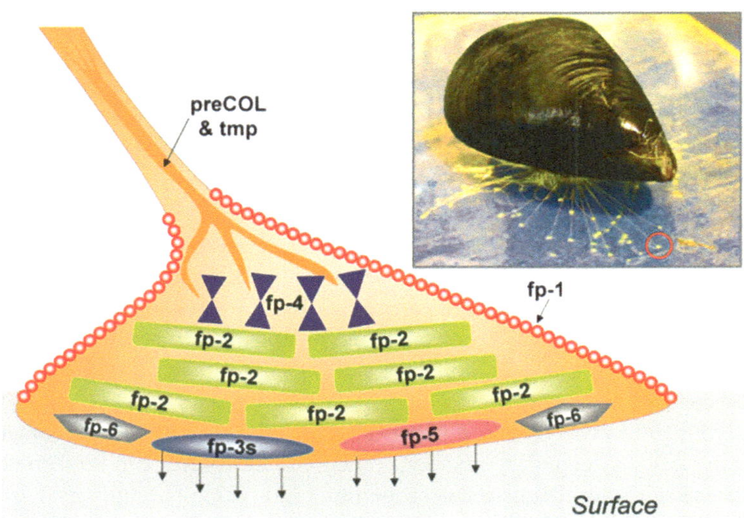

FIG. 10.12 Basic schematic of the byssal plaque and the location of thread proteins. The organization of the foot proteins is considered to be similar for all mussel species. *(Image from D.S. Hwang, H. Zeng, A. Masic, M.J. Harrington, J.N. Israelachvili, J.H. Waite, J. Biol. Chem., 285 (2010) 25850–25858.)*

mfp-2 from mica resulting from bridging adhesion, and eventually results from the mica surface mfp-3 bridge. In addition, mfp-3 has strong interactions with mfp-5,which is stronger than mfp-3 itself against mica. Whereas mfp-2 has no intrinsic interaction with mfp-1, -4, or -6. Due to the diversity of various proteins' interactions, the specific role of dopa is not clear-cut yet.

10.4 APPLICATIONS

Based on the chelating and reaction ability of catechol groups, the mussel-inspired adhesive materials are divided into three categories, including the synthetic DOPA-containing protein, polypeptide and its copolymer, DOPA modified polymers and polydopamine [91].

Artificial synthesis of polypeptide binding molecules containing DOPA is used to obtain a refined structure analogous to natural mussel protein, which is a more direct mimic of mussel adhesion protein. Anti-fouling sticking polypeptides prepared by Saxer et al. [115] can firmly adhere to the surface of TiO_2 and maintain stability under salt solution and high temperature conditions. Podsiadlo et al. [116] incorporated the laminated structure of the shellfish and the strong adhesion properties of the adherent proteins. They assembled the clay and DOPA-containing peptide layers to obtain layered nanocomposites. The introduction of a small amount of DOPA groups increased the strength and hardness of the films.

The introduction of catechol groups into the polymers helps them to endure adhesive pendants or terminal groups. Bilic et al. [117] synthesized hydrogel-like tissue adhesives containing catechol groups and formed non-destructive, non-toxic and effective adhesion between the adhesive and the adhered tissue during in vitro experiments. Brubaker et al. [118] prepared an adherent sol, which maintained its in vivo adhesion over one year. Inspired by the multi-scale micro-and nano-structure of gecko toes and mussel adhesion protein, Lee et al. [119] dipped and pulled the PDMS nano-column array in a catechol group-containing polymer solution to obtain a reusable nano adhesive (Fig. 10.13). The adhesion of this material is better than that of geckos in dry and wet conditions, but especially in water; in fact, about 15 times that of the gecko. On this basis, Glass et al. [63] adjusted the array size and the thickness of the polymer coating to enhance the wet adhesion of the composite surface. In the absence of the suction effect, the reversible adhesion at the macroscopic scale was strengthened.

Polydopamine can form a uniform cohesive film on the surface of many materials. After a secondary reaction, functional molecules can be formed on the surface with multi-functional properties [120–122]. The universal adherence and strong chelation ability of polydopamine can enhance the interfacial interaction between different materials, and further extend the two-dimensional surface adhesion to the three-dimensional structure [123]. Yu et al. [121] found that polydopamine membranes have pH-responsive permeability. To further

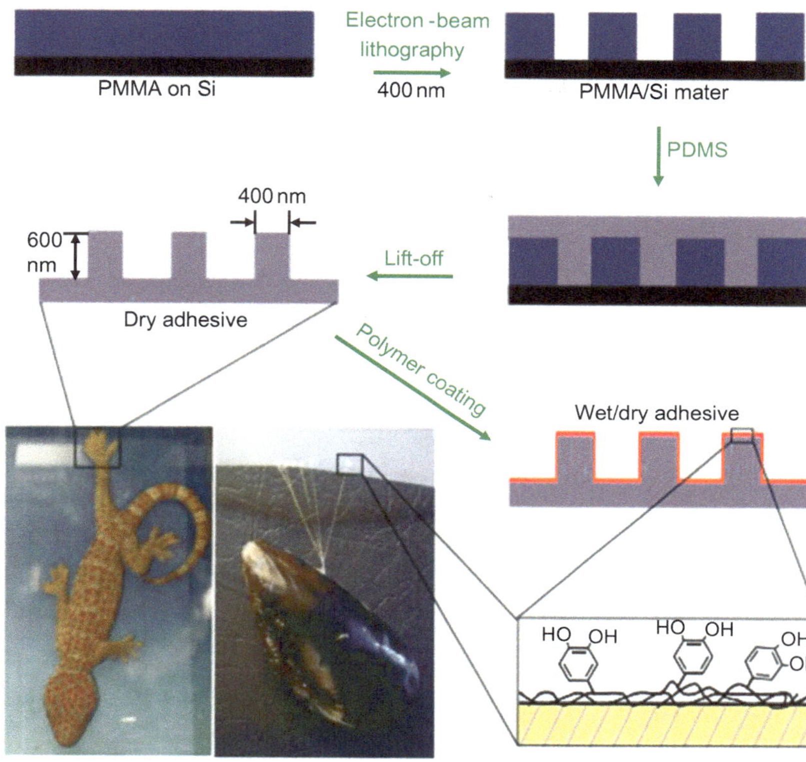

FIG. 10.13 Rational design and fabrication of wet/dry hybrid nano-adhesive. *(Image from H. Lee, B.P. Lee, P.B. Messersmith, Nature, 448 (2007) 338–341.)*

understand the adhesion mechanism, more and more mussel-inspired wet adhesive materials have been developed. Recently, B. Kollbe Ahn and colleagues [124] synthesized a kind of biomimetic primer based on mussel byssus proteins. Compared with commonly used primers (such as phosphates and siloxanes), primed molecules with catechol groups can deposit mineral surfaces with cross-linked polymethacrylate in just 1 nm while increasing, by nearly an order of magnitude, toughness, which is almost equivalent to that of natural mussels. Song pingan et al. [125] reported a simple biomimetic granular nano-structured polyvinyl alcohol (PVA) polymer film. PVA film embedded with uniform dispersion of sulfonated styrene-ethylene/butylene-styrene triblock copolymer (SSEBS) can be synthesized by simple solution mixing and casting methods. Compared with most engineering plastics, this bionic polymer film has an ultra-high toughness of 122 (±6.1) J/g, with a breaking strain at failure as high as 205%, and a tensile strength of up to 91.2 MPa (Fig. 10.14). This outstanding combination of properties is attributed to the unique nano-particulate phase-separated structure.

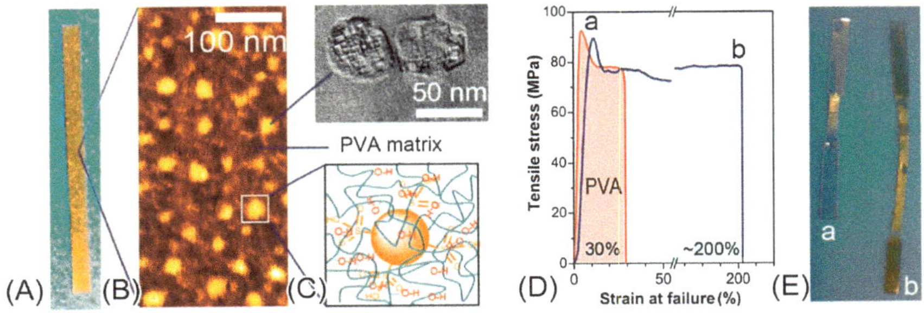

FIG. 10.14 (A) PVA/10 wt% SSEBS film. (B) AFM image of the PVA film showing interspersed granular SSEBS nano-domains, and SSEBS interacts with PVA via multiple hydrogen bonds. (C) TEM image for the phase-separation domains of SSEBS via self-assembly driven by the competitive hydrophilic-hydrophobic interactions. (D) Tensile stress-strain curves of (a) PVA and (b) PVA/10 wt% SSEBS, and (E) their corresponding digital images after mechanical failure. *(Image from P. Song, Z. Xu, M.S. Dargusch, Z.G. Chen, H. Wang, Q. Guo, Adv. Mater., (2017).)*

Inspired by the cuticle of marine mussel byssus, Filippidi et al. [126] synthesized new mussel-inspired elastomeric material to avoid an inherent compromise that links to dry, loosely cross-linked epoxy networks by sacrificing reversible iron-catechol links. These elastomeric materials provide excellent mechanical properties with higher rigidity, toughness, and ductility. By sacrificing iron-catechol bonds, the new material can be stretched by 50%. Once the pressure disappears, the molecular bonds can reform and recover. Compared with materials without these molecular bonds, the new material with molecular bonds can stretch by 770 times and increase tenacity by 58 times. This is mainly due to the synergistic effect of the cross-linked iron-catechol complex, which provides enhanced cross-linking density, and limits the chain-forming ionomer nanoclusters they formed. Zhao et al. [127] fabricated a material with strong adhesion under water, and its under-water adhesion can be reversible by controlling the interface temperature (Fig. 10.15). The design concept of the material originates from the high adhesive strength of mussel byssi and rapid failure of underwater adhesion. In order to obtain the material with high underwater adhesion, a submerged adhesion polymer was prepared by introducing a supramolecular host-guest, which is a bionic adhesive linear polymer prepared by the polymerization of adhesive dopamine molecules and hydrophobic monomers. A temperature-responsive polymer (poly-*N*-isopropylacrylamide) was also designed and fabricated to adjust the underwater adhesive strength. The polymer exhibited a highly hydrated state when intermolecular hydrogen bonds are formed with water molecules below its phase transition temperature, but a dehydrated state above its phase transition temperature. Therefore, based on the principle of underwater adhesive failure, this responsive molecule covered the surface of the adhesion molecule. The material can achieve underwater reversible adhesion by adjusting the temperature. These biomimetic polymer films are expected to be applied for tissue engineering and biomaterials such as artificial skin and tendons.

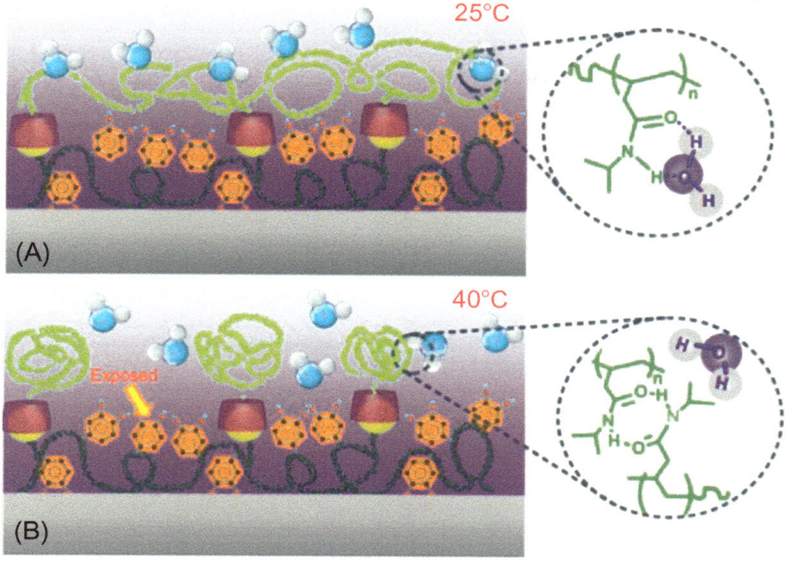

FIG. 10.15 Working mechanism. (A) Schematic drawing shows the screening of the interfacial adhesion when the local temperature of the adhesive is below LCST. In this case, the pNIPAM can easily form intermolecular hydrogen bonding with adjacent water molecules and the infused water layer transforms the pNIPAM side chains to a swelling layer. As a result of selective recognition between the host and guest copolymers, the adhesive moiety DOPA is spatially stabilized and confined underneath the swelling pNIPAM chains, screening the interfacial interaction of the adhesive moiety DOPA with target surface. (B) The collapse of pNIPAM-CD chains is also associated with the formation of numerous agglomerates during the phase transition. As a result, the adhesive group is exposed, leading to a pronounced interfacial adhesion. *(Image form Y. Zhao, Y. Wu, L. Wang, M. Zhang, X. Chen, M. Liu, J. Fan, J. Liu, F. Zhou, Z. Wang, Nat. Commun., 8 (2017) 2218.)*

10.5 CONCLUSIONS

In recent years, through the intensive study of biomaterials with excellent adhesion properties, many kinds of adhesive materials with superb properties have been manufactured. Inspired by the gecko toes, a series of dry adhesive materials with multi-scale and nano-scale structures have been synthetized. In addition, inspired by the underwater adhesive mechanism of mussels, many polymers with strong adhesion function have been designed and fabricated. These bionic adhesive materials have shown excellent application prospects in biology, medicine, new materials, chemistry and so on. This chapter summarizes the most important two types of biomass materials in the wet/dry adhesion, and the most recent cutting-edge applications and developments. In the field of dry adhesion, the gecko feet with a self-cleaning function has always been a hot spot. Based on the explanation of the gecko's self-cleaning and dry-adhesion mechanism, this chapter discusses some potential applications and future development prospects. For wet adhesion, a great deal of work has been done with good results, but the progress in mechanistic and direct bionicism is not enough. In addition, the materials with integrated functions are more desirable. Thus, it is an inevitable tendency to synthesize materials with two or more properties.

ACKNOWLEDGMENT

We thank the National Nature Science Foundation of China (No. 51505501), Beijing Nova Program (Z171100001117058) for their support.

REFERENCES

[1] Q. Cheng, L. Jiang, Z. Tang, Acc. Chem. Res. 47 (2014) 1256.
[2] M.A. Meyers, P.Y. Chen, Y.M. Lin, Y. Seki, Prog. Mater. Sci. 53 (2008) 1–206.
[3] B. Bhushan, Philos. Trans. 367 (2009) 1445.
[4] P. Vukusic, J.R. Sambles, Nature 424 (2004) 852–855.
[5] K. Krieger, Science 305 (2004) 636–637.
[6] W. Yang, I.H. Chen, B. Gludovatz, E.A. Zimmermann, R.O. Ritchie, M.A. Meyers, Adv. Mater. 25 (2013) 31–48.
[7] R.O. Ritchie, Nat. Mater. 13 (2014) 435–437.
[8] F. Song, A.K. Soh, Y.L. Bai, Biomaterials 24 (2003) 3623–3631.
[9] W.M. Kier, A.M. Smith, Integr. Comp. Biol. 42 (2002) 1146.
[10] L. Feng, S. Li, Y. Li, H. Li, L. Zhang, J. Zhai, Y. Song, B. Liu, L. Jiang, D. Zhu, Adv. Mater. 14 (2002) 1857–1860.
[11] X. Jin, S. Yang, Z. Li, K. Liu, L. Jiang, SCIENCE CHINA Chem. (2012) 1–7.
[12] K. Liu, L. Jiang, Nano Today 6 (2011) 155–175.
[13] J.-S. Kwak, T.-W. Kim, Int. J. Precis. Eng. Manuf. 11 (2010) 171–186.
[14] K. Autumn, M. Sitti, Y.A. Liang, A.M. Peattie, W.R. Hansen, S. Sponberg, T.W. Kenny, R. Fearing, J.N. Israelachvili, R.J. Full, Proc. Natl. Acad. Sci. 99 (2002) 12252–12256.
[15] K. Liu, J. Du, J. Wu, L. Jiang, Nanoscale 4 (2012) 768–772.
[16] R. Hatt, K. Versailles, Moisture Impacts on Coal Handling and Heat Rate as Published by World Coal, Coal Combustion, Inc., Versailles, KY, 1997.
[17] J.J. Wilker, Angew. Chem. Int. Ed. 49 (2010) 8076–8078.
[18] K. Liu, M. Zhang, J. Zhai, J. Wang, L. Jiang, Appl. Phys. Lett. 92 (2008) 183103.
[19] W.R. Hansen, K. Autumn, Proc. Natl. Acad. Sci. U. S. A. 102 (2005) 385–389.
[20] K. Autumn, N. Gravish, Philos. Trans. R. Soc. Lond. A Math. Phys. Eng. Sci. 366 (2008) 1575–1590.
[21] H. Lee, J. Rho, P.B. Messersmith, Adv. Mater. 21 (2009) 431–434.
[22] J. Yu, W. Wei, E. Danner, R.K. Ashley, J.N. Israelachvili, J.H. Waite, Nat. Chem. Biol. 7 (2011) 588–590.
[23] J.H. Waite, Biol. Rev. 58 (1983) 209–231.
[24] F. Xia, L. Jiang, Adv. Mater. 20 (2008) 2842–2858.
[25] K. Autumn, Y.A. Liang, S.T. Hsieh, W. Zesch, W.P. Chan, T.W. Kenny, R. Fearing, R.J. Full, Nature 405 (2000) 681–685.
[26] A.M. Peattie, C. Majidi, A. Corder, R.J. Full, J. R. Soc. Interface 4 (2007) 1071.
[27] M.D. Bartlett, A.B. Croll, D.R. King, B.M. Paret, D.J. Irschick, A.J. Crosby, Adv. Mater. 24 (2012) 1078.

[28] B. Chen, P.D. Wu, H. Gao, Proc. Math. Phys. Eng. Sci. 464 (2008) 1639–1652.
[29] P. Loskill, J. Puthoff, M. Wilkinson, K. Mecke, K. Jacobs, K. Autumn, J. R. Soc. Interface 10 (2013).
[30] M.S. Prowse, M. Wilkinson, J.B. Puthoff, G. Mayer, K. Autumn, Acta Biomater. 7 (2011) 733–738.
[31] M. Zhou, N. Pesika, H. Zeng, W. Jin, X. Zhang, Y. Meng, S. Wen, T. Yu, J. R. Soc. Interface 9 (2012) 2424–2436.
[32] A.G. Kluge, Hamadryad 26 (2001).
[33] D. Han, K. Zhou, A.M. Bauer, Biol. J. Linn. Soc. 83 (2004) 353–368.
[34] R. Ruibal, V. Ernst, J. Morphol. 117 (1965) 271–293.
[35] U. Hiller, Zoomorphology 62 (1968) 307–362.
[36] D.J. Irschick, C.C. Austin, K. Petren, R.N. Fisher, J.B. Losos, O. Ellers, Biol. J. Linn. Soc. 59 (1996) 21–35.
[37] H.H. Schleich, W. Kästle, Amphibia-Reptilia 7 (1986) 141–166.
[38] K. Autumn, P.H. Niewiarowski, J.B. Puthoff, Annu. Rev. Ecol. Evol. Syst. 45 (2014) 445–470.
[39] A.P. Russell, J. Zool. 176 (1975) 437–476.
[40] E. Williams, J. Peterson, Science 215 (1982) 1509–1511.
[41] B. Bhushan, Introduction to Tribology, John Wiley & Sons, 2013.
[42] B. Bhushan, Nanotribology and Nanomechanics—An Introduction, Springer, Berlin Heidelberg, New York, 2005.
[43] K. Johnson, K. Kendall, A. Roberts, Proc. R. Soc. Lond. A Math. Phys. Eng. Sci. (1971) 301–313.
[44] E. Arzt, S. Gorb, R. Spolenak, Proc. Natl. Acad. Sci. 100 (2003) 10603–10606.
[45] G. Huber, S.N. Gorb, R. Spolenak, E. Arzt, Biol. Lett. 1 (2005) 2–4.
[46] K. Kendall, J. Phys. D. Appl. Phys. 8 (1975) 1449.
[47] R. Spolenak, S. Gorb, H. Gao, E. Arzt, Proc. R. Soc. Lond. A Math. Phys. Eng. Sci. (2005) 305–319.
[48] N. Rizzo, K. Gardner, D.e.a. Walls, N. Keiper-Hrynko, T. Ganzke, D. Hallahan, J. R. Soc. Interface 3 (2006) 441–451.
[49] N.S. Pesika, H. Zeng, K. Kristiansen, B. Zhao, Y. Tian, K. Autumn, J. Israelachvili, J. Phys. Condens. Matter 21 (2009).
[50] G. Huber, H. Mantz, R. Spolenak, K. Mecke, K. Jacobs, S.N. Gorb, E. Arzt, Proc. Natl. Acad. Sci. U. S. A. 102 (2005) 16293–16296.
[51] P.H. Niewiarowski, S. Lopez, L. Ge, E. Hagan, A. Dhinojwala, PLoS One 3 (2008).
[52] W. Sun, P. Neuzil, T.S. Kustandi, S. Oh, V.D. Samper, Biophys. J. 89 (2005) L14–L17.
[53] J.B. Puthoff, M.S. Prowse, M. Wilkinson, K. Autumn, J. Exp. Biol. 213 (2010) 3699–3704.
[54] J. Davies, S. Haq, T. Hawke, J. Sargent, Int. J. Adhes. Adhes. 29 (2009) 380–390.
[55] H.E. Jeong, J.-K. Lee, H.N. Kim, S.H. Moon, K.Y. Suh, Proc. Natl. Acad. Sci. 106 (2009) 5639–5644.
[56] T.I. Kim, H.E. Jeong, K.Y. Suh, H.H. Lee, Adv. Mater. 21 (2009) 2276–2281.
[57] T.-i. Kim, C. Pang, K.Y. Suh, Langmuir 25 (2009) 8879–8882.
[58] H. Yoon, H.E. Jeong, T.-i. Kim, T.J. Kang, D. Tahk, K. Char, K.Y. Suh, Nano Today 4 (2009) 385–392.
[59] M.P. Murphy, B. Aksak, M. Sitti, Small 5 (2009) 170–175.
[60] S. Kim, M. Sitti, T. Xie, X. Xiao, Soft Matter 5 (2009) 3689–3693.
[61] B. Aksak, M.P. Murphy, M. Sitti, IEEE International Conference on Robotics and Automation, (2008), pp. 3058–3063.
[62] E. Cheung, M. Sitti, Langmuir 25 (2009) 6613.

[63] P. Glass, H. Chung, N.R. Washburn, M. Sitti, Langmuir 25 (2009) 6607.
[64] C. Greiner, E. Arzt, A. Del Campo, Adv. Mater. 21 (2010) 479–482.
[65] H.E. Jeong, S.H. Lee, P. Kim, K.Y. Suh, Colloids Surf. A Physicochem. Eng. Asp. 313–314 (2008) 359–364.
[66] M.P. Murphy, S. Kim, M. Sitti, ACS Appl. Mater. Interfaces 1 (2009) 849.
[67] L. Qu, L. Dai, Adv. Mater. 19 (2010) 3844–3849.
[68] L. Qu, L. Dai, M. Stone, Z. Xia, Z.L. Wang, Science 322 (2008) 238.
[69] S. Sethi, L. Ge, L. Ci, P.M. Ajayan, A. Dhinojwala, Nano Lett. 8 (2008) 822.
[70] M. Xu, D.N. Futaba, M. Yumura, K. Hata, ACS Nano 6 (2012) 5837.
[71] M. Sitti, R.S. Fearing, J. Adhes. Sci. Technol. 17 (2003) 1055–1073.
[72] H.E. Jeong, S.H. Lee, P. Kim, K.Y. Suh, Nano Lett. 6 (2006) 1508–1513.
[73] Y. Zhang, S. Qu, X. Cheng, X. Gao, X. Guo, J. Bionic Eng. 13 (2016) 132–142.
[74] D. Sameoto, C. Menon, J. Micromech. Microeng. 19 (2009) 115002.
[75] A. Parness, D. Soto, N. Esparza, N. Gravish, M. Wilkinson, K. Autumn, M. Cutkosky, J. R. Soc. Interface (2009). rsif. 2009.0048.
[76] K. DR, B. MD, G. CA, I. DJ, C. AJ, Adv. Mater. (Deerfield Beach, Fla.) 26 (2014) 4345.
[77] S. Gorb, M. Varenberg, A. Peressadko, J. Tuma, J. R. Soc. Interface 4 (2007) 271.
[78] A.Y. Ho, L.P. Yeo, Y.C. Lam, I. Rodríguez, ACS Nano 5 (2011) 1897–1906.
[79] T.S. Kustandi, V.D. Samper, D.K. Yi, W.S. Ng, P. Neuzil, W. Sun, Adv. Funct. Mater. 17 (2007) 2211–2218.
[80] D. Sameoto, C. Menon, Smart Mater. Struct. 19 (2010) 103001.
[81] A.K. Geim, S. Dubonos, I. Grigorieva, K. Novoselov, A. Zhukov, S.Y. Shapoval, Nat. Mater. 2 (2003) 461–463.
[82] J. Lee, R.S. Fearing, Langmuir 24 (2008) 10587–10591.
[83] A.G. Gillies, J. Puthoff, M.J. Cohen, K. Autumn, R.S. Fearing, ACS Appl. Mater. Interfaces 5 (2013) 6081–6088.
[84] Y. Mengüç, M. Röhrig, U. Abusomwan, H. Hölscher, M. Sitti, J. R. Soc. Interface 11 (2014).
[85] S. Hu, S. Lopez, P.H. Niewiarowski, Z. Xia, J. R. Soc. Interface 9 (2012) 2781–2790.
[86] Q. Xu, Y. Wan, T.S. Hu, T.X. Liu, D. Tao, P.H. Niewiarowski, Y. Tian, Y. Liu, L. Dai, Y. Yang, Nat. Commun. 6 (2015) 8949.
[87] C.G. Hurlburt, S.W. Hurlburt, Mar. Fish. Rev. 37 (1975) 10–18.
[88] M.W. Denny, B. Gaylord, Annu. Rev. Mar. Sci. 2 (2010) 89–114.
[89] J.H. Waite, The formation of mussel byssus: anatomy of a natural manufacturing process, in: Structure, Cellular Synthesis and Assembly of Biopolymers, Springer, 1992, pp. 27–54.
[90] B.P. Lee, P.B. Messersmith, J.N. Israelachvili, J.H. Waite, Annu. Rev. Mater. Res. 41 (2011) 99–132.
[91] H.G. Silverman, F.F. Roberto, Mar. Biotechnol. 9 (2007) 661–681.
[92] C. Sun, J.H. Waite, J. Biol. Chem. 280 (2005) 39332–39336.
[93] N. Holten-Andersen, J. Waite, J. Dent. Res. 87 (2008) 701–709.
[94] S. Haemers, The adhesive of the Blue Mussel: controlling cross-linking, ISBN: 9040724024. Delft, TU Delft, Delft University of Technology 2003. https://repository.tudelft.nl/islandora/object/uuid:bacef753-895b-48e9-92f4-587c518f1618?collection=research
[95] S.W. Taylor, D.B. Chase, M.H. Emptage, M.J. Nelson, J.H. Waite, Inorg. Chem. 35 (1996) 7572–7577.
[96] N. Holten-Andersen, H. Zhao, J.H. Waite, Biochemistry 48 (2009) 2752–2759.
[97] N. Holten-Andersen, G.E. Fantner, S. Hohlbauch, J.H. Waite, F.W. Zok, Nat. Mater. 6 (2007) 669–672.
[98] K. Inoue, Y. Takeuchi, D. Miki, S. Odo, J. Biol. Chem. 270 (1995) 6698–6701.
[99] L.M. Rzepecki, K.M. Hansen, J.H. Waite, Biol. Bull. 183 (1992) 123–137.

[100] R.Y. Floriolli, J. von Langen, J.H. Waite, Mar. Biotechnol. 2 (2000) 352–363.
[101] V.V. Papov, T.V. Diamond, K. Biemann, J.H. Waite, J. Biol. Chem. 270 (1995) 20183–20192.
[102] H. Zhao, N.B. Robertson, S.A. Jewhurst, J.H. Waite, J. Biol. Chem. 281 (2006) 11090–11096.
[103] H. Zhao, J.H. Waite, Biochemistry 45 (2006) 14223–14231.
[104] J.H. Waite, X. Qin, Biochemistry 40 (2001) 2887–2893.
[105] H. Zhao, J.H. Waite, J. Biol. Chem. 281 (2006) 26150–26158.
[106] D.S. Hwang, H.J. Yoo, J.H. Jun, W.K. Moon, H.J. Cha, Appl. Environ. Microbiol. 70 (2004) 3352–3359.
[107] J. Yu, Antioxidant is a key factor in mussel protein adhesion, in: Adhesive Interactions of Mussel Foot Proteins, Springer, 2014, , pp. 31–41.
[108] M.J. Harrington, J.H. Waite, J. Exp. Biol. 210 (2007) 4307–4318.
[109] X.X. Qin, K.J. Coyne, J.H. Waite, J. Biol. Chem. 272 (1997) 32623.
[110] X.X. Qin, J.H. Waite, Proc. Natl. Acad. Sci. U. S. A. 95 (1998) 10517.
[111] X. Qin, J.H. Waite, J. Exp. Biol. 198 (1995) 633–644.
[112] J.H. Waite, X.-X. Qin, K.J. Coyne, Matrix Biol. 17 (1998) 93–106.
[113] J. Yu, W. Wei, E. Danner, J.N. Israelachvili, J.H. Waite, Adv. Mater. 23 (2011) 2362–2366.
[114] D.S. Hwang, H. Zeng, A. Masic, M.J. Harrington, J.N. Israelachvili, J.H. Waite, J. Biol. Chem. 285 (2010) 25850–25858.
[115] S. Saxer, C. Portmann, S. Tosatti, K. Gademann, S. Zurcher, M. Textor, Macromolecules 43 (2009) 1050–1060.
[116] P. Podsiadlo, Z. Liu, D. Paterson, P.B. Messersmith, N.A. Kotov, Adv. Mater. 19 (2007) 949–955.
[117] G. Bilic, C. Brubaker, P.B. Messersmith, A.S. Mallik, T.M. Quinn, C. Haller, E. Done, L. Gucciardo, S.M. Zeisberger, R. Zimmermann, Am. J. Obstet. Gynecol. 202 (2010). 85. e81-85. e89.
[118] C.E. Brubaker, H. Kissler, L.-J. Wang, D.B. Kaufman, P.B. Messersmith, Biomaterials 31 (2010) 420–427.
[119] H. Lee, B.P. Lee, P.B. Messersmith, Nature 448 (2007) 338–341.
[120] H. Lee, S.M. Dellatore, W.M. Miller, P.B. Messersmith, Science 318 (2007) 426–430.
[121] B. Yu, J. Liu, S. Liu, F. Zhou, Chem. Commun. 46 (2010) 5900–5902.
[122] S.H. Yang, S.M. Kang, K.-B. Lee, T.D. Chung, H. Lee, I.S. Choi, J. Am. Chem. Soc. 133 (2011) 2795–2797.
[123] W.-H. Zhou, C.-H. Lu, X.-C. Guo, F.-R. Chen, H.-H. Yang, X.-R. Wang, J. Mater. Chem. 20 (2010) 880–883.
[124] S. Seo, D.W. Lee, J.S. Ahn, K. Cunha, E. Filippidi, S.W. Ju, E. Shin, B.S. Kim, Z.A. Levine, R.D. Lins, Adv. Mater. 29 (2017).
[125] P. Song, Z. Xu, M.S. Dargusch, Z.G. Chen, H. Wang, Q. Guo, Adv. Mater. (2017)
[126] E. Filippidi, T.R. Cristiani, C.D. Eisenbach, J.H. Waite, J.N. Israelachvili, B.K. Ahn, M.T. Valentine, Science 358 (2017) 502–505.
[127] Y. Zhao, Y. Wu, L. Wang, M. Zhang, X. Chen, M. Liu, J. Fan, J. Liu, F. Zhou, Z. Wang, Nat. Commun. 8 (2017) 2218.

Chapter 11

Flame Retardancy of Wood-Polymeric Composites

Xing Yang* **and Wei Zhang**[†]
Viance LLC, Charlotte, NC, United States, †Sustainable Biomaterials Department, Virginia Tech, Blacksburg, VA, United States

Chapter Outline

ABBREVIATIONS

ABS	acrylonitrile butadiene styrene
AHP	aluminum hypophosphite
APP	ammonium polyphosphate
ATH	aluminum trihydroxide
BA	boric acid
BX	borax
CB	carbon black
DAP	diammonium phosphate
EG	expandable graphite

Polymer-Based Multifunctional Nanocomposites and Their Applications
https://doi.org/10.1016/B978-0-12-815067-2.00011-1

EHC	effective heat of combustion
FR	flame retardants
HDPE	high-density PE
HRR	heat release rate
LDPE	low-density PE
LOI	limited oxygen index
MAP	monoammonium phosphate
M-APP	modified APP
MD	manganese dioxide
MEL	melamine
MEL	melamine (MEL)
MH	magnesium hydroxide
MLR	mass loss rate
MMT	montmorillonite
MP	melamine phosphate
MPP	melamine polyphosphate
O-MH	ordinary MH
PE	polyethylene
PER	pentaerythritol
Ph-LDH	phytic acid modified layered double hydroxide
p-HRR	peak of heat release rate
PP	polypropylene
PS	polystyrene
PVC	polyvinyl chloride
RP	red phosphorus
TB	PP/LDPE/HDPE ternary blends
TD	stannic oxide
THR	total heat release
TSR	total smoke release
TTI	time to ignition
U-MH	ultrafine MH
WF	wood flours
WPC	wood-polymeric composite
ZB	zinc borate

11.1 INTRODUCTION

Wood-polymeric composite (WPC) is composed of wood flours (WF) as fillers and polymer as a matrix. Wood fibers, as an alternative to synthetic fillers, provide the composite with unique advantages such as low cost, ease of processing, abundant availability, environmental sustainability, and biodegradability [1]. Polypropylene (PP) and polyethylene (PE) are the most widely used polymeric matrixes. Acrylonitrile butadiene styrene (ABS) and polyvinyl chloride (PVC) have also been used to process the composite. Because of WPC's unique features, it has been widely used in residential construction, transportation, and production of furniture. Serious safety concerns related to fire risk have arisen in many of these applications. Abundant studies have focused on the enhancement of flame retardancy of WPC. Because the studies involved different polymeric matrices, weight ratios of WF/polymer matrix/flame retardants (FR), and testing methods, the primary goal of this chapter is to describe them according to their different flame retardant systems, and provide the organized structure for comparison.

11.2 APPLYING METHODS OF FLAME RETARDANTS

In the literature, fire retardants are introduced into WPC by two methods. One method that is used more frequently is to mix fire retardants with wood fibers and polymers, as well as other additives, together to process the composites. Research on this method will be discussed in more detail in the following sections.

The other method is to pretreat wood fibers with liquid flame retardants by impregnation first, and then produce the composite with treated fibers and polymers using the conventional process. Seefeldt et al. impregnated a salt of a phosphoric-acid derivate (Disflamoll TP LXS 51064) into wood fibers, and produced the Wood-PP composites. With 10 wt% Disflamoll TP LXS 51064 actives, the composite provided a 15% increase of TTI, and 10% and 8% decreases of pHRR and THR, compared with WF/PP (60/40) [2]. Hamalainen et al. pre-treated wood fibers with two different phosphate-based fire-retardant solutions (MCF and HR-Prof), and melamine formaldehyde resin (EXPH 522) solution [3]. With a 32%–37% weight gain in the treated fibers, WF/PP/HR-Prof (64/30/32) performed the best with a 13% increase of TTI, and 44% and 39% decreases of pHRR and THR, followed by WF/PP/MCF (64/30/32) with a 2% increase of TTI, and 30% and 24% decreases of pHRR and THR, compared with WF/PP (64/30). Cavdar et al. impregnated wood fibers with boric acid (BA), borax (BX), and a mixture of BA/BX (50/50), and studied the treatment effects on flame retardancy of WF/HDPE composites [4]. BA, BX, and BA/BX treatments all provided similar improvements of flame retardancy of composites, with a 14% increase of LOI and 39%–45% decrease of horizontal burning rate, compared with the pristine WF/HDPE.

11.3 TESTING METHODS FOR FLAMMABILITY OF WPC

The main regulatory fire test for WPC is ASTM E84, or UL723. However, the burning tunnel in the test is 24′, which requires approximately of 4 m^2 of materials, so it is not practical for new product development or experimental research. Among the numerous flammability tests for material screening or performance evaluation, the cone calorimeter, limited oxygen index (LOI), UL-94, and horizontal burning tests are the most common methods used to evaluate flame retardancy of WPC. Because those methods have been reviewed previously, this chapter will only provide a general introduction to them [5–8].

11.3.1 Cone Calorimeter

Cone calorimeter is the most widely used instrument to study fire behavior of materials because it provides abundant information with relatively small-size samples. The method has been standardized in ASTM E1354 and ISO 5660. The basic principle is to measure the decreasing oxygen concentration in the combustion gases of a sample ($100 \times 100 \times 4\,mm^3$) subjected to a given heat flux (i.e., 10–100 kW/m^2). The heat flux is typically set at 50 kW/m^2 to simulate the ignition burner heat fluxes in ASTM 84 or UL723 test. Cone calorimeter tests reviewed in this chapter all used the 50 kW/m^2 heat fluxes. The test provides various flammability parameters, including heat release rate (HRR), total heat release (THR), time to ignition (TTI), mass loss rate (MLR), total smoke release (TSR), and effective heat of combustion (EHC). Because the peak of heat release rate (p-HRR), total heat release (THR), and time to ignition (TTI) were provided by most of the studies in the open literature, it is convenient to simply focus on these parameters in the following review.

11.3.2 Limiting Oxygen Index

The Limiting Oxygen Index (LOI) test is a simple method that is widely used to quickly screen materials based on flammability. It has been standardized in ASTM E2863 and ISO 4589. LOI is defined as the minimal O_2 concentration if the O_2/N_2 mixture that maintains flame combustion of the sample for 3 min, or consumes a length of 5 cm of the sample. The sample is placed in a vertical position, and the top is ignited with a burner. Generally, low LOI indicates highly flammability of materials, while less flammable materials have high LOI.

11.3.3 UL-94V

UL-94V is the most commonly used one in the set of UL94 tests approved by the Underwriters Laboratories. It is mainly used to test the flammability of plastic materials for parts in devices and appliances. The ignitability and flame spread of vertical bulk materials are measured as exposed to a small flame. The materials can be classified into V-0, V-1, and V-2 ratings, ordered from less flammable to highly flammable.

11.3.4 Horizontal Burning Test

The horizontal burning test has been used in some studies in the reviewed literature. The method has been standardized with ASTM D635. The specimen is placed horizontally, and one end of the specimen is ignited by a burner.

A burning time from the first reference mark (25 mm from the end) to the second reference mark (100 mm from the end) was recorded. Burning rates of the materials are calculated.

11.4 OPEN LITERATURE OF WPC WITH FLAME RETARDANTS

Generally, common flame retardants can be fit into halogenated and non-halogenated categories. Halogenated flame retardants typically including brominated and chlorinated molecules can reduce heat efficiently in the gas phase during combustion, and inhibit burning. However, halogenated flame retardants are often toxic, and release toxic or corrosive gases during burning; therefore, they have been gradually phased out due to environmental and regulatory concerns. Because a majority of the work has been done to find halogen-free flame retardants in WPC processing, this review only focuses on this category, which can be broken down into subcategories such as phosphorus-based, boron-based, graphite-based, filler-based, and synergy flame retardants.

11.4.1 WPC With Phosphorus-Based Flame Retardants

Phosphorus-based flame retardants have been widely used in WPC processing, including ammonium polyphosphate (APP), diammonium phosphate (DAP), monoammonium phosphate (MAP), melamine phosphate (MP), melamine polyphosphate (MPP), aluminum hypophosphite (AHP), and red phosphorus (RP). Phosphoric acid can act in the condensed phase by enhancing charring, but also in the gas phase, by reducing flammable radicals [9].

11.4.1.1 Ammonium Polyphosphate

Ammonium polyphosphate (APP) is composed of polyphosphoric acid and ammonia in the chains. It is reported to act mainly in the condensed phase to promote char formation with acid catalysis; but also in some cases dilute the flammable decomposition products with the release of non-flammable carbon dioxide in the gas phase. Because the peak of heat release rate (p-HRR), total heat release (THR), and time to ignition (TTI) are provided by most of the studies in the open literature, it is convenient to simply focus on these parameters in the following review. The results of cone calorimeter test in the open literature regarding WF/PP/APP and WF/other polymer/APP are shown in Tables 11.1 and 11.2 respectively. Regarding TTI, there are no consistent positive or negative effects with APP in WPC. For wood-PP composites, the literature reported that adding APP resulted in around a 30% increase of TTI [10–12]. Wang et al. observed over a 50% increase [13]. Nikolaeva et al. found no change in TTI [14]. However, Kalali et al. and Arao et al. reported a 29% and 5% reduction of TTI, respectively [15,16]. For the wood-HDPE system, the increase of TTI was reported; however, the same group of authors found the reduction of TTI in another report [17,18]. Stark et al. replaced 10% of PE

TABLE 11.1 Results of Cone Calorimeter Test Regarding APP in WF/PP

Sample	Wt. Ratio	TTI (s)	ΔTTI (%)	pHHR (kW/m²)	ΔpHHR (%)	THR (MJ/m²)	ΔTHR (%)	Ref.
WF/PP	40/58	19		505		253		[12]
WF/PP/APP	40/48/10	24	29	342	−32	225	−11	[12]
WF/PP/APP	50/38/10	25	31	316	−37	196	−23	[12]
WF/PP/APP	60/28/10	23	23	279	−45	151	−40	[12]
WF/PP	50/46.7	35		585		139		[15]
WF/PP/APP	50/35.7/11	25	−29	503	−14	133	−4	[15]
WF/PP	55/40			340		128		[21]
WF/PP/APP	50/30/15			244	−28	111	−13	[21]
WF/PP	55/40			343		283		[22]
WF/PP/APP 422	50/30/15			214	−38	235	−17	[22]
WF/PP/APP 760	50/30/15			248	−28	253	−11	[22]
WF/PP	60/40	21		359		87		[23]
WF/PP/APP	60/40/20	22	5	208	−42	68	−22	[23]

WF/PP	64/30	22		442				[14]
WF/PP/APP	44/30/20	22	0	328	−26			[14]
WF/PP/APP	34/30/30	25	14	322	−27			[14]
WF/PP	40/60	13		696		89		[10]
WF/PP/APP	40/60/25	17	31	391	−44	71	−20	[10]
WF/PP	40/55	15		1003		170		[13]
WF/PP/APP	28/37/30	23	53	560	−44	124	−27	[13]
WF/PP	40/60	15		758		123		[24]
WF/PP/APP	28/42/30	17	13	383	−49	77	−37	[24]
WF/PP	60/32.5	12		296		53		[11]
WF/PP/APP	40.5/22/30	16	33	176	−41	34	−36	[11]
WF/PP/ETA-APP	40.5/22/30	17	42	202	−32	39	−26	[11]
WF/PP	50/46.7	21		563		93		[16]
WF/PP/APP	50/36.7/10	20	−5	312	−45	79	−15	[16]

Note: Δ% *calculated according to the change% between FR/WPC and the pristine WPC.*

TABLE 11.2 Results of Cone Calorimeter Test Regarding APP in WF/Non-PP

Sample	Wt. Ratio	TTI (s)	Δ TTI (%)	pHHR (kW/m^2)	Δ pHHR (%)	THR (MJ/m^2)	Δ THR (%)	Ref.
WF/HDPE	30/70	23		471		117		[17]
WF/HDPE/APP	30/56/14	33	43	355	−25	109	−7	[17]
WF/HDPE	30/70	28		194		71		[18]
WF/HDPE/APP	30/70/10	19	−32	113	−42	62	−13	[18]
WF/LDPE	30/70	37		432		98		[20]
WF/LDPE/APP	30/40/30	91	146	162	−63	69	−30	[20]
WF/PE	50/45	25		505		373		[19]
WF/PE/APP	50/35/10	22	−10	310	−39	255	−32	[19]
WF/ABS	40/60			351				[25]
WF/ABS/APP	32/48/20			222	−37			[25]

Note: *Δ % calculated according to the change% between FR/WPC and the pristine WPC.*

by APP and found a 10% decrease of TTI [19]. For wood-LDPE composite, Altun et al. found a 146% increase of TTI [20]. Overall, there is no clear correlation between TTI and wood/polymer ratio or percent of added APP.

The peak of heat release rate (p-HRR) and total heat release (THR) are two important parameters to evaluate the flammability of materials. All the related literature shows the reduction of pHRR and THR in different wood/polymer systems. Ayrilmis et al. reported the reductions of pHRR and THR as wt% of PP decreased and wt% of wood fiber increased with the addition of 10% APP [12]. Generally, as APP ranged from 10% to 20%, the pHRR reduction ranged from 14% to 38%, compared with the pristine WPC. When more than 25% APP was added, pHRR reduction reached more than 40%. Kalali et al. reported WF/PP/APP (50/35.7/11) with the weight ratio of 50/35.7/11 produced 14% and 4% decreases in pHRR and THR, respectively, when compared with the WF/PP (50/46.7) with the weight ratio of 50/46.7 [15]. Seefeldt et al. reported that adding 15% APP in to a WF/PP (50/30) system resulted in 28% and 13% reductions of pHHR and THR, respectively, compared with the WF/PP (55/40) [21]. Naumann et al. studied the effects of two types of APP [22]. They found that 15% APP 422 provided WF/PP (50/30) system 38% and 17% of deceases in pHRR and THR, respectively; while 15% APP 760 resulted in 28% and 11% of reductions in PHHR and THR, compared with WF/PP (55/40). Yu et al. found that adding 20% APP decreased the 42% and 22% of pHRR and THR, respectively [23]. Nikolaeva et al. reported that WF/PP/APP (44/30/20) and WF/PP/APP (34/30/30) systems resulted in similar reductions (26%–27%) of pHRR compared with WF/PP (64/30) [14]. Wang et al. reported that adding 25% APP to WF/PP (40/60) decreased pHHR and THR by 44% and 20%, respectively [10]. They also found that WF/PP/APP (28/37/30) caused a 44% and 27% decrease of pHHR and THR, compared with WF/PP (40/55), and WF/PP/APP (28/42/30) caused a 49% and 37% reduction of pHHR and THR, compared with WF/PP (40/60) [13,24]. Guan et al. found that WF/PP/APP (40.5/22/30) provided 41% and 36% decreases of pHRR and THR [11]. The report also showed that ethanolamine modified APP performed similarly with 32% and 36% decreases of pHRR and THR. Arao et al. presented a different finding; that replacing 10% PP from WF/PP (50/46.7) by APP provided 45% and 15% decreases of pHRR and THR [16]. For a WF/HDPE system, Pan et al. reported that adding 10% APP decreased 42% and 13% of pHRR and THR regarding WF/HDPE (30/70), while they found WF/HDPE/APP (30/56/14) only produced 25% and 7% decreases of pHRR and THR [17,18].

Altun et al. replaced 30% of LDPE by APP and found 63% and 30% decreases of pHRR and THR compared with WF/LDPE (30/70) [20]. Stark et al. replaced 10% of PE by APP and found 39% and 32% decreases of pHRR and THR compared with WF/PE (50/45) [19]. Adding 20% APP into the WF/ABS system was also found to improve the flammability demonstrated by a 37% reduction of pHRR [25].

The Limiting Oxygen Index (LOI), UL94V, and horizontal burning test also have been used. The results are summarized in Table 11.3. Kalali et al. reported

TABLE 11.3 Results of LOI, UL-94V, and Horizontal Burning Tests Regarding APP in WPC

Sample	Wt. Ratio	LOI	ΔLOI (%)	UL-94V	Burning Rate (mm/min)	Ref.
WF/PP	50/46.7	19		Fail		[15]
WF/PP/APP	50/35.7/11	23	22	Fail		[15]
WF/PP	40/60	21				[10]
WF/PP/APP	40/60/25	24	12			[10]
WF/PP	60/40	22		Fail		[26]
WF/PP/APP	60/40/25	27	20	V-0		[26]
WF/PP	40/60	20				[24]
WF/PP/APP	28/42/30	24	18			[24]
WF/PP	60/32.5	19		Fail		[11]
WF/PP/APP	40.5/22/30	29	53	V-0		[11]
WF/LDPE	30/70	18		Fail		[20]
WF/LDPE/APP	30/40/30	24	38	Fail		[20]
WF/ABS	40/60	20		Fail		[25]
WF/ABS/APP	32/48/20	25	24	Fail		[25]
WF/HDPE/PVC	30/50/18			Fail		[27]
WF/HDPE/PVC/APP	30/50/18/20			Fail		[27]
WF/PP	50/46.7				32.4	[16]
WF/PP/APP	50/36.7/10				Self-extinguish	[16]
WF/HDPE	30/67				21	[28]
WF/HDPE/APP	30/67/25				12	[28]
WF/HDPE	49.5/38.6				144	[29]
WF/HDPE/APP	45.5/35.5/9.1				Self-extinguish	[29]

Note: *Δ% calculated according to the change% between FR/WPC and the pristine WPC.*

WF/PP/APP (50/35.7/11) produced a 22% increase of LOI compared with WF/PP (50/46.7), but they failed the UL-94V test [15]. By adding 25% APP, Wang et al. reported it increased 12% of LOI for WF/PP (40/60), while Bai et al. found that it increased 12% of LOI for WF/PP (60/40) and achieved the V-0 rating for the UL-94V test [10,26]. By adding 30% APP, Wang et al. reported it increased 18% of LOI for WF/PP (40/60), while Guan et al. found that it increased 53% of LOI for WF/PP (60/32.5) and achieved the V-0 rating for the UL-94V test [11,24]. WF/LDPE/APP (30/40/30), WF/HDPE/PVC/APP (30/50/18/20), and WF/ABS/APP (32/48/20) all failed the UL-94V test [20,25,27]. Pan et al. found WF/HDPE/APP (30/56/14) provided an 8% increase of LOI [17]. Altun et al. replaced 30% of LDPE by APP and found a 38% increase of LOI, compared with WF/LDPE (30/70) [20]. Adding 20% APP into the WF/ABS system increased LOI by 24%, compared with WF/ABS (40/60) [25]. With the horizontal burning test, WF/PP/APP (50/36.7/10), WF/HDPE/APP (30/67/25), and WF/HDPE/APP (45.5/35.5/9.1) reduced the burning rates, compared with the pristine composites [16,28,29].

11.4.1.2 Diammonium Phosphate

The results of burning tests in the open literature regarding DAP are shown in Table 11.4. Ayrilmis et al. investigated the flammability of WF/PP/DAP by varying the percent of DAP. WF/PP/DAP (34/54/12) was found to be the best, with a 6% increase of TTI, and 15% and 10% reductions of pHRR and THR [30]. Akbulut et al. studied the flammability of WF/HDPE/DAP by varying the percent of DAP. WF/HDPE/DAP (31/51/12) performed the best with a 10% increase of TTI, and 19% and 9% reductions of pHRR and THR [31]. Chindaprasirt et al. evaluated the flammability of the WF/PS (20/80) composite by adding different percentages of DAP [32]. Twenty percent DAP resulted in a 17% increase of LOI, and provided the self-distinguishing performance during the horizontal burning test. Generally speaking, DAP is not as effective as APP in improving flame retardancy of wood-polymeric composites.

11.4.1.3 Melamine Polyphosphate

The results of burning tests in the open literature regarding MPP are shown in Table 11.5. Ayrilmis et al. investigated the flammability of WF/PP/MPP by varying the wt% of MPP. WF/PP/MPP (34/54/12) performed the best, with an 8% increase of TTI, and 19% and 10% reductions of pHRR and THR [30]. Arao et al. presented a different finding; replacing 10% PP from WF/PP (50/46.7) by MPP provided a 37% decrease of pHRR, but a 5% decrease of TTI, and 6% increase of THR [16]. Akbulut et al. studied the flammability of WF/HDPE/MPP by varying the percent of MPP. WF/HDPE/MPP (31/51/12) was found to be the best, with an 11% increase of TTI, and 21% and 11% reductions of pHRR and THR [31]. Li et al. reported that adding 35% MPP into WF/HDPE (40/60) provided a 3% increase of TTI, and 52% and 56% decreases of pHRR and THR [33]. Also, this system obtained a 10% increase of LOI, and achieved a V-2 rating during the UL-94V test. The effectiveness of MPP showed by the open literature is comparable to DAP, but still lower than APP.

TABLE 11.4 Results of Cone Calorimeter, LOI, and Horizontal Burning Tests Regarding DAP in WPC

Sample	Wt. Ratio	TTI (s)	ΔTTI (%)	pHHR (kW/m^2)	ΔpHHR (%)	THR (MJ/m^2)	ΔTHR (%)	LOI	ΔLOI (%)	Burning rate (mm/min)	Ref.
WF/PP	40/60	19		535		319					[30]
WF/PP/DAP	38/58/4	20	2	499	−7	308	−3				[30]
WF/PP/DAP	36/56/8	19	−4	495	−7	299	−6				[30]
WF/PP/DAP	34/54/12	20	6	455	−15	286	−10				[30]
WF/HDPE	40/60	23		523		335					[31]
WF/HDPE/DAP	37/57/4	22	−2	466	−11	311	−7				[31]
WF/HDPE/DAP	34/54/8	26	13	446	−15	312	−7				[31]
WF/HDPE/DAP	31/51/12	25	10	425	−19	305	−9				[31]
WF/PS	20/80							18		44.6	[32]
WF/PS/DAP	20/80/10							20	8	26.1	[32]
WF/PS/DAP	20/80/20							21	17	<25	[32]
WF/PS/DAP	20/80/30							23	28	<25	[32]
WF/PS/DAP	20/80/40							23	30	<25	[32]

Note: Δ *% calculated according to the change% between FR/WPC and the pristine WPC.*

TABLE 11.5 Results of Cone Calorimeter, LOI, and UL-94V Tests Regarding MPP in WPC

Sample	Wt. Ratio	TTI (s)	ΔTTI (%)	pHHR (kW/m^2)	ΔpHHR (%)	THR (MJ/m^2)	ΔTHR (%)	LOI	ΔLOI (%)	UL-94V	Ref.
WF/PP	40/60	19		535		319					[30]
WF/PP/MPP	38/58/4	17	−11	543	1	318	0				[30]
WF/PP/MPP	36/56/8	20	1	470	−12	300	−6				[30]
WF/PP/MPP	34/54/12	21	8	434	−19	288	−10				[30]
WF/PP	50/46.7	21		563		93					[16]
WF/PP/MPP	50/36.7/10	20	−5	352	−37	99	6				[16]
WF/HDPE	40/60	23		523		335					[31]
WF/HDPE/MPP	37/57/4	25	8	459	−12	325	−3				[31]
WF/HDPE/MPP	34/54/8	26	15	448	−14	314	−6				[31]
WF/HDPE/MPP	31/51/12	25	11	411	−21	298	−11				[31]
WF/HDPE	40/60	35		422		102		25		Fail	[33]
WF/HDPE/MPP	40/60/35	36	3	202	−52	45	−56	27	10	V-2	[33]

Note: Δ% *calculated according to the change% between FR/WPC and the pristine WPC.*

11.4.1.4 Other Phosphorus-Based FRs

The results of burning tests in the open literature regarding other phosphorus based FRs are shown in Table 11.6. Li et al. reported that adding 35% aluminum hypophosphite (AHP) into WF/HDPE (40/60) provided a 9% increase of TTI, and 48% and 49% decreases of pHRR and THR [33]. Also, this system obtained an 8% increase of LOI, and achieved a V-2 rating during the UL-94V test. Seefeldt et al. presented that adding 5% of red phosphorus (RP) into a WF/PP (50/40) system resulted in 10% and 6% reductions of pHRR and THR, respectively [21]. Altun et al. replaced 5% of LDPE by RP, and found a 17% increase of LOI, compared with WF/LDPE (30/70), and a V-2 rating of the UL-94V test [20]. Stark et al. replaced 10% of PE with melamine phosphate (MP), and found 31% and 33% decreases of pHRR and THR, as well as a 2% reduction of TTI, compared with WF/PE (50/45) [19].

11.4.2 WPC With Boron-Based Flame Retardants

Boron-based flame retardants form a glassy layer as a barrier to volatile gases and heat transfer [1].

The open literature shows that adding zinc borate (ZB) into wood-polymeric composites improved the fire retardancy with the increased TTI and the deceases of pHRR and THR. The results of cone calorimeter test regarding ZB-based FRs are shown in Table 11.7. Ayrilmis et al. reported the increase of TTI, and the reductions of pHRR and THR as wt% of PP decreased, and wt% of wood fiber increased with the addition of 10% APP [12]. However, in another report, they presented that WF/PP/ZB (34/54/12) resulted in a 5% increase of TTI, with 11% and 1% reductions of PHHR and THR [30]. Because of the high flammability of PP, the difference of two reports might be ascribed to the different wt% of PP in the composites. Turku et al. found that replacing 10% of WF in WF/PP (50/43) by ZB provided a 6% increase of TTI, and 18% and 4% reductions of pHHR and THR [34]. When the wt% ZB increased to 30%, Nikolaeva et al. found that WF/PP/ZB (34/30/30) gave a 14% increase of TTI, and a 34% decrease of pHRR [14]. Stark et al. replaced 10% of PE by ZB, and found 37% and 32% decreases of pHRR and THR, as well as a 6% increase of TTI, compared with WF/PE (50/45) [19]. Akbulut et al. studied the flammability of WF/HDPE/ZB by varying the percent of ZB. WF/HDPE/ZB (31/51/12) was found to be the best, with a 24% increase of TTI, and 16% and 8% reductions of pHRR and THR [31]. When adding 6% ZB into WF/PVC (35/58), a 4% increase of TTI and 12% decrease of THR were reported [35]. However, pHRR increased 31%. It indicates that ZB is not a promising flame retardant for the WF/PVC system, which might be due to the "natural" fire-resistance of PVC [36].

Boric acid (BA)-borax (BX) mixture have been evaluated to be the potential flame retardants for WPC. The results of cone calorimeter test are shown in Table 11.8. Ayrilmis et al. investigated the effects of a boric acid (BA)-borax

TABLE 11.6 Results of Cone Calorimeter, LOI, and UL-94V Tests Regarding AHP, RP, and MP in WPC

Sample	Wt. Ratio	TTI (s)	ΔTTI (%)	pHHR (kW/m^2)	ΔpHHR (%)	THR (MJ/m^2)	ΔTHR (%)	LOI	ΔLOI (%)	UL-94V	Ref.
WF/HDPE	40/60	35		422		102		24.8		Fail	[33]
WF/HDPE/AHP	40/60/35	38	9	221	−48	52	−49	26.8	8	V-2	[33]
WF/PP	55/40			340		128					[21]
WF/PP/RP	50/40/5			307	−10	120	−6				[21]
WF/LDPE	30/70							17.5		Fail	[20]
WF/LDPE/RP	30/65/5							20.5	17	V-2	[20]
WF/PE	50/45	25		505		373					[19]
WF/PE/MP	50/35/10	24	−2	350	−31	250	−33				[19]

Note: Δ% *calculated according to the change% between FR/WPC and the pristine WPC.*

TABLE 11.7 Results of Cone Colorimeter Test Regarding ZB in WPC

Sample	Wt. Ratio	TTI (s)	ΔTTI (%)	pHHR (kW/m^2)	ΔpHHR (%)	THR (MJ/m^2)	ΔTHR (%)	Ref.
WF/PP	40/58	19		505		253		[12]
WF/PP/ZB	40/48/10	22	14	402	−20	237	−6	[12]
WF/PP/ZB	50/38/10	25	34	337	−33	216	−15	[12]
WF/PP/ZB	60/28/10	24	25	276	−45	184	−27	[12]
WF/PP	40/60	19		535		319		[30]
WF/PP/ZB	34/54/12	20	5	474	−11	315	−1	[30]
WF/PP	50/43	28		395		162		[34]
WF/PP/ZB	50/43/10	29	4	325	−18	155	−4	[34]
WF/PP	64/30	22		442				[14]
WF/PP/ZB	34/30/30	25	14	292	−34			[14]
WF/PE	50/45	25		505		373		[19]
WF/PE/ZB	50/35/10	26	6	320	−37	255	−32	[19]
WF/HDPE	40/60	25		477		273		[31]
WF/HDPE/ZB	37/57/4	25	2	465	−3	329	21	[31]
WF/HDPE/ZB	34/54/8	26	4	483	1	335	23	[31]
WF/HDPE/ZB	31/51/12	28	14	440	−8	309	13	[31]
WF/PVC	35/58	26		160		49		[35]
WF/PVC/ZB	35/58/6	27	4	209	31	43	−12	[35]

Note: Δ*% calculated according to the change% between FR/WPC and the pristine WPC.*

TABLE 11.8 Results of Cone Calorimeter Test Regarding BA-BX in WPC

Sample	Wt. Ratio	TTI (s)	ΔTTI (%)	pHHR (kW/m^2)	ΔpHHR (%)	THR (MJ/m^2)	ΔTHR (%)	Ref.
WF/PP	40/60	19		535		319		[30]
WF/PP/BA-BX	37/57/4	20	4	465	−13	318	0	[30]
WF/PP/BA-BX	34/54/8	20	5	507	−5	310	−3	[30]
WF/PP/BA-BX	31/51/12	22	16	495	−7	306	−4	[30]
WF/HDPE	40/60	26		516		319		[31]
WF/HDPE/BA-BX	37/57/4	25	−3	510	−1	320	0	[31]
WF/HDPE/BA-BX	34/54/8	27	2	494	−4	310	−3	[31]
WF/HDPE/BA-BX	31/51/12	26	1	458	−11	311	−3	[31]
WF/HDPE	56.4/36	37		331		116		[37]
WF/HDPE/BA-BX	56.4/36/12	41	11	281	−15	101	−13	[37]

Note: Δ% *calculated according to the change% between FR/WPC and the pristine WPC.*

(BX) mixture (weight ratio: 1/1) on the flammability of WF/PP. WF/PP/BA-BX (31/51/12) was found to be the best, with a 16% increase of TTI, and 7% and 4% reductions of PHHR and THR [30]. Akbulut et al. found that WF/HDPE/BA-BX (31/51/12) provided a 15% increase of TTI, and 12% and 7% reductions of pHHR and THR [31]. Wu et al. presented that WF/HDPE/BA-BX (56.4/36/12) provided an 11% increase of TTI, and 15% and 13% reductions of pHHR and THR [37]. Increasing the wt% of WF and decreasing wt% of HDPE reduced the heat release.

11.4.3 WPC With Metal Hydroxide Flame Retardants

Metal hydroxide decomposed under the heating by the endothermic reaction that consumes the heat energy. Also, water is release during the decomposition, which dilutes the flammable gases in the gas phase [1,38]. The results of cone calorimeter test in the open literatures regarding metal hydroxide FRs are shown in Table 11.9. Ayrilmis et al. reported that the reductions of pHRR and THR as wt% of PP decreased, and wt% of wood fiber increased, with the addition of 10% magnesium hydroxide (MH) [12]. WF/PP/MH (60/28/10) provided the best performance with 45% and 25% decreases of pHRR and THR, as well as a 55% increase of TTI, compared with WF/PP (40/58). WF/PE/MH (50/35/10) was reported to result in the 43% and 30% decreases of pHRR and THR, as well as a 27% increase of TTI, compared with WF/PE (50/45) [19]. Wu et al. investigated the size effects of MH on flame retardancy of WF/HDPE [39]. While 30% ordinary MH provided WF/HDPE/MH (28/42/30) 29% and 24% reductions of pHRR and THR, as well as a 53% increase of TTI, ultrafine performed better with 48% and 62% reductions of pHRR and THR, as well as a 161% increase of TTI compared with WF/HDPE (40/60). Zadeh et al. studied the flame retardancy of PP/LDPE/HDPE ternary blends containing date palm fibers by varying the weight percentage of MH [40]. The flame retardance of the composite was found to have the correlation with the percentage of MH positively. Forty percent of MH performed the best, with 47% and 28% reductions of pHRR and THR, as well as a 47% increase of TTI, compared with the pristine composite.

Aluminum trihydroxide (ATH), another common flame retardant, has been studied by incorporating it into the WF/PP composite. Arao et al. found that WF/PP/ATH (50/36.7/10) provided 17% and 6% reductions of pHRR and THR, as well as a 19% increase of TTI, compared with WF/PP (50/46.7) [16]. Turku et al. reported that WF/PP/ATH (40/43/10) provided 17% and 6% reductions of pHRR and THR, as well as a 19% increase of TTI, compared with WF/PP (50/43) [34]. WF/PP/ATH (28/37/30) was reported to have 33% and 10% reductions of pHRR and THR, as well as a 60% increase of TTI, compared with WF/PP (40/55) [13].

TABLE 11.9 Results of Cone Calorimeter Test Regarding MH and ATH in WPC

Sample	Wt. Ratio	TTI (s)	ΔTTI (%)	pHHR (kW/m^2)	ΔpHHR (%)	THR (MJ/m^2)	ΔTHR (%)	Ref.
WF/PP	40/58	19		505		253		[12]
WF/PP/MH	40/48/10	25	33	407	−19	249	−2	[12]
WF/PP/MH	50/38/10	28	47	330	−35	212	−16	[12]
WF/PP/MH	60/28/10	29	55	277	−45	191	−25	[12]
WF/PE	50/45	25		505		373		[19]
WF/PE/MH	50/35/10	31	27	290	−43	260	−30	[19]
WF/HDPE	40/60	51		365		313		[39]
WF/HDPE/O-MH	28/42/30	78	53	258	−29	238	−24	[39]
WF/HDPE/U-MH	28/42/30	133	161	189	−48	118	−62	[39]
WF/TB	10/90	17		724		101		[40]
WF/TB/MH	10/90/10	22	29	736	2	95	−6	[40]
WF/TB/MH	10/90/20	21	24	539	−26	87	−14	[40]
WF/TB/MH	10/90/40	25	47	385	−47	73	−28	[40]

Continued

TABLE 11.9 Results of Cone Calorimeter Test Regarding MH and ATH in WPC—cont'd

Sample	Wt. Ratio	TTI (s)	ΔTTI (%)	pHHR (kW/m^2)	ΔpHHR (%)	THR (MJ/m^2)	ΔTHR (%)	Ref.
WF/PP	50/46.7	21		563		93		[16]
WF/PP/ATH	50/36.7/10	25	19	467	−17	99	6	[16]
WF/PP	50/43	28		395		162		[34]
WF/PP/ATH	40/43/10	30	7	336	−15	152	−6	[34]
WF/PP	40/55	15		1003		170		[13]
WF/PP/ATH	28/37/30	24	60	673	−33	136	−20	[13]

Note: *Δ % calculated according to the change% between FR/WPC and the pristine WPC.*

11.4.4 WPC With Graphite-Based Flame Retardants

Expandable graphite (EG) can expand up to 300 times its original volume when it is subjected to high temperatures, and create a heat insulating layer that enhances the fire retardancy of EG-based composites [1,38]. The results of burning tests in the open literatures regarding graphite-based FRs are shown in Table 11.10. Turku et al. replaced 10% of WF with EG in WF/PP (50/43), resulting in a 57% increase of TTI, and a 13% reduction of pHRR, but no change of THR [34]. Naumann et al. reported that WF/PP/EG (50/30/15) provided 55% and 49% reductions of pHRR and THR, while Seefeldt et al. found 73% and 1% reductions of pHRR and THR, compared with WF/PP (55/40) [21,22]. Yu et al. added 20% EG into WF/PP (60/40) and found the improved flame retardancy with 50% and 26% reductions of pHRR and THR, as well as a 52% increase of TTI [23]. When 25% EG was added, WF/PP/EG (60/40/25) provided 75%–77% and 53%–57% reductions of pHRR and THR [26,41]. Also, with a 32% increase of LOI, a V-0 rating of the UL-94V test was achieved [26]. Zheng et al. reported that WF/ABS/EG (32/48/20) provided a 50% decrease of pHRR and achieved a 54% increase of LOI, as well as a V-0 rating of the UL-94V test, compared with the pristine WF/ABS (40/60) [25]. Idumah et al. studied the flame retardancy of Kenaf/PP (20/75) with varying percentages of exfoliated graphite nanoplatelets (GNP). With only 3% GNP, the composite provided a 46% reduction of pHRR, and a 133% increase of TTI, and achieved a 93% increase of LOI and a V-0 rating of the UL-94V test [42].

11.4.5 WPC With Filler-Based Flame Retardants

Multiple types of fillers were studied as additives to improve the flame retardancy of wood-polymeric composites. However, their efficiency is limited. The results of burning tests in the open literatures regarding filler-based FRs are shown in Table 11.11. Naumann et al. found a 15% commercial product, Struktol SA 0832, provided 39% and 30% decreases of pHRR and THR [22]. Nikolaeva et al. studied the effects of talc, $CaCO_3$, soapstone, and two different types of wollastonite on the flame retardancy of WF/PP (44/30) [43]. Only 30%–40% talc provided promising effects, with a 52%–76% increase of TTI and a 28%–29% decrease of pHRR. All other fillers presented limited effects with less than 10% reduction of pHRR. The effects of TiO_2 particles on either WF/PP or WF/HDPE are also limited [17,34]. Liu et al. functionalized the lignin with nitrogen, phosphorus, and metal ions [44]. By replacing 15% of the WF with a functionalized lignin in WF/PP (20/76.5), 21% and 26% decreases of pHRR, and THR as well as a 33% increase of TTI, were found. The composite achieved a V-1 rating on the UL-94V test.

TABLE 11.10 Results of Cone Calorimeter, LOI, and UL-94V Tests Regarding EG in WPC

Sample	Wt. Ratio	TTI (s)	ΔTTI (%)	pHHR (kW/m^2)	ΔpHHR (%)	THR (MJ/m^2)	ΔTHR (%)	LOI	ΔLOI (%)	UL-94V	Ref.
WF/PP	50/43	28		395		162					[34]
WF/PP/EG	40/43/10	44	57	343	−13	162	0				[34]
WF/PP	55/40			340		128					[21]
WF/PP/EG	50/30/15			91	−73	127	−1				[21]
WF/PP	55/40			343		283					[22]
WF/PP/EG	50/30/15			156	−55	145	−49				[22]
WF/PP	60/40	21		359		87					[23]
WF/PP/EG	60/40/20	32	52	181	−50	64	−26				[23]
WF/PP	40/60	20		390		90		22		Fail	[26]
WF/PP/EG	40/60/25	24	20	99	−75	39	−57	29	32	V-0	[26]
WF/PP	60/40	28		358		102					[41]
WF/PP/EG	60/40/25	23	−18	84	−77	48	−53				[41]
WF/ABS	40/60			351				20		Fail	[25]
WF/ABS/EG	32/48/20			176	−50			31	54	V-0	[25]
Kenaf/PP	20/75	15		496		68		15		Fail	[42]
Kenaf/PP/EG	20/75/3	35	133	269	−46	100	47	29	93	V-0	[42]

Note: Δ% *calculated according to the change% between FR/WPC and the pristine WPC.*

TABLE 11.11 Results of Cone Calorimeter and UL-94V Tests Regarding Fillers in WPC

Sample	Wt. Ratio	TTI (s)	ΔTTI (%)	pHHR (kW/m^2)	ΔpHHR (%)	THR (MJ/m^2)	ΔTHR (%)	UL-94V	Ref.
WF/PP	55/40			343		283			[22]
WF/PP/Struktol SA 0832	50/30/15			210	−39	240	−15		[22]
WF/PP	64/30	22		444					[43]
WF/PP/talc	44/30/20	25	14	415	−7				[43]
WF/PP/talc	34/30/30	33	52	321	−28				[43]
WF/PP/talc	24/30/40	38	76	315	−29				[43]
WF/PP/$CaCO_3$	44/30/20	25	14	416	−6				[43]
WF/PP/waste $CaCO_3$	44/30/20	23	6	421	−5				[43]
WF/PP/Soapstone	44/30/20	24	12	437	−2				[43]
WF/PP/wollastonite 1	44/30/20	22	2	417	−6				[43]
WF/PP/wollastonite 2	44/30/20	29	33	432	−3				[43]
WF/HDPE	30/70	23		471		117			[17]
WF/HDPE/TiO_2	30/56/14	33	43	355	−25	109	−7		[17]
WF/HDPE	50/43	28		395		162			[34]
WF/HDPE/TiO_2	40/43/10	29	4	362	−8	154	−5		[34]
WF/PP	20/76.5	18		595		94		Fail	[44]
WF/PP/f-lignin	5/76.5/15	24	33	470	−21	70	−26	V-1	[44]

Note: Δ% *calculated according to the change% between FR/WPC and the pristine WPC.*

11.4.6 Synergy of Flame Retardants in WPC

Synergy effects of multiple flame retardants have been studied, with the goal of solving different problems, such as enhancement of flame retardancy, or reduction of overall loading of flame retardants, or improvement of compatibility in the composite. The intumescent system consisting of a dehydrating agent, char forming agent, and blowing agent shows the synergy effects [45]. A typical example is the case of the APP/pentaerythritol (PER)/Melamine (MEL) system. The results of burning tests in the open literatures regarding APP/PER/MEL are shown in Table 11.12. Wang et al. reported that adding PER had positive effects for the WF/PP/APP system [24]. Their study showed that WF/PP/APP/PER (28/42/22.5/7.5) was the best, with 59% and 35% decreases of pHRR and THR, as well as a 33% increase of TTI, while WF/PP/APP (28/42/30) provided 49% and 37% decreases of pHRR and THR, as well as a 13% increase of TTI, compared with the pristine WF/PP (40/60). Krehula et al. found that adding 7% PER helped the WF/HDPE/PVC/APP (30/50/18/20) that failed during the UL-94V test, achieving a V-0 rating [27]. Adding 3%–5% SiO_2 and $CaCO_3$ showed different effects, with V-0 and failure of the UL-94V test, respectively. WF/PE/APP/PER/MEL (37.5/37.5/14.3/7.1/3.6) resulted in an 87% increase of LOI compared with WF/PE (50/50) [46]. Ren et al. found WF/PP/APP/PER/MEL (18/57/15/5/5) provided the 37% reduction of pHRR, as well as an 8% increase of TTI; and the system had an 83% increase of LOI, and achieved a V-1 rating on the UL-94V test [47]. The authors also studied the effects of zinc borate (ZB), montmorillonite (MMT), manganese dioxide (MD), and stannic oxide (TD) on the flammability of the system. While all of them showed the synergy effects with APP/PER/MEL, MMT performed the best, with a 66% decrease of pHRR, as well as 251% increase of TTI. The composite with MMT also provided a 95% increase of LOI, and achieved a V-0 rating on the UL-94V test.

Filler also offers potential synergy effects for APP. The results of burning tests in the open literatures regarding APP/filler synergy are shown in Table 11.13. Pan et al. studied the effects of different ratios of APP/SiO_2 on the flame retardancy of WF/HDPE [17]. The results showed that WF/HDPE/APP/SiO_2 (30/56/8/6) was the best, with 44% and 18% decreases of pHRR and THR, as well as a 78% increase of TTI, while WF/HDPE/APP (30/56/14) provided 25% and 7% decreases of pHRR and THR, as well as a 43% increase of TTI, compared with the pristine WF/HDPE (30/70). Katancic et al. found that WF/HDPE/APP/SiO_2 (30/70/20/5) had 44% and 45% decreases of pHRR and THR, as well as a 28% increase of LOI, compared with WF/HDPE (30/70) [48]. However, $CaCO_3$ was not as effective as SiO_2, with 29% and 39% decreases of pHRR and THR, as well as a 22% increase of LOI. Wang et al. studied the synergy effects of two different zeolites (4A and 13X) with APP on the flame retardancy of WF/PP [10]. The results showed that WF/PP/APP/4A (40/60/25/2) provided 53% and 22% decreases of pHRR and THR, as well as a 38% increase of TTI. WF/PP/APP/13X (40/60/25/2) had 50%

TABLE 11.12 Results of Cone Calorimeter, LOI, and UL-94V Tests Regarding APP/PER in WPC

Sample	Wt. Ratio	TTI (s)	ΔTTI (%)	pHHR (kW/m^2)	ΔpHHR (%)	THR (MJ/m^2)	ΔTHR (%)	LOI	ΔLOI (%)	UL-94V	Ref.
WF/PP	40/60	15		758		123					[24]
WF/PP/APP	28/42/30	17	13	383	−49	77	−37				[24]
WF/PP/APP/PER	28/42/22.5/7.5	20	33	312	−59	80	−35				[24]
WF/HDPE/PVC	30/50/18									Fail	[27]
WF/HDPE/PVC/APP	30/50/18/20									Fail	[27]
WF/HDPE/PVC/APP/PER	30/50/18/20/7									V-0	[27]
WF/PE	50/50							18.1			[46]
WF/PE/APP/PER/MEL	37.5/37.5/14.3/7.1/3.6							33.9	87		[46]
WF/PP	25/75	37		436				18.1		Fail	[47]
WF/PP/APP/PER/MEL	18/57/15/5/5	40	8	273	−37			33.2	83	V-1	[47]
WF/PP/APP/PER/MEL/ZB	18/57/12/4/4/5	49	32	135	−69			37.7	108	V-0	[47]
WF/PP/APP/PER/MEL/MMT	18/57/12/4/4/5	130	251	147	−66			35.3	95	V-0	[47]
WF/PP/APP/PER/MEL/MD	18/57/12/4/4/5	43	16	202	−54			38.1	110	V-0	[47]
WF/PP/APP/PER/MEL/TD	18/57/12/4/4/5	47	27	128	−71			34.5	91	V-0	[47]

Note: *Δ% calculated according to the change% between FR/WPC and the pristine WPC.*

TABLE 11.13 Results of Cone Calorimeter and LOI Tests Regarding APP/Fillers in WPC

Sample	Wt. Ratio	TTI (s)	ΔTTI (%)	pHHR (kW/m^2)	ΔpHHR (%)	THR (MJ/m^2)	ΔTHR (%)	LOI	ΔLOI (%)	Ref.
WF/HDPE	30/70	23		471		117				[17]
WF/HDPE/APP	30/56/14	33	43	355	−25	109	−7			[17]
WF/HDPE/APP/SiO_2	30/56/12/2	36	57	297	−37	101	−14			[17]
WF/HDPE/APP/SiO_2	30/56/10/4	37	61	356	−24	101	−14			[17]
WF/HDPE/APP/SiO_2	30/56/8/6	41	78	262	−44	96	−18			[17]
WF/HDPE	30/70			917		34		18		[48]
WF/HDPE/APP/SiO_2	30/70/20/5			517	−44	19	−45	23	28	[48]
WF/HDPE/APP/ $CaCO_3$	30/70/20/5			650	−29	20	−39	22	22	[48]
WF/PP	40/60	13		696		89				
WF/PP/APP	40/60/25	17	31	391	−44	71	−20			[10]
WF/PP/APP/4A	40/60/25/2	18	38	325	−53	69	−22			[10]
WF/PP/APP/13X	40/60/25/2	18	38	351	−50	68	−24			[10]

TABLE 11.13 Results of Cone Calorimeter and LOI Tests Regarding APP/Fillers in WPC—cont'd

Sample	Wt. Ratio	TTI (s)	ΔTTI (%)	pHHR (kW/m^2)	ΔpHHR (%)	THR (MJ/m^2)	ΔTHR (%)	LOI	ΔLOI (%)	Ref.
WF/PP	50/46.7	35		585		139				[15]
WF/PP/APP	50/35.7/11	25	−29	503	−14	133	−4			[15]
WF/PP/APP/Ph-LDH	50/35.7/10.5/0.5	22	−37	301	−49	88	−37			[15]
WF/PP/APP/Ph-LDH	50/35.7/10/1	25	−29	285	−51	91	−35			[15]
WF/PP/APP/Ph-LDH	50/35.7/9.5/1.5	25	−29	280	−52	103	−26			[15]
WF/PP/APP/Ph-LDH	50/35.7/9/2	23	−34	282	−52	83	−40			[15]

Note: *Δ% calculated according to the change% between FR/WPC and the pristine WPC.*

and 24% decreases of pHRR and THR, as well as a 38% increase of TTI. However, WF/PP/APP (30/56/14) showed 44% and 20% decreases of pHRR and THR, as well as a 31% increase of TTI, compared with the pristine WF/PP (30/70). Kalali et al. investigated the synergy effect between APP and phytic acid modified layered double hydroxide (Ph-LDH) on the flame retardancy of WF/PP [15]. The results showed that WF/PP/APP/Ph-LDH (50/35.7/9/2) was the best, with 52% and 40% decreases of pHRR and THR, as well as a 34% increase of TTI; while WF/PP/APP (50/35.7/11) provided 14% and 4% decreases of pHRR and THR, as well as a 29% increase of TTI, compared with the pristine WF/PP (50/46.7). The potential synergy between APP and EG has been investigated. The results of cone calorimeter test in the open literatures regarding APP/EG synergy are shown in Table 11.14. Zheng et al. found that WF/ABS/APP/EG (32/48/7.5/12.5) provided a 46% decrease of pHRR, while WF/ABS/APP (32/48/20) provided a 37% decrease of pHRR, compared with the pristine WF/ABS (40/60) [25]. Yu et al. reported that WF/PP/APP/EG/carbon black(CB) (60/40/5/15/3) had 53% and 44% decreases of pHRR and THR, as well as a 43% increase of TTI; while WF/PP/APP (60/40/20) provided 42% and 22% decreases of pHRR and THR, as well as a 5% increase of TTI, compared with the pristine WF/PP (60/40) [23]. Bai et al. replaced EG by APP and a char forming agent (CFA) prepared in their lab, and found that WF/PP/APP/EG/CFA (60/40/12/10/3) had 77% and 62% decreases of pHRR and THR, as well as a 50% increase of TTI; while WF/PP/EG (60/40/25) provided 75% and 57% decreases of pHRR and THR, as well as a 20% increase of TTI compared with the pristine WF/PP (60/40) [26]. Guo et al. studied the synergy effects of EG and APP modified with 3-(methylacryloxyl) propyltrimethoxy silane (M-APP) on the flammability of WF/PP [41]. The results showed that the reductions of pHRR and THR increased, but TTI decreased, as the EG part increased and the M-APP part decreased. Kurt et al. studied the effects of boron compound synergists with APP on the burning rates of wood/HDPE polymer composites [28]. WF/HDPE/APP/BA/BX (30/67/22/1.59/1.41) performed the best with a 55% decrease of the burning rate, compared with the 43% reduction for WF/HDPE/APP (30/67/25).

The studies on other combinations of different FRs have been summarized in Table 11.15. Altun et al. found that there is no significant difference between WF/LDPE/APP (30/40/30) and WF/LDPE/APP/RP (30/40/25/5) from cone calorimeter burning results [20]. However, WF/LDPE/APP/RP (30/40/25/5) provided a 55% increase of LOI, and achieved a V-0 rating on the UL-94V test, while WF/LDPE/APP (30/40/30) had a 38% increase of LOI, and failed the UL-94V test, compared with WF/LDPE (30/70). Li et al. studied the synergistic effect of melamine polyphosphate (MPP) and aluminum hypophosphite (AHP) on flame retardancy of HDPE/wood flour composites [33]. WF/HDPE/MPP/AHP (40/60/21/14) performed better than either WF/HDPE/MPP (40/60/35) or WF/HDPE/AHP (40/60/35), based on the TTI, pHRR, and THR results.

TABLE 11.14 Results of Cone Calorimeter Test Regarding APP/EG in WPC

Sample	Wt. Ratio	TTI (s)	ΔTTI (%)	pHHR (kW/m^2)	ΔpHHR (%)	THR (MJ/m^2)	ΔTHR (%)	Ref.
WF/ABS	40/60			351				[25]
WF/ABS/APP	32/48/20			222	−37			[25]
WF/ABS/APP/EG	32/48/12.5/7.5			188	−46			[25]
WF/PP	60/40	21		359		87		[23]
WF/PP/APP	60/40/20	22	5	208	−42	68	−22	[23]
WF/PP/APP/EG/CB	60/40/5/15/3	30	43	170	−53	49	−44	[23]
WF/PP	60/40	20		390		90		[26]
WF/PP/EG	60/40/25	24	20	99	−75	39	−57	[26]
WF/PP/EG/APP/CFA	60/40/10/12/3	30	50	90	−77	34	−62	[26]
WF/PP	60/40	28		358		102		[41]
WF/PP/M-APP	60/40/25	34	21	174	−51	85	−17	[41]
WF/PP/M-APP/EG	60/40/24/1	33	18	154	−57	74	−27	[41]
WF/PP/M-APP/EG	60/40/22.5/2.5	34	21	144	−60	82	−20	[41]
WF/PP/M-APP/EG	60/40/20/5	32	14	124	−65	76	−25	[41]
WF/PP/M-APP/EG	60/40/17.5/7.5	28	0	104	−71	67	−34	[41]
WF/PP/M-APP/EG	60/40/15/10	27	−4	90	−75	56	−45	[41]
WF/PP/M-APP/EG	60/40/12.5/12.5	27	−4	89	−75	54	−47	[41]

Note: Δ% *calculated according to the change% between FR/WPC and the pristine WPC.*

TABLE 11.15 Results of Cone Calorimeter, LOI, UL-94V, and Horizontal Burning Tests Regarding APP/BA/BX, APP/RP, and MPP/AHP in WPC

Sample	Wt. Ratio	TTI (s)	ΔTTI (%)	pHHR (kW/m^2)	ΔpHHR (%)	THR (MJ/m^2)	ΔTHR (%)	LOI	ΔLOI (%)	UL-94V	Burning rate (mm/min)	Ref.
WF/HDPE	30/67										21	[28]
WF/HDPE/APP	30/67/25										12	[28]
WF/HDPE/APP/BA/BX	30/67/22/ 1.59/1.41										9.5	[28]
WF/LDPE	30/70	37		432		98		17.5		Fail		[20]
WF/LDPE/APP	30/40/30	91	146	162	−63	69	−30	24.2	38	Fail		[20]
WF/LDPE/APP/RP	30/40/25/5	74	100	172	−60	72	−27	27.2	55	V-0		[20]
WF/HDPE	40/60	35		422		102						[33]
WF/HDPE/MPP	40/60/35	36	3	202	−52	45	−56					[33]
WF/HDPE/AHP	40/60/35	38	9	221	−48	52	−49					[33]
WF/HDPE/MPP/AHP	40/60/21/14	43	23	166	−61	43	−58					[33]

Note: *Δ% calculated according to the change% between FR/WPC and the pristine WPC.*

11.5 CONCLUSION AND OUTLOOK

Literature from the past decade has shown a variety of strategies for fire retardancy enhancement of wood-polymeric composites. The synergy effects in multiple flame retardants are essential to improved performance. APP-based synergic systems such as APP/PER/MEL and APP/PER/EG demonstrated the most promising effects. As the technology has advanced, a variety of techniques, especially the cone calorimeter method, have provided a much better understanding of burning behavior. However, it is still a big challenge to correlate the test results among different testing methods, especially regulatory fire tests. Even different regulatory fire tests emphasize various parameters, such as ASTM E84 (UL723) for flame spread, and UL-94 for dripping. Therefore, understanding the required regulatory fire test in each specific market will help provide the right direction for flammability improvements as outlined in the new research.

REFERENCES

[1] M. Nikolaeva, T. Kärki, A review of fire retardant processes and chemistry, with discussion of the case of wood-plastic composites, Baltic For 17 (2) (2011) 314–326.

[2] H. Seefeldt, U. Braun, A new flame retardant for wood materials tested in wood-plastic composites, Macromol. Mater. Eng. 297 (8) (2012) 814–820.

[3] K. Hämäläinen, T. Kärki, Effects of wood flour modification on the fire retardancy of wood–plastic composites, Eur. J. Wood Wood Prod. 72 (6) (2014) 703–711.

[4] A.D. Cavdar, F. Mengeloğlu, K. Karakus, Effect of boric acid and borax on mechanical, fire and thermal properties of wood flour filled high density polyethylene composites, Measurement 60 (2015) 6–12.

[5] F. Laoutid, et al., New prospects in flame retardant polymer materials: from fundamentals to nanocomposites, Mater. Sci. Eng. R. Rep. 63 (3) (2009) 100–125.

[6] M.E. Mngomezulu, M.J. John, Thermoset-Cellulose Nanocomposites: Flammability Characteristics, Wiley, 2017.

[7] R.M. Rowell, M.A. Dietenberger, Thermal properties, combustion, and fire retardancy of wood, in: Handbook of Wood Chemistry and Wood Composites, 2013, , pp. 127–150.

[8] Seefeldt, H., Flame Retardancy of Wood-Plastic Composites. PhD diss., Technischen Universität Berlin, 2012.

[9] B. Schartel, Phosphorus-based flame retardancy mechanisms—old hat or a starting point for future development? Materials 3 (10) (2010) 4710–4745.

[10] W. Wang, et al., Synergistic effect of synthetic zeolites on flame-retardant wood-flour/polypropylene composites, Constr. Build. Mater. 79 (2015) 337–344.

[11] Y.-H. Guan, et al., An effective way to flame-retard biocomposite with ethanolamine modified ammonium polyphosphate and its flame retardant mechanisms, Ind. Eng. Chem. Res. 54 (13) (2015) 3524–3531.

[12] N. Ayrilmis, et al., Effects of fire retardants on physical, mechanical, and fire properties of flat-pressed WPCs, Eur. J. Wood Wood Prod. 70 (1-3) (2012) 215–224.

[13] W. Wang, et al., Effect of ammonium polyphosphate to aluminum hydroxide mass ratio on the properties of wood-flour/polypropylene composites, Polymers 9 (11) (2017) 615.

[14] M. Nikolaeva, T. Kärki, Reaction-to-fire properties of wood–polypropylene composites containing different fire retardants, Fire. Technol 51 (1) (2015) 53–65.

[15] E.N. Kalali, et al., Flame-retardant wood polymer composites (WPCs) as potential fire safe bio-based materials for building products: Preparation, flammability and mechanical properties, Fire Saf. J. (2017)

[16] Y. Arao, et al., Improvement on fire retardancy of wood flour/polypropylene composites using various fire retardants, Polym. Degrad. Stab. 100 (2014) 79–85.

[17] M. Pan, et al., Synergistic effect of nano silicon dioxide and ammonium polyphosphate on flame retardancy of wood fiber–polyethylene composites, Compos. A: Appl. Sci. Manuf. 66 (2014) 128–134.

[18] M. Pan, C. Mei, Y. Song, A novel fire retardant affects fire performance and mechanical properties of wood flour-high density polyethylene composites, Bioresources 7 (2) (2012) 1760–1770.

[19] N.M. Stark, et al., Evaluation of various fire retardants for use in wood flour–polyethylene composites, Polym. Degrad. Stab. 95 (9) (2010) 1903–1910.

[20] Y. Altun, M. Doğan, E. Bayramlı, The effect of red phosphorus on the fire properties of intumescent pine wood flour–LDPE composites, Fire Mater. 40 (5) (2016) 697–703.

[21] H. Seefeldt, U. Braun, M.H. Wagner, Residue stabilization in the fire retardancy of wood–plastic composites: combination of ammonium polyphosphate, expandable graphite, and red phosphorus, Macromol. Chem. Phys. 213 (22) (2012) 2370–2377.

[22] A. Naumann, et al., Material resistance of flame retarded wood-plastic composites against fire and fungal decay, Polym. Degrad. Stab. 97 (7) (2012) 1189–1196.

[23] F. Yu, et al., Expandable graphite's versatility and synergy with carbon black and ammonium polyphosphate in improving antistatic and fire-retardant properties of wood flour/polypropylene composites, Polym. Compos. 38 (4) (2017) 767–773.

[24] W. Wang, et al., Effect of pentaerythritol on the properties of wood-flour/polypropylene/ammonium polyphosphate composite system, Bioresources 10 (4) (2015) 6917–6927.

[25] J. Zheng, et al., Flame-retardant properties of acrylonitrile–butadiene–styrene/wood flour composites filled with expandable graphite and ammonium polyphosphate, J. Appl. Polym. Sci. 131 (10) (2014).

[26] G. Bai, C. Guo, L. Li, Synergistic effect of intumescent flame retardant and expandable graphite on mechanical and flame-retardant properties of wood flour-polypropylene composites, Constr. Build. Mater. 50 (2014) 148–153.

[27] L. Kratofil Krehula, et al., Study of fire retardancy and thermal and mechanical properties of HDPE-wood composites, J. Wood Chem. Technol. 35 (6) (2015) 412–423.

[28] R. Kurt, F. Mengeloglu, H. Meric, The effects of boron compounds synergists with ammonium polyphosphate on mechanical properties and burning rates of wood-HDPE polymer composites, Eur. J. Wood Wood Prod. 70 (1–3) (2012) 177–182.

[29] M. Garcia, et al., Wood–plastics composites with better fire retardancy and durability performance, Compos. A: Appl. Sci. Manuf. 40 (11) (2009) 1772–1776.

[30] N. Ayrilmis, et al., Effect of boron and phosphate compounds on physical, mechanical, and fire properties of wood–polypropylene composites, Constr. Build. Mater. 33 (2012) 63–69.

[31] T. Akbulut, et al., Effect of boron and phosphate compounds on thermal and fire properties of wood/HDPE composites. (2011).

[32] P. Chindaprasirt, et al., Properties of wood flour/expanded polystyrene waste composites modified with diammonium phosphate flame retardant, Polym. Compos. 36 (4) (2015) 604–612.

[33] L. Li, W. Guo, C. Guo, Synergistic effect of melamine polyphosphate and aluminum hypophosphite on mechanical properties and flame retardancy of HDPE/wood flour composites, Wood Sci. Technol. 51 (3) (2017) 493–506.

[34] I. Turku, M. Nikolaeva, T. Kärki, The effect of fire retardants on the flammability, mechanical properties, and wettability of co-extruded PP-based wood-plastic composites, Bioresources 9 (1) (2014) 1539–1551.

[35] Y. Fang, et al., Effect of zinc borate and wood flour on thermal degradation and fire retardancy of polyvinyl chloride (PVC) composites, J. Anal. Appl. Pyrolysis. 100 (2013) 230–236.
[36] A.A. Klyosov, Wood-Plastic Composites, John Wiley & Sons, 2007.
[37] G.-F. Wu, M. Xu, Effects of boron compounds on the mechanical and fire properties of wood-chitosan and high-density polyethylene composites, Bioresources 9 (3) (2014) 4173–4193.
[38] S. Chapple, R. Anandjiwala, Flammability of natural fiber-reinforced composites and strategies for fire retardancy: a review, J. Thermoplast. Compos. Mater. 23 (6) (2010) 871–893.
[39] Z. Wu, et al., The effect of ultrafine magnesium hydroxide on the tensile properties and flame retardancy of wood plastic composites, J. Nanomater. 2014 (2014) 196.
[40] K.M. Zadeh, D. Ponnamma, M.A.A. Al-Maadeed, Date palm fibre filled recycled ternary polymer blend composites with enhanced flame retardancy, Polym. Test. (2017).
[41] C. Guo, L. Zhou, J. Lv, Effects of expandable graphite and modified ammonium polyphosphate on the flame-retardant and mechanical properties of wood flour-polypropylene composites, Polym. Polym. Compos. 21 (7) (2013) 449.
[42] C.I. Idumah, A. Hassan, S. Bourbigot, Influence of exfoliated graphene nanoplatelets on flame retardancy of kenaf flour polypropylene hybrid nanocomposites, J. Anal. Appl. Pyrolysis 123 (2017) 65–72.
[43] M. Nikolaeva, T. Kärki, Influence of mineral fillers on the fire retardant properties of wood-polypropylene composites, Fire Mater. 37 (8) (2013) 612–620.
[44] L. Liu, et al., Fabrication of green lignin-based flame retardants for enhancing the thermal and fire retardancy properties of polypropylene/wood composites, ACS Sustain. Chem. Eng. 4 (4) (2016) 2422–2431.
[45] S. Bourbigot, S. Duquesne, Fire retardant polymers: recent developments and opportunities, J. Mater. Chem. 17 (22) (2007) 2283–2300.
[46] E.Y. Dong, Y.L. Ren, Y.S. Jin, Evaluation of flame retardants on wood-flour/PE composites, in: Advanced Materials Research, Trans Tech Publications, 2013.
[47] Y. Ren, et al., Evaluation of intumescent fire retardants and synergistic agents for use in wood flour/recycled polypropylene composites, Constr. Build. Mater. 76 (2015) 273–278.
[48] Z. Katančić, et al., Effect of modified nanofillers on fire retarded high-density polyethylene/wood composites, J. Compos. Mater. 48 (30) (2014) 3771–3783.

Index

Note: Page numbers followed by *f* indicate figures, *t* indicate tables, and *s* indicate schemes.

D

N

T

U

V

W

X

Y

Z